The Biochemical Basis of Neuropharmacology

EIGHTH EDITION

JACK R. COOPER, PH.D.
Emeritus Professor of Pharmacology
Yale University School of Medicine

FLOYD E. BLOOM, M.D.
Chairman, Department of Neuropharmacology
The Scripps Research Institute

ROBERT H. ROTH, PH.D.
Professor of Pharmacology and Psychiatry
Yale University School of Medicine

OXFORD
UNIVERSITY PRESS
2003

This book is dedicated to the memory of
Nicholas J. Giarman, colleague and dear friend

OXFORD
UNIVERSITY PRESS

Oxford New York
Auckland Bangkok Buenos Aires Cape Town Chennai
Dar es Salaam Delhi Hong Kong Istanbul Karachi Kolkata
Kuala Lumpur Madrid Melbourne Mexico City Mumbai
Nairobi São Paulo Shanghai Taipei Tokyo Toronto

Published by Oxford University Press, Inc.
198 Madison Avenue, New York, New York, 10016
http://www.oup-usa.org

Library of Congress Cataloging-in-Publication Data
Cooper, Jack R., 1924–
The biochemical basis of neuropharmacology /
Jack R. Cooper, Floyd E. Bloom, Robert H. Roth.—8th ed.
p.; cm. Includes bibliographical references and index.
ISBN 0-19-514007-9 (cloth)—ISBN 0-19-514008-7 (pbk.)
1. Neurochemistry. 2. Neuropharmacology.
I. Bloom, Floyd E. II. Roth, Robert H., 1939–III. Title.
[DNLM: 1. Neuropharmacology. 2. Nerve Tissue—chemistry.
3. Neurotransmitters—physiology.
QV 76.5 C777b 2003] QP356.3 .C66 2003 615'.78–dc21 2002025192

9 8 7 6 5 4 3 2 1

Printed in the United States of America
on acid-free paper

Preface to the Eighth Edition

We must confess it came as somewhat of a shock when it was pointed out that since the first edition of this book was published in 1970, we have attempted to educate a whole generation of neuroscience-oriented individuals. While we modestly lower our eyes at this accomplishment, we feel a bit frustrated these days because neuroscience has exploded so dramatically in the past 30 years that it is becoming increasingly difficult to know what to cover and in what detail without swamping the reader with too much information. We have tried to compensate for the brevity of the discussions by providing appropriate references at the end of each chapter.

Aside from the usual updating of material, the major change in this edition is an extensive rewriting of the chapter on memory and learning, to emphasize that genes involved in behavior are not immutable but their expression can be modified by transcription factors. Thus, with respect to learning, that old question about which is more important, nature or nurture, genetics or environment, should be answered with the question, which leg is more important for walking, the left or the right?

J. R. C.
F. E. B.
R. H. R.

Contents

1. Introduction, 1
2. Cellular Foundations of Neuropharmacology, 7
3. Molecular Foundations of Neuropharmacology, 39
4. Receptors, 65
5. Modulation of Synaptic Transmission, 85
6. Amino Acid Transmitters, 105
7. Acetylcholine, 151
8. Norepinephrine and Epinephrine, 181
9. Dopamine, 225
10. Serotonin (5-Hydroxytryptamine), Histamine, and Adenosine, 271
11. Neuroactive Peptides, 321
12. Cellular Mechanisms in Learning and Memory, 357
13. Treating Neurological and Psychiatric Diseases, 373

Index, 401

The Biochemical Basis of Neuropharmacology

1

Introduction

Neuropharmacology can be defined simply as the study of drugs that affect nervous tissue. This, however, is not a practical definition since a great many drugs whose therapeutic value is extraneural can affect the nervous system. For example, the cardiotonic drug digitalis will not uncommonly produce central nervous system (CNS) effects ranging from blurred vision to disorientation. For our purposes, we must accordingly limit the scope of neuropharmacology to those drugs specifically employed to affect the nervous system. The domain of neuropharmacology would thus include psychotropic drugs that affect mood and behavior, anesthetics, sedatives, hypnotics, narcotics, anticonvulsants, analgesics, and a variety of drugs that affect the autonomic nervous system.

Since, with few exceptions, the precise molecular mechanism of action of these drugs is unknown and recitations of their absorption, metabolism, therapeutic indications, and toxic liability can be found in most textbooks of pharmacology, we have chosen to take a different approach to the subject. We will concentrate on the biochemistry and physiology of nervous tissue, emphasizing neurotransmitters, and will introduce the neuropharmacological agents where their action is related to the subject under discussion. Thus, a discussion of lysergic acid diethylamide (LSD) is included in the chapter on serotonin, and a suggested mechanism of action of the antipsychotic drugs is found in Chapters 9 and 13.

It is not difficult to justify this focus on either real or proposed neurotransmitters since the drugs act at junctions rather than on the events that occur with axonal conduction or within the cell body. Except for local anesthetics, which interact with axonal membranes, all neuropharmacological agents whose mechanisms of action are to some extent documented seem to be involved primarily with synaptic events. This finding appears to be quite logical in view of the regulatory mechanisms in the transmission of nerve impulses. The extent to which a neuron is depolarized or hyperpolarized will depend largely on its excitatory and inhibitory synaptic inputs, and these inputs must obviously involve neurotransmitters, neuromodulators, or neurohormones. What is enormously difficult to comprehend is the contrast between the action of a drug on a simple neuron, which causes it either to fire or not to fire, and the wide diversity of CNS effects, including subtle changes in mood and behavior, which that same drug will induce. As will become clearer in subsequent chapters, at the molecular level, an explanation of the action of a drug is often possible; at the cellular level, an explanation is sometimes possible; but at the behavioral level, our ignorance is abysmal. There is no reason to assume, for example, that a drug that inhibits the firing of a particular neuron will therefore produce a depressive state in an animal: there may be dozens of unknown intermediary reactions involving transmitters and modulators between the demonstration of the action of a drug on a neuronal system and the ultimate effect on behavior.

However, the fact that there are compounds with a specific chemical structure to control a given pathological condition is an exciting experimental finding since it suggests an approach that the neuropharmacologist can use to clarify normal as well as abnormal brain chemistry and physiology. For instance, the use of drugs that affect the adrenergic nervous system has uncovered basic and hitherto unknown neural properties, such as the uptake, storage, and release of the biogenic amines. Recognition of the analogy between curare poisoning in animals and myasthenia gravis in humans led to the understanding of the cholinergic neuromuscular transmission problem in myasthenia gravis and to subsequent treatment with anticholinesterases.

We have already referred to neuroactive agents involved in synaptic transmission as neurotransmitters, neuromodulators, and neurohormones, so definitions are now in order. Although we can define these terms in a strict, rigid fashion, it will be apparent, as noted later, that it is an exercise in futility to apply these definitions to a neuroactive agent as a classification unless one understands its activity and specifies its locus of action. Briefly, the traditional definition of a *neurotransmitter* states that the compound must be synthesized and released presynaptically; it must mimic the action of the endogenous compound that is released on nerve stimulation; and, where possible, a pharmacological identity is required where drugs that either potentiate or block postsynaptic responses to the endogenously released agent also act identically to the suspected neurotransmitter that is administered. Conventionally, based on studies of acetylcholine at the neuromuscular junction, transmitter action was thought to be a brief and highly restricted point-to-point process. If one takes the word *modulation* literally, then a *neuromodulator* has no intrinsic activity but is active only in the face of ongoing synaptic activity, where it can modulate transmission either pre- or postsynaptically. In many instances, however, a modulating agent does produce changes in conductance or membrane potential. Typically, modulatory effects involve a second-messenger system. A *neurohormone* has intrinsic activity; can be released from both neuronal and nonneuronal cells; and, most important to the definition, travels in the circulation to act at a site distant from its release site. Just how far a neurohormone has to travel before it loses its neurotransmitter status and becomes a neurohormone has never been determined.

We stated earlier that, while we could define these terms, it would be of little use to pigeonhole known neuroactive compounds until the site of action and the activity of the agent were specified. For example, dopamine is a certified neurotransmitter in the striatum, yet it is released from the hypothalamus and travels through the hypophyseal circulation to the pituitary, where it inhibits the release of prolactin. Here, it obviously fits the definition of a neurohormone. Similarly, serotonin is a neurotransmitter in the raphe nuclei, yet at the facial motor nucleus it acts primarily as a neuromodulator and secondarily as a transmitter. Most peptides, with their multiple activities in the brain and gut, are generally considered to be neuromodulators, yet substance P fulfills the criteria of a transmitter at sensory afferents to the dorsal horn of the spinal cord. In sum, the plethora of exceptions to the aforementioned definitions of *transmitter*, *modulator*, and *hormone* has generated confusion in the literature. Better to describe the activity of a neuroactive agent at a specified site rather than attempt to give a profitless definition.

The multidisciplinary aspects of pharmacology in general are particularly relevant in the field of neuropharmacology, where a pure neurophysiologist or neurochemist would be severely handicapped in elucidating drug action at the molecular level. Since neuropharmacology is not a specific discipline with its own technology, the neuropharmacologist should be aware of the methodologies that are available for the total dissection of a biological problem at all levels of resolution from the molecular to the behavioral.

Until relatively recently, medicinal chemists felt fortunate if they could produce a few dozen potential therapeutic agents a year. However, in the past several years, technologies utilizing combinatorial chemistry and high-throughput screening have produced libraries that can yield over 5000 potential drugs a week. It is therefore not a problem these days to develop neuropharmacological agents. A major problem is finding ways to get these agents, small organic molecules as well as peptides and proteins, into the brain and into the correct location. Further, in the case of gene replacement, it is necessary to ensure that the gene is not only expressed but also programmed to make just the right amount of protein. To date, a number of novel strategies have been employed to circumvent the blood–brain barrier (see Chapter 2) and to prevent enzymatic destruction of the agent before it gets to the brain. An example of one technique is the administration of levodopa (L-DOPA) plus a peripherally acting DOPA decarboxylase inhibitor. Unlike dopamine, DOPA via the neutral amino acid transport system can penetrate the brain, where it will be decarboxylated to dopamine, the neuroactive agent that is depleted in parkinsonism (see Chapter 9). In addition to this prodrug approach, other techniques involve encapsulating the drug with lipids or biodegradable polymers, coupling the drug to a molecule that possesses a specific transport mechanism, utilizing retrograde transport to deliver trophic factors to specific sites in the brain, or transiently opening the blood–brain barrier (e.g., injecting a hyperosmotic solution or a bradykinin analog). Gene transfer is a considerably more difficult technology, which is still developing. Virus-based vectors, such as adenovirus, herpes simplex virus, or *Salmonella*, that have been rendered nonpathogenic along with a cloned gene can be directly injected into the target cell. This is the in vivo approach. In the ex vivo approach to gene therapy, cells are removed from the patient, cultured, modified with a vector and the cloned gene, and then injected back into the patient. In both instances, delivery of the gene to the appropriate site may still be problematic.

In science, one measures something. One must know what to measure, where to measure it, and how to measure it. This sounds rather obvious, but the student should be aware that, particularly in the neural sciences, these seemingly simple tasks can be enormously difficult. For example, suppose one were interested in elucidating the presumed biochemical aberration in schizophrenia. What would one measure? Adenosine triphosphate? Glucose? Ascorbic acid? Unfortunately, this problem had been zealously investigated early on by people who measured everything they could think of, generally in the blood, in their search for differences between normal individuals and schizophrenics. As could be predicted, the problem was not solved. (It may be assumed, however, that these studies produced a large population of anemic schizophrenics with all this bloodletting.) In recent times, it has been demonstrated that antipsychotic drugs block a dopamine receptor (see Chapter 9). Although this biochemical reaction takes place immediately in test tubes containing brain tissue, patients who are given antipsychotic medication do not show beneficial effects for about 2 weeks. The inference, therefore, is that the drug itself and its biochemical reaction do not produce the ameliorative effect; rather, it is the adaptation of the brain to the presence of the drug that is beneficial. The question then is what is this adaptation; the answer is that we still do not know what to measure (but see Chapter 13).

Deciding where to measure something in neuroscience is complicated by the heterogeneity of nervous tissue: in general, unless one has a particular axon to grind, it is preferable to use peripheral nerves rather than the CNS. Suburban neurochemists (they work on the peripheral nervous system) have an easier time than their CNS counterparts since it is a question not only of which region of the brain to use for the test preparation but also of which of the multitude of cell types within each area to choose. If a project involved a study of amino acid transport in nervous tissue, for example, would one use isolated nerve-ending particles (synaptosomes), glial cells, neuronal cell bodies in culture, a myelinated axon, or a ganglion cell? Up to the present time, most investigators have used cortical brain slices, but the obvious disadvantage of this preparation is that one has no idea which cellular organelle takes up the amino acid.

How to measure something is a surprisingly easy question to answer, at least if one is dealing with simple molecules. With the recent advances in microseparation techniques and in fluorometric, radiometric, and immunological assays, there is virtually nothing that cannot be measured with a high degree of specificity and sensitivity. In this regard, one should be careful not to overlook the classic bioassay, which tends to be scorned by young investigators but is in fact largely responsible for striking progress in our knowledge of both the prostaglandins and the opiate receptors with their peptide agonists. The major problem is with macromolecules. How can neuronal membranes be quantified, for example, if extraneuronal constituents are an invariable contaminant and if markers to identify unequivocally a cellular constituent are often lacking? The quantitative and spatial measurement of receptors utilizing autoradiography is also a key problem (see Chapter 4).

This discussion of measurement is meant to point out that what appears on the surface to be the simplest part of research can in fact be very difficult. It is vital that students learn not to accept data without a critical appraisal of the procedures that were employed to obtain the results. Current trends in the neural sciences that are related to neuropharmacology include identifying subclasses of ion channels, utilizing molecular genetics to uncover genes whose expression is activated or suppressed by exposure to drugs or peptides, neural cartography (the mapping of transmitters and neuroactive peptides in the CNS), searching for toxins with specific effects on conduction or transmission, cloning and characterizing receptors and ion channels, and identifying trophic factors involved in synaptogenesis and neuronal regulation. It can also easily be predicted that within the next few years an intensive search will be undertaken to explain the function of the thousands of receptor subtypes. Clearly, neuropharmacological agents will be invaluable probes in this search.

In this introductory chapter, we have flitted over a number of topics relevant to neuropharmacology. Now it is time to get down to serious business.

Selected References

Nestler, E. J., S. E. Hyman, and R. C. Malenka (2001). *Molecular Neuropharmacology*. McGraw-Hill, New York.

Siegel, G. J., B. W. Agranoff, R. W. Albers, S. K. Fisher, and M. D. Uhler (1999). *Basic Neurochemistry*, 6th ed. Lippincott-Raven, Philadelphia.

2

Cellular Foundations of Neuropharmacology

As we begin to consider the particular problems that underlie the analysis of drug actions in the central nervous system, it may be asked "Just what is so special about nervous tissue?" Nerve cells have two special properties that distinguish them from all other cells in the body. First, they can conduct bioelectric signals for long distances without any loss of signal strength. Second, they possess specific intercellular connections with other nerve cells and with innervated tissues such as muscles and glands. These connections determine the types of information a neuron can receive and the range of responses it can yield in return.

Cytology of the Nerve Cell

We do not need the high resolution of the electron microscope to identify the characteristic structural features of the nerve cell. The classic studies of the Spanish Nobel Prize–winning cytologist Santiago Ramón y Cajal demonstrated the heterogeneous size and shape of neurons as individual cells. An inescapable rule of neurocytology and neuroanatomy is that structures have several synonymous names. So, for example, we find that the body of the nerve cell is also called the *soma* and the *perikaryon*—literally, "the part that surrounds the nucleus." A fundamental scheme classifies nerve cells by the number of cytoplasmic processes they possess. In the simplest case, the perikaryon has one process, called an *axon.* The best example of this cell type is the sensory neuron, whose perikarya occur in groups in the sensory or dorsal root ganglia. In this case, the axon conducts the signal, which was generated by the sensory receptor in the skin or other organs, centrally through the dorsal root into the spinal cord or cranial nerve nuclei. At the next step of complexity, we find neurons possessing two processes: the bipolar nerve cells. The sensory receptor nerve cells of the retina, the olfactory mucosa, and the auditory nerve are of this form, as is a class of small nerve cells of the brain known as granule cells.

All other nerve cells tend to fall into the class known as multipolar nerve cells. These cells possess only one axon or efferent-conducting process, which may be short or long, be branched or straight, and possess a recurrent or collateral branch that feeds back onto the same type of nerve cell from which the axon arises. Their main differences relate to the extent and size of the receptive field of the neuron, termed the *dendrites* or *dendritic tree*. In silver-stained preparations for the light microscope, the branches of dendrites look like trees in wintertime, although they may be long and smooth, be short and complex, or bear short spines like a cactus. It is on these dendritic branches as well as on the cell body where the termination of axons from other neurons makes the specialized interneuronal communication point known as the synapse.

Regardless of their shape and size, neurons have very characteristic cytoplasmic organelles. Neuronal cytoplasm in the perikaryon and dendrites, but not in the axons, is rich in rough and smooth endoplasmic reticula, connoting an emphasis on their secretory activity. In all structural compartments, neurons are rich in mitochondria, connoting their dependence on high rates of adenosine triphosphate (ATP) generation. In addition, neuronal processes, the axons and dendrites, exhibit prominent microtubules to maintain their polarized shapes.

The Synapse

The characteristic specialized contact zone that has been presumptively identified as the site of functional interneuronal communication is the *synapse*. It contains special organelles. As the axon approaches the site of its termination, it exhibits structural features not found more proximally. Most striking is the occurrence of dilated regions of the axon (*varicosities*), within which are clustered large numbers of microvesicles (*synaptic vesicles*). Synaptic vesicles tend to be spherical in shape, with diameters varying between 400 and 1200 Å. Depending on the type of fixation used, the shape and staining properties of the vesicles can be related to their neurotransmitter content. The nerve endings also exhibit mitochondria but not microtubules unless the varicosity is a "preterminal" region of an axon as it extends toward its terminal target. One or more of these varicosities may form a specialized contact with one or more dendritic branches before the ultimate termination. Such endings are known as *en passant* terminals. In this sense, the term *nerve terminal*, or *nerve ending*, connotes a functional transmitting site rather than the end of the axon.

Electron micrographs of synaptic regions in the central nervous system reveal a specialized contact zone between the axonal nerve ending and the postsynaptic structure (Fig. 2–1). Cell types arising from the embryonic ectoderm often have such specialized intercellular contact zones, which are generally presumed to maintain the structural integrity of the cells within a layer. In the nervous system, the specialized contact zone at synapses has been viewed as the site of active chemical transmission and response. This conjecture posed substantial controversy in the era before any of the molecules associated with presynaptic transmitter release or postsynaptic transmitter response had been characterized. However, the controversy now seems to have ended with the

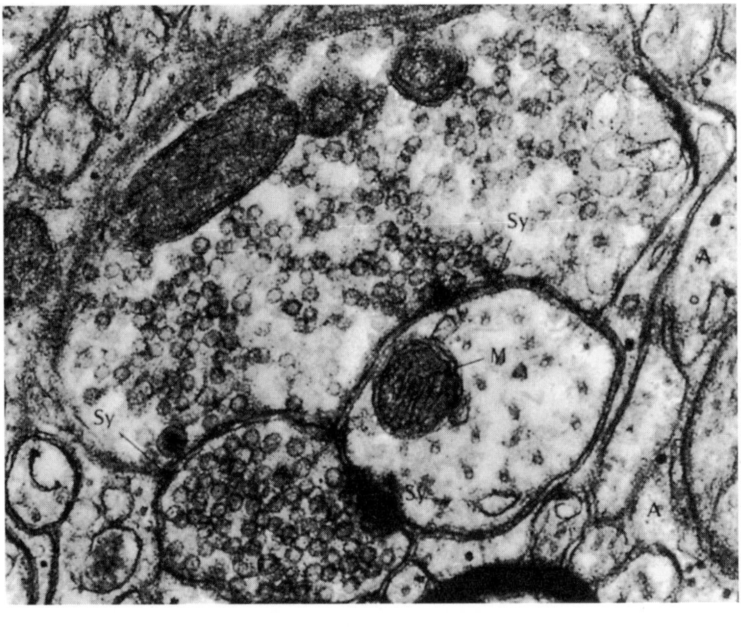

FIGURE 2–1. High-power view of two nerve terminals contacting a small dendritic spine containing one round mitochondrion (M). At this magnification, the synaptic vesicles can be seen clearly, as can the zones of specialized contact (Sy). Astrocyte processes containing glycogen (A) can be seen. Note that the larger nerve terminal makes a specialized contact on the small terminal (axoaxonic) as well as on the dendrite (axodendritic). ×12,000.

identification of many of the specific proteins that have precise functional properties in and on the surface of the synaptic vesicles and at sites along the inside of the presynaptic membrane associated with vesicle docking and release. Similarly, the direct chemical identification at discrete sites within the postsynaptic surface of the specialized contacts of receptor proteins and other proteins capable of modifying the response to neurotransmitters has given near consensus to the functional inference of a synaptic active zone. In some neurons, especially the single-process and small-granule cell types, the dendrite may also be structurally specialized to store and release transmitter.

Glia

A second element in the maintenance of the neuron's integrity depends on a type of cell known as *neuroglia*. There are two main types of neuroglia, together termed *macroglia*.

The first is called the *fibrous astrocyte*, a descriptive term based on its starlike shape when viewed in the light microscope and on the fibrous nature of its cytoplasmic organelles, which can be seen with both light and electron microscopy. Astrocytes are found mainly in regions of axons and dendrites; they tend to surround and closely contact the adventitial surface of blood vessels. Functions such as insulation (between conducting surfaces) and organization (to surround and separate functional units of nerve endings and dendrites) have been empirically attributed to the astrocyte, mainly on the basis of its structural characteristics. However, studies of carbohydrate-metabolizing properties of astrocytes in culture have demonstrated their capacity to accumulate glucose, synthesize glycogen, and provide two-carbon energy substrates to neurons. Furthermore, the link between glucose uptake into astrocytes is also activity-dependent and regulated by extracellular cations, as well as by at least some neurotransmitters. These findings have led to the suggestion that the regional localization of glucose uptake, as demonstrated in the human brain by positron emission tomography and in the brain of other species by 2-deoxyglucose autoradiography, reflects primarily the uptake of glucose by astrocytes and not by neurons.

The second major type of neuroglia is known as the *oligodendrocyte*. It is called the *satellite cell* when it occurs close to nerve cell bodies and the *Schwann cell* when it occurs in the peripheral nervous system. The cytoplasm of the oligodendrocyte is characterized by rough endoplasmic reticulum, but its most prominent characteristic is the enclosure of concentric layers of its own surface membrane around the axon. These concentric layers come together so closely that the oligodendrocyte cytoplasm is completely squeezed out and the original internal surfaces of the membrane become fused, presenting the ringlike appearance of the myelin sheath in cross section. Along the course of an axon, which may be many centimeters, many oligodendrocytes are required to constitute its myelin sheath. At the boundary between adjacent portions of the axon covered by separate oligodendrocytes, there is an uncovered axonal portion known as the node of Ranvier.

Many central axons and certain elements of the peripheral autonomic nervous system do not possess myelin sheaths. Even these axons, however, are not bare or exposed directly to the extracellular fluid; rather, they are enclosed within single invaginations of the astrocyte surface membrane. Because of this close relationship between the conducting portions of the nerve cell, its axon, and the astrocyte, it is easy to see the origin of the proposition that the astrocyte may contribute to the nurture of the nerve cell.

There is yet a third nonneuronal cell class in the brain, termed *microglia*, that is gaining increasing importance as a pharmacological target. Microglia are of mesodermal origin and related to macrophages and monocytes. Some microglia reside within the brain, often around blood vessels. During events that cause tissue necrosis, such as stroke, trauma, or infection, however, macrophages enter the brain and secrete chemical signals (cytokines, see Chapter 11) to recruit lymphocytes and leukocytes to seal off and repair the tissue damage.

Brain Permeability Barriers

While the unique cytological characteristics of neurons and glia are sufficient to establish the complex intercellular relationships of the brain, there is yet another histophysiological concept to consider. Numerous chemical substances pass from the bloodstream into the brain at rates that are far slower than for entry into all other organs in the body. There are similar slow rates of transport between the cerebrospinal fluid and the brain, although there is no good standard in other organs against which to compare this latter movement.

These permeability barriers appear to be the end result of numerous contributing factors that present diffusional obstacles to chemicals on the basis of molecular size, charge, solubility, and specific carrier systems. The difficulty has not been in establishing the existence of these barriers but in determining their mechanisms. When the relatively small protein horseradish peroxidase (molecular weight = 43,000 daltons), is injected intravenously into mice, its eventual location within the tissue can be demonstrated histochemically with the electron microscope. As opposed to the easy transvascular movement of this substance across muscle capillaries, in the brain the peroxidase molecule hardly penetrates the continuous layer of vascular endothelial cells at all. The endothelial cells of brain capillaries differ from those of other tissues in that the intercellular zones of membrane apposition are much more highly developed in the brain and virtually continuous along all surfaces of these cells. Furthermore, cerebral vascular endothelial cells lack pinocytotic vesicles, considered to be the transvascular carrier systems of both large and small molecules in other tissues. Recent studies suggest that a very reduced transcytosis may occur in brain capillary endothelial cells.

Since the enzyme marker can barely go through or between the endothelial cells, an operationally defined barrier exists. Whether the same barrier is also applicable to highly charged lipophobic small molecules cannot be determined from these observations. As neuropharmacologists, what concerns us more are the factors that retard the entrance of these smaller molecules, such as norepinephrine and serotonin, their amino acid precursors, or drugs that affect the metabolism of these and other neurotransmitters. Charged molecules can, however, diffuse widely through the extracellular spaces of the brain when permitted entry via the cerebrospinal fluid.

Astrocytes are currently thought to elicit expression of the proteins that constitute the blood–brain barrier in cerebral capillary endothelial cells. Recent research suggests that a model blood–brain barrier system may be attainable under culture conditions. So long as central astrocytes are viable, brain fragments experimentally transplanted to vascular beds, such as the anterior chamber of the eye, that would normally lack a blood–brain barrier, nevertheless retain a functional barrier when revascularized. This suggests that the properties of the barrier reside not in the endothelial cells themselves but in some functional response to the adjacent central astrocytes.

Substances that have difficulty entering the brain, in general, also have difficulty leaving it. Thus, when monoamines are increased in concentration by blocking their catabolism (see Chapter 9), high levels of amine persist until the inhibiting agents are metabolized or excreted. One such excretory route is the acid-transport system, by which the choroid plexus and/or brain parenchymal cells actively secrete acid catabolites, as well as drugs such as penicillin or zidovudine, which one might like to keep in. This step can be blocked by the drug probenecid, resulting in increased brain and cerebrospinal fluid amine catabolite and drug levels.

The choroid plexus may also be an interface between the peripheral vascular system and the immune response system, where antigens for which immunosurveillance is required can be recognized and immune responses mounted. Lymphocytes may normally "wander" through the brain in search of immune targets, although how they can slip through the endothelial cell barriers without reducing their normal barrier functions remains unclear.

Since the precise nature of these barriers still cannot be formulated, students would be wise to avoid the "great wall of China" concept and lean toward the possibility of a series of variously placed, progressively selective filtration sites that discriminate substances on the basis of several molecular characteristics. With lipid-soluble, weak electrolytes—a characteristic of most centrally acting drugs—transport occurs by a process of passive diffusion. Thus, a drug will penetrate the endothelial cell only in the undissociated form and at a rate consonant with its lipid solubility and its pK_a (negative logarithm of acid ionization constant).

Specialized sets of neurons known as the circumventricular organs exist within discrete sites along the linings of the cerebroventricles but are functionally on the blood side of the blood–brain barrier. Considered to be "windows" through which the normally excluded central milieu can monitor the components of the bloodstream, these neurons can communicate directly with neurons well within the enclosure of the blood–brain barrier.

Bioelectric Properties of the Nerve Cell

Given these structural details, we can now turn to the second striking feature of nerve cells, namely, their bioelectric property. However, even for this introductory presentation, we must understand certain basic concepts of the physical phenomena of electricity in order to have a working knowledge of the bioelectric characteristics of living cells.

The initial concept is that of a difference in potential existing within a charged field, as occurs when charged particles are separated and prevented from randomly redistributing themselves. When a potential difference exists, the amount of charge per unit of time that will flow between the two sites (i.e., *current flow*) depends on the resistance separating them. If the resistance tends to 0, no net current will flow since no potential difference can exist in the absence of a measurable resistance. If the resistance is extremely high, only a minimal current will flow; and that will be proportional to the electromotive force or potential difference between the two sites. The relationship between voltage, current, and resistance is Ohm's law: $V = I \cdot R$.

When we come to measuring the electrical properties of living cells, these basic physical laws apply but with one exception. The pioneer electrobiologists, who did their work before the discovery and definition of the electron, developed a convention for the flow of charges based not on the electrons but on the flow of positive charges. Therefore, since in biological systems the flow of charges is not carried by electrons but by ions, the direction of flow is expressed in terms of the movement of positive charges. To analyze the electrical potentials of a living system, we use small electrodes (a microprobe for detecting current flow or potential), electronic amplifiers for increasing the size of the current or potential, and oscilloscopes or polygraphs for displaying the potentials observed against a time base.

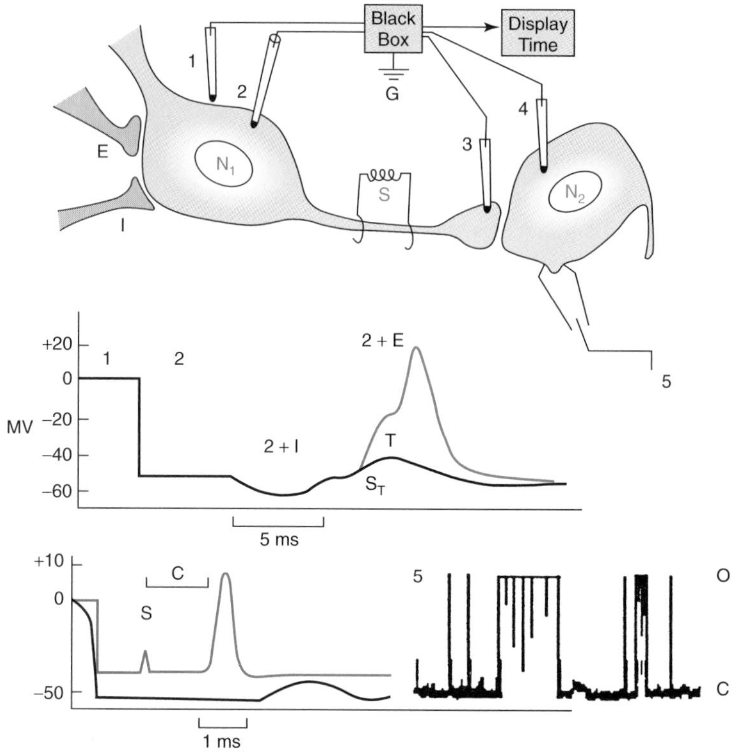

Figure 2–2. At the *top* is shown a hypothetical neuron (N_1) receiving a single excitatory pathway (E) and a single inhibitory pathway (I). A stimulating electrode (S) has been placed on the nerve cell's axon; microelectrode 1 is extracellular to nerve cell 1, while microelectrode 2 is in the cell body, and microelectrode 3 is in its nerve terminal. Microelectrode 4 records from within postsynaptic cell 2. The potentials and current, recorded by each of these electrodes, are compared through a "black box" of electronics with a distant extracellular grounded electrode and displayed on an oscilloscope screen. When the cell is resting and the electrode is on the outside of the cell, no potential difference is observed (1). In the resting state, electrode 2 records a steady potential difference between inside and outside of approximately −50 mV (2). While recording from electrode 2 and stimulating the inhibitory pathway, the membrane potential is hyperpolarized during the inhibitory postsynaptic potential (2 + I). When recording from electrode 2 and stimulating the excitatory pathway, a subthreshold stimulus (S_T) produces an excitatory postsynaptic potential indicated by a brief depolarization of the resting membrane potential (2 + E). When the excitatory effects are sufficient to reach threshold (T), an action potential is generated which reverses the inside negativity to inside positivity (2 + E). On the lower scale, potentials recorded by electrodes 3 (blue line) and 4 (black line) are compared on the same time base following axonal stimulation of nerve cell 1, which is assumed to be excitatory. The point of stimulus is seen as an electrical artifact at point S. The action potential generated at the nerve terminal occurs after a finite lag period due to the conduction time (C) of the axon between the stimulating

Membrane Potentials

If we take two electrodes and place them on the outside of a living cell or tissue, we will find little, if any, difference in potential. However, if we injure a cell so as to break its membrane or insert one ultrafine electrode across the otherwise intact membrane, we will find a potential difference such that the inside of the cell is 50 mV or more negative with respect to the extracellular electrode (Fig. 2–2). This transmembrane potential difference has been found in almost all types of living cell in which it has been sought; such a membrane is said to be *electrically polarized.* By passing negative ions into the cell through the microelectrode (or extracting cations), the inside can be made more negative (*hyperpolarized*). If positive current is applied to the inside of the cell, the transmembrane potential difference is decreased and the potential is said to be *depolarized.* The potential difference across the membrane of most living cells can be accounted for by the relative distribution of the intracellular and extracellular ions.

The extracellular fluid is particularly rich in sodium (Na) and relatively low in potassium (K). Inside the cell, the cytoplasm is relatively high in K content and very low in Na. While the membrane of the cell permits K ions (K^+) to flow back and forth with relative freedom, it resists the movement of Na ions (Na^+) from the extracellular fluid to the inside of the cell. Since K^+ can cross the membrane, they tend to flow along the concentration gradient, which is highest inside the cell. K^+ diffusion out of the cell leaves a relative negative charge behind, owing to the negative charges of the macromolecular proteins. As the negative charge inside the cell begins to build, the further diffusion of K^+ from inside to outside is retarded. Eventually, an equilibrium point will be reached that is proportional to certain physical constants and to the relative concentrations of intracellular and extracellular K^+ and chloride (Cl^-) ions. These concepts of ionic diffusion potentials across

---

electrode and the nerve terminal. The action potential in the nerve ending does not directly influence postsynaptic cell 2 until after the transmitter has been liberated and can react with nerve cell 2's membrane, causing the excitatory postsynaptic potential. The time between the beginning of the action potential recorded by microelectrode 3 and the excitatory postsynaptic potential recorded by electrode 4 is the time required for excitation secretion coupling in the nerve terminal and the liberation of sufficient transmitter to produce effects on nerve cell 2. Electrode 5 is a patch clamp electrode attached to neuron 2; it indicates the effects of transmitters acting through receptors located elsewhere on the neuron's surface, mediating channel opening (O) and closing (C) events through intracellular second messengers.

semipermeable membranes apply generally not only to nerve and muscle but also to blood, glandular, and other cells large enough to have their transmembrane potential measured.

Membrane Ion Pumps

When the nerve cell or muscle fiber can be impaled by electrodes to record transmembrane potential, the relation between the membrane potential and external K^+ concentration can be directly tested by exchanging the extracellular fluid for artificial solutions of varying K concentration. When this experiment is performed on muscle cells, we find that the membrane potential bears a linear relationship to the external K concentration at normal to high K concentrations but that it deviates from this linear relationship when the external K concentration is less than normal. To account for this discrepancy, we must reexamine an earlier statement. While the plasma membranes of nerve and muscle cells and other types of polarized cell are relatively impermeable to the flow of Na ions along the high concentration gradient from extracellular to intracellular, they are not completely impermeable.

Radioisotope experiments can establish that a certain amount of Na "leaks" into the resting cell from outside. The amount of measurable Na entering the cell is sufficient to double the intracellular Na concentration in approximately 1 hour if there is not some opposing process to maintain the relatively low intracellular Na concentration. The process that continuously maintains the low intracellular Na concentration is known as *active Na transport* or, colloquially, as the *sodium pump*. This pump mechanism ejects Na from the inside of the cell against the high concentration and electrical gradients forcing it in. However, the pump does not handle Na exclusively but requires the presence of extracellular K^+. Thus, when a Na ion is ejected from the cell, a K ion is incorporated into the cell, giving the mechanism yet another apt name, the Na–K ATPase exchanger.

When the external K^+ concentration is near normal, the transmembrane potential, which is based mainly on K concentration differences, behaves as if there were actually more extracellular K than really exists. This is because the Na–K exchange mechanism elevates the amount of K coming into the cell. Remember that K permeability is relatively high and that K^+ tends to diffuse out of the cell because of its concentration gradient but to diffuse into the cell because of charge attraction. Therefore, two factors operate to drive K^+ into the cell in the presence of relatively low external K^+ concentration: *(1)* the electrical gradient across the membrane and *(2)* the Na–K pump mechanism. The latter system could be considered electrogenic since at low external K^+ concentrations it modifies the electrical status of the muscle membrane. Other metabolic pumps simply exchange cationic species across the membrane and are nonelectrogenic. The relative electrogenicity of a pump may depend on the ratio of the exchange cations (i.e., 1:1, 2:2, or 3:2). The pump is immediately dependent on metabolic energy and can be blocked by several metabolic poisons, such as dinitrophenol and the rapid-acting cardiac glycoside ouabain. Astrocytes have such pumps too.

The Uniqueness of Nerves

All that we have said regarding the transmembrane ionic distributions applies equally to the red blood cell or glia as to the neuron. Thus, possession of a transmembrane potential difference is not sufficient to account for the neuron's bioelectric properties. However, applying depolarizing currents across the membrane can bring out the essential difference between the red blood cell and the nerve cell. When the red blood cell membrane is depolarized, the difference in potential across the cell passively follows the imposed polarization. However, when a nerve cell membrane, such as the giant axon of an invertebrate, is depolarized from a resting value of approximately −70 mV to approximately −10 to −15 mV, an explosive self-limiting process occurs, namely, the action potential. In the *action potential*, the transmembrane potential is reduced not merely to 0 but beyond 0 so that the inside of the membrane now becomes positive with respect to the outside. This overshoot may extend for 10 to 30 mV in the positive direction. Because of this explosive response to an electrical depolarization, the nerve membrane is said to be "electrically excitable," and the resultant voltage polarity shift is the action potential. While astrocytes can also show variable membrane potentials, they do not exhibit this excitability.

Analysis of Action Potentials

In an elegant series of pioneering experiments that are now classic, Hodgkin, Huxley, and Katz analyzed the various ionic steps responsible for the action potential. When the cell begins to depolarize in response to stimulation, the current flow across the membrane is carried by K. As the membrane becomes more depolarized, the resistance to Na decreases (i.e., Na conductance increases) and more Na enters the cell along its electrical and concentration gradients. As Na enters, the membrane becomes more and more depolarized, which further increases the conductance to Na and thus depolarizes the membrane more and faster. Such conductance changes are "voltage-dependent." This self-perpetuating process continues, driven by the flow of Na ions moving toward their equilibrium distribution, which should be proportional to the original extracellular and intracellular concentrations of Na.

However, the peak of the action potential does not attain the equilibrium potential predicted on the basis of transmembrane Na concentrations because of a second phase of events. The voltage-dependent increase in Na^+ conductance and the consequent depolarization also activate a voltage-dependent K^+ conductance, and K flow then increases along its concentration gradient from inside to outside the cell. This process restricts the height of the reversal potential since it tends to maintain the inside negativity of the cell and begins to reduce the membrane conductance to Na, thus making the action potential a self-limiting phenomenon. In most nerve axons, the action potential lasts for approximately 0.2 to 0.5 milliseconds, depending on the type of fiber and the temperature at which it is measured.

Once the axon has been sufficiently depolarized to reach threshold for an action potential, the wave of activity travels at a rate proportional to the diameter of the axons (through which the bioelectric currents will flow). In large axons, the rate is further accelerated by the insulation provided by myelin sheaths, restricting the flow of transmembrane currents to the opening at the nodes of Ranvier. Therefore, instead of the action potential propagating from minutely contiguous sites of the membrane, the action potential in the myelinated axon leaps from node to node. This saltatory conduction is consequently much more rapid.

The threshold level for an all-or-none action potential is also inversely proportional to the diameter of the axon: large myelinated axons respond to low values of imposed stimulating current, whereas fine and unmyelinated axons require much greater depolarizing currents. Local anesthetics block activation of Na conductance, preventing depolarization.

Once the threshold has been reached, a complete action potential will not develop if it occurs too quickly after the preceding action potential, an interval termed the *refractory phase*. This phase varies for different types of excitable nerve and muscle cell and appears to be related to the activation process that increases Na conductance, a phenomenon that has a finite cycling period. That is, the membrane cannot be reactivated before a finite interval of time has occurred. K conductance increases with the action potential and lasts slightly longer than the activation of Na conductance. This results in a prolonged phase of after-hyperpolarization due to the continued redistribution of K from inside to outside the membrane. If the axonal membrane is artificially maintained at a transmembrane potential equal to the K^+ equilibrium potential, no after-hyperpolarization can be seen.

Ion Channels

The experiments of Hodgkin and Huxley defined the kinetics of cation movement during nerve membrane excitation without constraint on the mechanisms accounting for the movement of ions through the membrane. The discovery that drugs can selectively block cation movement and that Na^+ permeability (blocked by tetrodotoxin) can be separated from K^+ permeability (blocked by tetraethylammonium [TEA]) made more detailed analysis of ion movement mechanisms feasible. Membrane physiologists now agree that there are several ion-specific pathways that form separate and independent "channels" for passive movement of Na^+, K^+, Ca^{2+}, and Cl^-. Thanks to several spectacular advances in the molecular biology of ion channels, it is becoming clear that these conductance mechanisms are actually quite complex arrays of interacting proteins. In some cases, the channel proteins open and close in an all-or-nothing fashion on time scales of 0.1 to 10 milliseconds to provide aqueous channels through the plasma membrane that ions can traverse. More precisely, channel macromolecules can exist in several interconvertible conformations, only one of which permits ion movement. The conformational shifts from one form to another are sensitive to the bioelectric fields operating on the membrane; by facilitating or retarding the conformational shifts, the ion channels are "gated." In this concept, Ca^{2+} acts at the membrane surface to alter permeability only by virtue of the effect its charge has on the electric fields of otherwise fixed (mainly negative) organic charges. The altered fields in turn can gate the channels because a part of the channel protein is able to sense the field and, thus, modulate the conformational shifts that open or close the gate. When ions flow across the membrane, the ionic current changes the membrane potential and other membrane properties.

From a variety of experimental methods, including those that can sample single ion channels in cultured neurons and other excitable cells, a large number of specific ion channels have been described. The current terminology recognizes four types of ion channel: *(1)* nongated, passive ion channels, previously referred to as "leakage channels," which are continuously open; *(2)* voltage-gated (i.e., voltage-sensitive) channels, in which channel opening and closing is affected by the membrane potential inside of the cell; *(3)* chemically gated channels, the opening and closing of which is affected by receptors on the external plasma membrane, such as those affected by drugs and other transmitters; and *(4)* ion-gated channels, whose opening and closing is affected by shifts in intracellular ion concentrations. Ion-gated channels are often also sensitive to membrane potential and to external regulatory receptors, and chemically gated channels are often also voltage-sensitive. These various modes of interaction provide an extremely rich spectrum of responses, thus greatly complicating what were once simple rules of excitability and ion conductance regulation. Additional forms of more complex ion flow regulation provide the means by which neurons communicate to their target cells through junctional transmission. As will be seen below, advances in the molecular biology of the ion channels have revealed important principles of their structure and function.

Junctional Transmission

While these ionic mechanisms appear to account adequately for the phenomena occurring in the propagation of an action potential down an axon, they do not per se explain what happens when the action potential reaches the nerve ending. At the nerve ending, the membrane of the axon is separated from the membrane of the postjunctional nerve cell, muscle, or gland by an intercellular space of 50 to 200 Å (Fig. 2–1). When an electrode can be placed in both the terminal axon and the postsynaptic cell, depolarization of the nerve terminal does not result in a direct instantaneous shift in the transmembrane potential of the postsynaptic element, except in cases where the connected cells are electronically coupled. With this exception, the junctional site seldom exhibits direct electrical excitability like the axon.

Postsynaptic Potentials

With the advent of microelectrode techniques for recording the transmembrane potential of nerve cells in vivo, it became possible to determine the effects of stimulation of nerve pathways that had previously been shown to cause either excitation or inhibition of synaptic transmission. From such studies, Eccles (1964) observed that subthreshold excitatory stimuli would produce postsynaptic potentials with time durations of 2 to 20 milliseconds. The excitatory postsynaptic potentials algebraically accumulate with other contemporary excitatory and inhibitory postsynaptic potentials. Most importantly, the duration of these postsynaptic potentials is longer than can be accounted for on the basis of electrical activity in the preterminal axon or on the electronic conductive properties of the postsynaptic membrane (Fig. 2–2). This latter observation combined with the fact that synaptic sites are not directly electrically excitable provides conclusive evidence that central synaptic transmission must be chemical: the prolonged time course is compatible with a rapidly released chemical transmitter whose time course of action is terminated by local enzymes, diffusion, and reuptake by the nerve ending.

By such experiments, it was possible to work out the basic ionic mechanisms for inhibitory and excitatory postsynaptic potentials. When an ideal excitatory pathway is stimulated, the presynaptic element liberates an excitatory transmitter, which activates ionic conductance of the postsynaptic membrane. This response leads to an increase in one or more transmembrane ionic conductances, depolarizing the membrane toward the Na equilibrium potential. In the resting state, as has already been discussed, the membrane resides near the K equilibrium potential. If the depolarization reaches the threshold for activating adjacent voltage-dependent conductances, an all-or-none action potential (spike) will be triggered. For many neurons, the axon hillock has the lowest spike threshold. If the resultant depolarization is insufficient to reach threshold, the cell can still discharge if additional excitatory postsynaptic potentials summate to threshold.

The postsynaptic potential resulting from the stimulation of an ideal inhibitory pathway to the postsynaptic cell has been explained in terms of the fact that an inhibitory transmitter selectively activates channels for Cl^- or K^+, resulting in diffusion of ions and hyperpolarization of the membrane. This counterbalances the excitatory postsynaptic potentials.

Because the sites of synaptic or junctional transmission are electrically inexcitable, the postsynaptic membrane potential can be maintained at various levels by applying current through intracellular electrodes and changing the intracellular concentrations of various ions. By such maneuvers, it is possible to poise the membrane at or near the so-called equilibrium potentials for each of the ionic species and to determine the ionic species whose equilibrium potential corresponds to the conductance change caused by the synaptic transmitter. This is the most molecular test for the identification of synaptic transmitter substance actions. (However, certain objections can be raised to this test in terms of those nerve endings making junctional contacts on distal portions of the dendritic tree. Here, the postsynaptic potentials may be incompletely transmitted to the cell body, where the recording electrode is placed.)

On re-reading the above section, note the use of the term *ideal.* It is generally considered that classically acting neurotransmitters produce their effects on receptor-coupled ion conductances that are voltage-independent; that is, the receptor will alter the coupled ion channel regardless of the membrane potential at the moment. Nevertheless, many nonclassic transmitters seem to operate on receptors coupled to voltage-sensitive mechanisms. Transmitters whose receptors are associated with intracellular second-messenger systems (e.g., activation of cyclic nucleotide synthesis) frequently produce these more complex forms of interaction (see Chapter 5). Similarly, many neuropeptides appear to affect certain of their target cells by modifying responses to other transmitters, while not showing any direct shifts in membrane potential or conductance when tested for actions on their own. For example, the β-adrenergic actions of locus ceruleus neurons on their central targets produce excitability changes that depend on which other afferent systems are activated synchronously (see Conditional Actions of Transmitters, below).

Two other aspects of ionic mechanisms bear some mention. First, despite our preoccupation with action potentials and their modification, Bullock has pointed out that the most numerous central nervous system neurons, the small single-process type of granule-like cell, may conduct its neuronal business within its restricted small spatial domain with no need ever to fire a spike. Second, those neurons that do fire spikes may sometimes do so unconventionally, using an influx of Ca^{2+} ions (voltage-sensitive Ca conductance) rather than Na^+. This Ca spike may represent a mechanism to transmit activity from the cell body out to the dendritic system and may play a functional role in those neurons whose dendrites can also release transmitter, such as the catecholamine cell body nuclei. Thus, the simplified ideal version of ionic mechanisms may be only one of many regulatory mechanisms between connected cells. In addition, through the use of ion-sensitive fluorescent dyes, such as fura-2, it has been possible to demonstrate that astrocytes in long-term cell cultures not only can exhibit dynamic ionic responses very much like those of neurons but also can do so in waves of coordinated activity. The student is advised to maintain an appreciative awareness of these potentially complex interaction systems. In the following section, we examine some of the less classic synaptic events and their advantageous properties.

During the 1980s, it became fashionable to accept transmitter response data only when they were recorded from stable neurons in vitro. In addition to slices of mammalian brains, other preferred systems included "model" neurons (i.e., neuronally derived cell lines, neuron–glia hybrid cells, and endocrine cells of normal or tumor origin previously exploited in lieu of real neurons). As a result, many experimental findings previously obtained with great difficulty from intact living brains were neglected in deference to the new, simpler preparations. Still beyond the pale, given the frenzy to work in vitro, is the reexamination of these transmitter-related effects in the intact nervous system, where their role in specific circuitry operations and interactions could actually be evaluated under living conditions.

The transductive mechanisms now considered to be part of the normal repertoire of regulatory processes include a variety of transmitter-regulated (transmitter-dependent) and voltage-sensitive ionic conductances that were formerly regarded as unusual for the mammalian central nervous system despite their nearly simultaneous demonstration in invertebrate ganglia, mammalian autonomic ganglia, and selected central mammalian neurons. Most of these unusual transmitter-regulated conductance mechanisms relate to an unexpectedly large number of distinctive ionic conductances for Ca^{2+} and K^+, with more modest expansions in the channels for Cl^- or Na^+ currents.

Given the onslaught of molecular biological characterizations for ion channels (see Chapter 3), it is important for the student to recognize that within a specific functional category of ion conductance (i.e., Na^+, K^+, Ca^+), there are subtypes of functional responses that are ligand-specific and that may be carried out by more than one ion channel protein (ionophore) complex. These precisely defined channel proteins can now be examined in intimate molecular detail to dissect how drugs, toxins, and imposed voltages can alter the excitability of a neuron.

Calcium Channels • Among the multiple voltage-sensitive Ca^{2+} conductances described in neurons, three are most consistent. The first is a transient, low-threshold Ca^{2+} conductance (T). This Ca^{2+} conductance is inactive at resting membrane potentials but is "deinactivated" by modest hyperpolarizations, providing a feature of oscillatory behavior. It is most frequently inhibited by Cd^{2+} or Co^{2+} and, in some cases, by Ni, Mg, or Mn as well; it can be activated by Ba^{2+}. The second Ca^{2+} conductance channel is a slowly inactivating, high-threshold Ca^{2+} conductance (L) seen mainly in nerve terminals. The third is a transient, high-threshold Ca conductance (N) observed in the soma and dendrites of large neurons in the neocortex, olfactory cortex, and hippocampal formation. The latter are blocked by Mn^{2+}, Co^{2+}, and Cd^{2+} and activated by Ba^{2+} and TEA; these responses may be inhibited functionally by endogenous purinergic receptors. The N-type Ca^{2+} channels have also been well studied in sympathetic neurons, where they are regulated through three separate transductive pathways, each of which may be engaged by different neurotransmitters and their specific intracellular mechanisms (see Chapter 5).

Potassium Channels • At least three types of K channel have been described in central neurons: *(1)* the "A" or "A-like" fast, transient K conductances inhibited by 4-aminopyridine, Ba^{2+}, or Co^{2+}; *(2)* the so-called anomalous rectifying K conductances (see below), of which the M current (closed by cholinergic muscarinic receptors) is one example; *(3)* the Ca-activated K conductances, blocked by Co, Mn, Cd, and some neurotransmitters (see Siggins and Gruol, 1986, and Hille, 1992). Although most data on these K^+ channel effects are pharmacological, the properties can clearly regulate cell firing and response patterns in distinct manners. By closing the M current, muscarinic receptors transduce cholinergic signals into more effective depolarization, once partial depolarization brings this channel into play. Somatostatin can oppose this effect, forcing the M channel to open. The latter effect may be mediated intracellularly by second messengers derived from arachidonic acid metabolism (see Chapter 5). By blocking the Ca-activated K channel of central neurons, the transmitter receptors for β-adrenergic agonists, 5-hydroxytryptamine, histamine, and corticotropin-releasing hormone enhance the ability of responsive neurons to follow long depolarizing pulses, thereby generating longer trains of spikes per afferent impulse.

Depending on the specific cells in which they were recognized (even Ca-activated K channels exist in many glands and completely nonneural cell types) and the conditions and possible inhibitors that may have been evaluated, as many as 12 different K conductances have been proposed. For example, many K channels are linked to second-messenger mediation (e.g., the channels activated by $GABA_B$ receptors and by D_2 dopamine receptors on rat substantia nigra neurons). With patch-clamp analyses of single K channels in locus ceruleus neurons acutely isolated from 1- to 7-day-old rats, opioids (at μ receptors), somatostatin, and α_2-adrenergic agonists seem to open a K channel that is not voltage-dependent and is directly regulated by receptor occupancy through a G protein (see Chapter 4) but with no known intervening second messenger.

Other Ion-Specific Channels • A "persistent" Na^+ conductance was first observed in cerebellar Purkinje neurons and later in hippocampal pyramidal neurons, as well as neurons throughout the neuraxis. This conductance provides long-lasting but low-amplitude depolarization, which does not lead directly to neuronal firing but, rather, provides a bias from which the conventional fast Na channels can produce full spike initiation. Persistent Na channels are typically blocked by tetrodotoxin and activated by TEA.

Slow Postsynaptic Potentials

Most of the postsynaptic potentials described by Eccles (1964) were relatively short, usually 20 milliseconds or less, and appeared to result from passive changes in ionic conductance. Postsynaptic potentials of slow onset and several seconds' duration have been described (Fig. 2–2), both of a hyperpolarizing nature and of a depolarizing nature. While such prolonged postsynaptic potentials could be the result of either prolonged release of transmitter or persistence of the transmitter at postsynaptic receptor sites, there is substantial support for the possibility that slow postsynaptic potentials could also be caused by other forms of synaptic communication. Many of these slow synaptic potentials are not accompanied by the expected increase in transmembrane ionic conductances but instead by increased transmembrane ionic impedance. Although multiple hypothetical explanations have been offered for such responses, the actual molecular mechanisms remain obscure. Among the more promising leads, transmitters such as the catecholamines can activate the synthesis of cyclic nucleotides, which in turn can activate intraneuronal protein kinases, which can phosphorylate specific membrane proteins (see Chapter 5). The phosphorylation of a membrane-mounted ion channel protein would be expected to alter its ionic permeability, and perhaps such changes lie at the root of the membrane effects of several types of neurotransmitter (also see Chapter 12).

Conditional Actions of Transmitters

Frequently, transmitters produce novel actions unlike those of classically conceived transmitters. These unconventional actions suggest that broader definitions are useful for conceptualizing the range of regulatory signals involved in interneuronal communication and for examining transmitter actions. For example, when the β-adrenergic effects of locus ceruleus stimulation are examined, target cell responses no longer adhere to standard concepts of inhibition. Rather, they appear to fit better the designation of "biasing" or "enabling." The latter indicates that the enabling transmitter (in this example, norepinephrine) can enhance or amplify the effectiveness of other transmitter actions converging on the common target neurons during the time period of the enabling circuit's activity (see Chapter 8). These β-adrenergic actions can enhance either excitatory or inhibitory afferents, a general effect referred to as *enabling* or, more ambiguously, *modulatory*. Some pharmacological actions of neuropeptides have been described as having the opposite effect, or *disenabling* (e.g., the effects of opioid peptides on the excitatory actions of sensory transmitters within the spinal cord) (see Chapter 11). This story is more complex (surprised?) because neuropeptides coexist with amino acid and amine transmitters.

To reexplore the issue of time course on the more complex interactions, it may be useful to speak of conditional and unconditional actions. *Unconditional actions* are those that a given transmitter evokes by itself (i.e., in the absence of other transmitters acting on the common target cell). *Conditional actions*, occurring either pre- or postsynaptically, include, but would not be limited to, the type of enhancement that is subsumed by enabling. In such a conditional interaction, each transmitter would act at its own pre- or postsynaptic transmitter receptor and interact on that target cell when both transmitters occupy their receptors simultaneously.

Thus, there are abundant circuits, abundant transmitters, and, for each of these, many classes of chemically coupled systems that can transduce the effects of active transmitter receptors. These receptors can operate either actively or passively, conditionally or unconditionally, over a wide range of time through nonspecific, dependent, or independent metabolic events. Clearly, neurons have a broad but finite and as yet incompletely characterized repertoire of molecular responses that messenger molecules (transmitters, hormones, and drugs) can elicit. The power of the chemical vocabulary of such components is their combinatorial capacity to act conditionally and coordinately and to integrate the temporal and spatial domains within the nervous system.

Transmitter Secretion

We have already seen that the cellular machinery of the neuron suggests that it functions as a secretory cell. Secretion of synaptic transmitters is the activity-locked expression of neuronal activity induced by depolarization of the nerve terminal. Recently, it has been possible to separate the excitation–secretion coupling process of the presynaptic terminal into at least two distinct phases. This has been made possible through analysis of the action of the puffer fish poison tetrodotoxin, which blocks the electrical excitation of the axon but does not block the release of transmitter substance from the depolarized nerve terminal. The best of these experiments have been performed in the giant synaptic junctions of the squid stellate ganglion, in which the nerve terminals are large enough to be impaled by recording and stimulating microelectrodes and with recording from the postsynaptic and presynaptic neurons. In this case, when tetrodotoxin blocks conduction of action potentials down the axon, electrical depolarization of the presynaptic terminal still results in the appearance of an excitatory postsynaptic potential in the ganglion neuron. Since tetrodotoxin selectively blocks voltage-dependent Na^{+} conductance, the excitation secretion must be coupled more closely to other ions. Present evidence strongly favors the view that voltage-sensitive Ca^{2+} conductance is required for transmitter secretion. Thus, the spike-generating and conducting events rest on voltage-dependent ion conductance changes, while synaptic events rest on voltage-independent or voltage-sensitive conductance.

Biochemical, ultrastructural, and physiological experiments have led to the concept that transmitter molecules are stored within vesicles in the nerve terminal and that the Ca-dependent excitation–secretion coupling within the depolarized nerve terminal requires the transient exchange of vesicular contents into the synaptic cleft. It is unclear whether the vesicle simply undergoes rapid fusion with the presynaptic specialized membrane to allow the transmitter stored in the vesicle to diffuse out or whether the process of exocytotic release simultaneously requires insertion of the vesicle membrane into the synaptic plasma membrane, reappearing later by the reverse process, namely, endocytosis. Information on the lipid and protein components of the two types of membrane once suggested that long-term fusion–endocytosis cycles were unlikely, but more recent data are compatible with either fusion release or contact release. In noradrenergic vesicles, for example, the transmitter is stored in very high concentrations in ternary complexes involving ATP, Ca, and possibly additional lipids or lipoproteins. Unfortunately, neurochemically homogeneous vesicles from central synapses have never been completely purified, and therefore all such analyses remain somewhat open to interpretation. For other molecules under active consideration as neurotransmitters, storage within brain synaptic vesicles has been extremely difficult to document chemically. The difficulties arise from the fact that homogenization of the brain to prepare synaptosomes disrupts both structural and functional integrity, and under these conditions the failure to demonstrate that amino-acid transmitters are stored in vesicles is rationalized as uncontrollable leakage. With electron microscopy and autoradiography, however, sites accumulating transmitters for which there is a high-affinity, energy-dependent uptake process can be demonstrated. Under these conditions, authentic "synaptic terminals" are identified, but glial processes are also labeled. The vesicle story is further discussed in Chapter 8.

In some cases, release of the transmitter can be modulated "presynaptically" by the neuron's own transmitter (autoreceptors). Autoreceptors are conceived to be receptors that are generally distributed over the surface of a neuron and are sensitive to the transmitter secreted by that neuron. In the case of the central dopamine-secreting neurons, such receptors have been related to the release of the transmitter and to its synthesis. Such effects seem to be achieved through receptor mechanisms different from those by which the same transmitter molecule acts postsynaptically. Presynaptic release may also be modified (by receptors other than auto-receptors), by coreleased neuropeptides, or by the effects of transmitters released by other neurons in the vicinity of the terminal or the cell body.

Analysis of Membrane Actions of Drugs and Transmitters in Vitro

The development of methods for the nearly complete functional maintenance of central neurons in vitro for several hours, such as tissue slice preparations, and for several days to weeks in single-cell or explant culture systems has led to a proliferation of additional electrophysiological methods to examine transmitter and drug action. In slice preparations, neuronal targets can be readily localized by inspection and intracellular electrodes can be inserted into suitably large neurons under visual control, while additional stimulating electrodes may activate sources of afferent circuitry within the slice. Transmitters and drugs can then be applied to the whole slice by superfusion within oxygenated buffered salt solutions or more locally by micropressure pulse or iontophoretic application methods. In many cases, excellent intracellular recordings can be obtained for long periods of time because there are no annoying respiratory, cardiac, or other movements to dislodge the electrode.

When long-term intracellular recordings can be obtained, two additional sources of information on transmitter actions can be analyzed. In voltage-clamp analysis, the experimenter inserts one or two electrodes into the cell and by injecting current holds the membrane potential of the neuron at a constant value. The cell is usually poised at a membrane potential more negative than resting in order to prevent spontaneous spikes. Transmitter action is monitored by the amount of current required to keep the membrane potential constant and thus measures transmembrane current flow directly. However, since the membrane potential stays constant, any of the nonlinear properties of that neuron's response that could occur when sufficient depolarization has occurred will be prevented.

The clamp can also be quickly changed to a new level of membrane potential, and the neuron's responses to this shift provide the basis for a pharmacological dissection (e.g., with ion channel blockers or ion substitutions) of the degree to which the effects of a transmitter can be explained as ion dependent or voltage dependent. Many of the actions ascribed to neuropeptides and some of those ascribed to monoamines fit the concept of voltage dependent, since they are modest effects at best at resting membrane potential levels, but they emerge as more substantive effects when the responding neuron is depolarized or hyperpolarized by other convergent transmitters.

Another method, termed noise analysis, also examines ion channel activity more directly than the standard in vivo methods. This method assumes that ion channels are either open or closed and that they switch instantaneously, and do so independently, between the two conditions. Using intracellular electrodes, the fluctuations in membrane potential (or, if voltage clamping is used, the fluctuations in membrane current flow) are analyzed statistically to infer the conductance of individual types of channel and the mean time they are open in the absence or presence of the transmitter to be analyzed.

When the neurons considered to be the appropriate targets of a specific transmitter can be maintained in long-term tissue culture, patch-clamp analysis, the current superstar of membrane action analysis methods, can be applied. This method offers the ability to study the behavior of single-ion channels under conditions of almost unbelievable precision. Special "fire-polished" microelectrodes are placed on the neuron's surface, and a slight vacuum is applied to the pipette to attain a very tight junction with the exposed surface of the neuronal membrane, thus requiring near-nude neurons for best application. The resulting cell–electrode junction will have such a high electrical resistance (gigohms) that the patch of enclosed membrane within the microelectrode's tip will be essentially isolated from the rest of the cell. Current flowing within that patch can then be analyzed independently of the responses of the rest of the neuron. With state-of-the-art, low-noise amplifiers, current flow through individual channels can be monitored and transmitter actions evaluated in terms of open time, amplitudes (number of channels opened), and closing times.

In addition, with clever micromanipulations, patch clamps can be done in three configurations. In the *cell-attached* mode, the pipette is sealed to the intact cell and measurements are made with no further physical disruption. However, further application of slight vacuum allows the patch of enclosed membrane to be removed from the cell but with enclosed ion channels still viable and responsive. In the *inside-out* patch, the previously intracellular surface will be on the exterior of the sealed membrane patch and simulations of changes in intracellular ions or, for example, catalysts of protein phosphorylations can examine the ion channel for regulation. It is also possible to demonstrate that, with clever handling and further negative pressure before pulling the membrane patch off the cell, an *outside-out* patch can be obtained. Here, the original patch is ruptured, the perimeter remains attached, and then the surrounding external membrane segments reseal once they are excised from the cell surface. By placing the outside surface into solutions with differing doses of transmitter, drug, or ion channel toxin, it is possible to analyze very discrete, single-channel pharmacology.

An Approach to Neuropharmacological Analysis

The business of analyzing bioelectrical potentials can be very complicated, even when restricted to changes in single neurons or to small portions of contiguous neurons; but if we restrict our examination of centrally active drugs to the effects on single cells, we can ask rather precise questions. For example, does drug X act on resting membrane potential or resistance, on an electrogenic pump, or on the Na- or K-activation phase of the action potential? Or does it block or modulate the effects of junctional transmission between two specific groups of cells?

To employ the modern powerful and precise electrophysiological tools, we must determine first the most likely target neuron to study. Earlier neuropharmacologists were forced to rely on much coarser tools, such as the electroencephalogram or sensory evoked potentials measured by electrodes placed on the scalp. Those methods, at their best, could measure the population response of a group of neurons to a drug, something that single-unit analysis can do only after many single recordings are collated. Macroelectrode methods are receiving increased attention again, since, as noninvasive methods, they can be used to examine drug actions clinically.

Approaches

If, as modern-day neuropharmacologists, we are chiefly concerned with uncovering the mechanisms of action of drugs in the brain, there are several avenues along which we can organize our attack. We could choose to examine the way in which drugs influence the perception of sensory signals by higher integrative centers of the brain. This is compatible with a single-neuron and ionic conductance types of analysis, directed, say, at how drugs affect inhibitory postsynaptic potentials. Drugs that cause convulsions, such as strychnine, have been analyzed in this respect; but all types of inhibitory postsynaptic potential are not affected by strychnine.

A second basic approach would be to use both macroelectrodes and microelectrodes to compare the drug responses of single units and populations of units in the same brain region. This approach is clearly limited, however, unless we understand the intimate functional relations between the multiple types of cell found even within one region of the brain.

A third approach is also possible. We could choose to separate the effects of drugs between those affecting the generation of action potential and its propagation and those acting on junctional transmission. For this type of analysis, we must identify the chemical synaptic transmitter for the junctions to be studied. Many of the interpretative problems already alluded to can be attacked through this approach. Thus, as might be expected, there is likely to be more than one type of excitatory and inhibitory transmitter substance, and a convulsant drug may affect the response to one type of inhibitory transmitter without affecting another. Moreover, a drug may have specific regional effects in the brain if it affects a unique synaptic transmitter there. In fact, using this approach, it may be possible to find drug effects not directly reflected in electrical activity at all but related more to the catabolic or anabolic systems maintaining the required functional levels of transmitter. We conclude this chapter by considering the techniques for identifying the synaptic transmitter for particular synaptic connections. The chapters that follow are organized to present in detail our current understanding of putative central neurotransmitter substances.

Identification of Synaptic Transmitters

How then do we identify the substance released by nerve endings? The entire concept of chemical junctional transmission arose from the classic experiments of Otto Loewi, who demonstrated chemical transmission by transferring the ventricular fluid of a stimulated frog heart onto a nonstimulated frog heart, thereby showing that the effects of the nerve stimulus on the first heart were reproduced by the chemical activity of the solution flowing onto the second heart. Since the phenomenon of chemical transmission originated from studies of peripheral autonomic organs, these peripheral junctions have become convenient model systems for central neuropharmacological analysis.

Certain interdependent criteria have been developed to identify junctional transmitters. By common-sense analysis, one would suspect that the most important criterion would be that a substance suspected of being a junctional transmitter must be demonstrated to be released from the prejunctional nerve endings when the nerve fibers are selectively stimulated. This criterion was relatively easily satisfied for isolated autonomic organs in which only one or, at most, two nerve trunks enter the tissue and the whole system can be isolated in an organ bath. In the central nervous system, however, satisfaction of this criterion presumes *(1)* that the proper nerve trunk or set of nerve axons can be selectively stimulated and *(2)* that release of the transmitter can be detected in the amounts released by single nerve endings after one action potential. This last subcriterion is necessary since we wish to restrict our analysis to the first set of activated nerve endings and not to examine the substances released by the secondary and tertiary interneurons in the chain, some of which might reside quite close to the primary endings. For many years, satisfaction of the second criterion was almost impossible because collection devices were so large as to injure the brain and detection methods were insufficiently sensitive. However, newer technologies, such as in vivo microdialysis, tissue voltammetry, and antibody-coated carbon filaments, have been applied to detect release effectively.

Localization

Because it is difficult, if not impossible, to identify the substance released from single nerve endings by selective stimulation, the next best evidence would be to prove that a suspected synaptic transmitter resides in the presynaptic terminal of our selected nerve pathway. Normally, we would expect the enzymes for synthesizing and catabolizing this substance also to be in the vicinity of this nerve ending, if not actually part of the nerve-ending cellular machinery. In the case of neurons secreting peptides or simple amino acid transmitters, however, these metabolic requirements may need further consideration. To document the presence of neurotransmitter, several types of specific cytochemical method for both light microscopy and electron microscopy have been developed. More commonly employed is the biochemical population approach, which analyzes the regional concentrations of suspected synaptic transmitter substances. However, presence per se indicates neither releasability nor neuroeffectiveness (e.g., acetylcholine in the nerve-free placenta or serotonin in the enterochromaffin cell). Although it has generally been considered that a neuron makes only one transmitter and secretes that same substance everywhere that synaptic release occurs, neuropeptide exceptions to this rule have become common.

Synaptic Mimicry: Drug Injections

A third criterion arising from peripheral autonomic nervous system analysis is that the suspected exogenous substance mimics the action of the transmitter released by nerve stimulation. In most pharmacological studies of the nervous system, drugs are administered intravascularly or onto one of the external or internal surfaces of the brain. The substances could also be directly injected into a given region of the brain, although the resultant structural damage would have to be controlled and the target verified histologically. Analysis of the effects of drugs given by each of these various gross routes of administration is quite complex.

We know that diffusional barriers selectively retard entry from the bloodstream of many types of molecule into the brain. These barriers have been demonstrated for most of the suspected central synaptic agents. In addition, we suspect that extracellular catabolic enzymes could destroy the transmitter as it diffuses to the postulated site of action. A further complicating aspect of these gross methods of administration is that the interval of time from the administration of the agent to the recording of the response is usually quite long (several seconds to several minutes) in comparison with the intervals required for junctional transmission (milliseconds). The delay in response further reduces the likelihood of detecting the primary site of action on one of a chain of neurons.

Microelectrophoresis

The student will now realize how important it is to have methods of drug administration equal in sophistication to those with which the electrical phenomena are detected. The most practical micromethod of drug administration yet devised is based on the principle of electrophoresis. Micropipettes are constructed in which one or several barrels contain an ionized solution of the chemical substance under investigation. The substance is applied by appropriately directing the current flow. The microelectrophoretic technique, when applied with controls to rule out the effects of pH, electrical current, and diffusion of the drug to neighboring neurons, has overcome the major limitations of classic neuropharmacological techniques. An alternative especially effective for testing poorly ionized molecules, such as neuropeptides, is to fill the delivery capillary with lower concentrations and then to "puff" very small volumes onto the test neuron by air pressure. Frequently, a multiple-barreled electrode is constructed from which one records the spontaneous extracellular discharges of single neurons while other attached pipettes are utilized to apply drugs. One can also construct an intracellular microelectrode glued to an extracellular drug-containing multielectrode so that the transmembrane effects of these suspected transmitter agents can be compared with the effects of nerve pathway stimulation. The intracellular electrode can also be used to poise the relative polarization of the membrane and to allow us to detect whether the applied suspected transmitter and that released by nerve stimulation cause the membrane to approach identical ionic equilibrium potentials.

Considerable experimentation with this technique has made possible certain generalizations regarding the actions of each putative neurotransmitter that has been studied. These substances are reviewed in detail in each of the chapters that follows. However, it should be borne in mind that certain substances have more or less invariable actions; for example, γ-aminobutyrate and glycine always inhibit, while glutamate and aspartate always excite. Insofar as we know, these actions arise from increased membrane conductances to Na, K, Ca, or Cl in every case. Other substances have many kinds of effect, depending on the nature of the cell whose receptors are being tested. Thus, acetylcholine frequently excites but can also inhibit, and the receptors for either response can be nicotinic or muscarinic. Similarly, dopamine, norepinephrine, and serotonin almost always inhibit; but they have a few excitatory actions that are probably not completely artifactual.

Pharmacology of Synaptic Effects

The fourth criterion for identifying a synaptic transmitter requires identical pharmacological effects of drugs potentiating or blocking postsynaptic responses to both the neurally released and the administered samples. Because the pharmacological effects are often not identical (most "classic" blocking agents are extrapolated to the brain from effects on peripheral autonomic organs), this fourth criterion is often satisfied indirectly with a series of circumstantial pieces of data. Recently, with the advent of drugs that block the synthesis of specific transmitter agents, the pharmacology for certain families of transmitters has been improved.

Electrophysiological analysis of drug and transmitter actions in the central nervous system was traditionally accomplished in terms of single-cell activity in vivo. Four types of physiological response served as the major indices to compare exogenously applied transmitter candidates and drugs with the effects of endogenous transmitters: *(1)* spontaneous activity, *(2)* orthodromic synaptically evoked activity, *(3)* antidromic activity, and *(4)* relative responses to independently acting excitatory or inhibitory transmitter released from another experimental source, such as another barrel of a multibarrel pipette. These techniques have been most successful when applied to large neurons, whose selected afferent pathways can be stimulated and the transmembrane effects specifically analyzed. However, when the unit recording techniques are applied to the intact mammalian brain, visualization and selection of the neuron under investigation are almost impossible. This is only one of the reasons why ex vivo brain slice methods and short-term tissue culture preparations, analyzed with voltage clamping and patch clamping, have gained in popularity.

The Steps of Synaptic Transmission

Let us now conclude this chapter by briefly examining the mechanisms of presumed synaptic transmission for the mammalian central nervous system. Each step in such transmission constitutes one of the potential sites of central drug action (Fig. 2–3). A stimulus activates an all-or-none action potential in a spiking axon by depolarizing its transmembrane potential above the threshold level. The action potential propagates unattenuated to the nerve terminal, where ion fluxes activate a mobilization process leading to transmitter secretion and transmission to the postsynaptic cell. From companion biochemical experiments (to be described in the following chapters), the transmitter substance is believed to be stored within the microvesicles or synaptic vesicles seen in nerve endings by electron microscopy. In certain types of nerve junction, miniature postsynaptic potentials can be seen in the absence of conducted presynaptic action potentials. These miniature potentials have a quantal effect on the postsynaptic membrane in that occasional potentials are statistical multiples of the smallest measurable potentials. The biophysical quanta have been related to the synaptic vesicles, although the proof for this relationship remains circumstantial.

When the transmitter is released from its storage site by the presynaptic action potential, the effects on the postsynaptic cells cause either excitatory

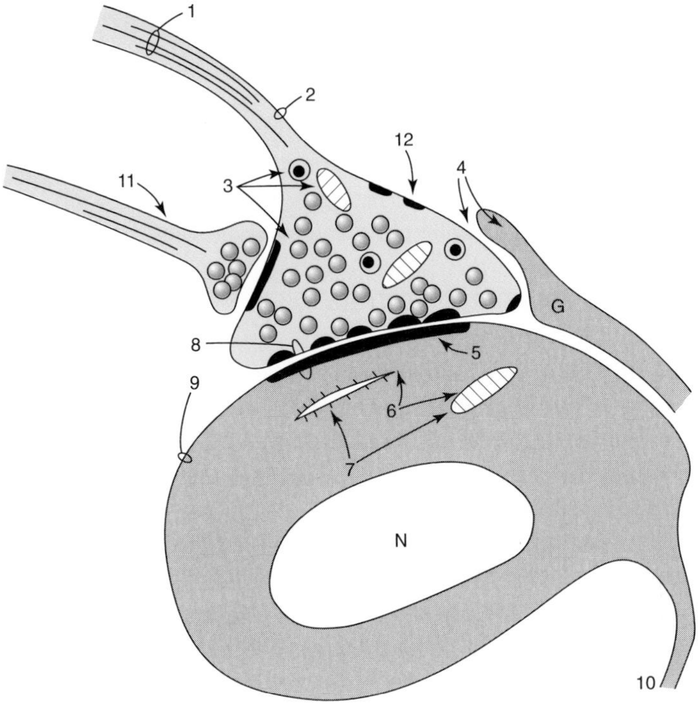

Figure 2–3. Twelve steps in the synaptic transmission process are indicated in this idealized synaptic connection. Step 1 is transport down the axon. Step 2 is the electrically excitable membrane of the axon. Step 3 involves the organelles and enzymes present in the nerve terminal for synthesizing, storing, and releasing the transmitter, as well as for the process of active reuptake. Step 4 includes the enzymes present in the extracellular space and within the glia for catabolizing excess transmitter released from nerve terminals. Step 5 is the postsynaptic receptor that triggers the response of the postsynaptic cell to the transmitter. Step 6 shows the organelles within the postsynaptic cells which respond to the receptor trigger. Step 7 is the interaction between genetic expression of the postsynaptic nerve cell and its influences on the cytoplasmic organelles that respond to transmitter action. Step 8 includes the possible "plastic" steps modifiable by events at the specialized synaptic contact zone. Step 9 includes the electrical portion of the nerve cell membrane that, in response to the various transmitters, is able to integrate the postsynaptic potentials and produces an action potential. Step 10 is the continuation of the information transmission by which the postsynaptic cell sends an action potential down its axon. Step 11, release of transmitter, is subjected to modification by a presynaptic (axoaxonic) synapse; in some cases, an analogous control can be achieved between dendritic elements. Step 12, release of the transmitter from a nerve terminal or secreting dendritic site, may be further subjected to modulation through autoreceptors that respond to the transmitter which the same secreting structure has released. Glia (G) can accumulate (4) released transmitters.

or inhibitory postsynaptic potentials, depending on the nature of the postsynaptic cell receptor for the particular transmitter agent. If sufficient excitatory postsynaptic potentials summate temporally from various inputs onto the cell, the postsynaptic cell will integrate these potentials and give off its own all-or-nothing action potential, which is then transmitted to each of its own axon terminals, and the process continues.

In trying to solve the problem of interneuronal chemical communication, it may be useful, nevertheless, to maintain an open mind with regard to three dimensions by which neuronal circuits can be characterized: *(1)* the spatial domain (those areas of the brain or peripheral receptive fields that feed onto a given cell and those areas into which that cell sends its efferent signals); *(2)* the temporal domain (the time spans over which the spatial signals are active); and *(3)* the functional domain (the mechanism by which the secreted transmitter substance operates on the receptive cell). When the receptive cell is closely coupled in time, space, and function to the secreting cell, almost everyone would agree that "real" synaptic actions have occurred. When the effects are long-lasting and widely distributed, however, many would prefer to call this action something else, even though the molecular agonist is stored in and released presynaptically from neurons onto the nerve cells they contact.

Selected References

Alexander, S. P. H and J. A. Peters (2000). Receptor and ion channel supplement. *Trends Pharmacol. Sci.* 21 (Suppl.), 1–120.

Armstrong, C. M. and B. Hille (1998). Voltage-gated ion channels and electrical excitability. *Neuron* 20, 371–380.

Aston-Jones, G. and G. R. Siggins (1994). Electrophysiology. In *Psychopharmacology: The Fourth Generation of Progress* (F. E. Bloom and D. J. Kupfer, eds.), Raven Press, New York, pp. 41–64.

Bloom, F. E. (2001). Neurohumoral transmission and the central nervous system. In *The Pharmacological Basis of Therapeutics*, 10th ed. (Hardman, J. G., Limbird, L. E. and Gilman, A. G., Eds.). McGraw-Hill, New York, pp. 293–320.

Bullock, T. H. (1980). Spikeless neurones: where do we go from here? *Soc. Exp. Biol. Semin. Ser.* 6, 269–284.

Catterall, W. A. (1993). Structure and function of voltage-gated ion channels. *Trends Neurosci.* 16, 500–506.

Eccles, J. C. (1964). *The Physiology of Synapses.* Academic Press, New York.

Hille, B. (1992). *Ionic Channels of Excitable Membranes*, 2nd ed. Sinauer Associates, Sunderland, MA.

Hodgkin, A. L. and A. F. Huxley (1952). Currents carried by sodium and potassium ions through the membrane of the giant axon of *Loligo*. *J. Physiol.* 116, 449.

Katz, B. (1966). *Nerve, Muscle and Synapse.* McGraw-Hill, New York.

Llinas, R. R. (1988). The intrinsic electrophysiological properties of mammalian neurons: insights into central nervous system function. *Science* 242, 1654–1660.

Loewi, O. (1921). Uber humorale Übertragbarkeit der Herznervenwirkung. *Pflugers Arch.* 189, 239.

Magistretti, P. J., L. Pellerin, and J.-L. Martin (1994). Brain energy metabolism: an integrated cellular perspective. In *Psychopharmacology: The Fourth Generation of Progress* (F. E. Bloom and D. J. Kupfer, eds.). Raven Press, New York, pp. 657–670.

Malenka, R. C. and R. A. Nicoll (1999). Long-term potentiation—a decade of progress? *Science* 251, 1870–1874.

Nicoll, R. A. (1988). The coupling of neurotransmitters to ion channels in the brain. *Science* 241, 545–553.

Siggins, G. R., and D. L. Gruol (1986). Synaptic mechanisms in the vertebrate central nervous system. In *Handbook of Physiology. Intrinsic Regulatory Systems of the Brain*, vol. IV (F. E. Bloom, ed.). American Physiological Society, Bethesda, MD, pp. 1–114.

Zigmond, M. J., F. E. Bloom, S. C. Landis, J. L. Roberts, and L. R. Squire, eds. (1999). *Fundamental Neuroscience*, Academic Press, San Diego.

3

Molecular Foundations of Neuropharmacology

Complete understanding of the basis for a drug's actions on the brain requires knowledge of all the molecules involved. However, until the last decade or so, most of the molecules involved in drug actions on the nervous system were recognized by their actions rather than their precise molecular structures or cellular compartmentalization. Thanks to advances in molecular biology, a growing number of these critical molecules can be specified in highly accurate terms to the level of their atomic structure. Nevertheless, as neuropharmacologists, the terms of reference remain very much the same. A drug is said to act *selectively* when it elicits responses from discrete populations of cells that possess "drug-recognizing" macromolecules, or receptors. Most drug receptors involve sites where neurotransmitters act. Some resemble, at the molecular level, specific molecular features of a neurotransmitter. However, drugs may also act by regulating intracellular enzymes critical for normal transmitter synthesis or breakdown or removal from the extracellular spaces of the brain. Receptors recognize drugs for a variety of reasons, which will be explored in subsequent chapters. Once having made that recognition, the activated receptor usually interacts with other molecules to alter membrane properties or intracellular metabolism. These cellular changes in turn regulate the interactions between cells in circuits. These circuit changes regulate the performance of functional systems (like the sensory, motor, or vegetative control systems) and eventually the behavior of the whole organism.

Thus, understanding the actions of drugs on the function of the brain, whether it be in terms of single cells or behavior, is a multilevel, multifaceted process that begins with and builds upon the concept of molecular interactions. Even beginning students of drug action on the nervous system will probably accept this statement as a reasonable hypothetical principle. In practice, however, this principle is severely compromised because most of the molecules in a very complex organ like the brain remain unknown.

When Watson and Crick deduced the three-dimensional, double-helical structure of DNA in 1953, the implications for the coding and replication of genetic information were recognized but could not be experimentally tested. Almost 25 years of effort were required before the new biological technology was launched. During this interval, it became clear how to combine genes and gene fragments from multicellular organisms with those of viruses, fungi, and bacteria to produce new genetic instructions and novel gene products. At last, the concept of a *gene* and its cellular *product* attained concrete form. Almost immediately, neuroscientists, who are always ready to exploit new technologies, began to apply these methods to the brain.

The power of molecular biological methods is realized from several related but independent developments: *(1)* the ability to clone genetic information (i.e., to isolate a selected segment and accurately reproduce it in large amounts), *(2)* the ability to determine the nucleic acid sequence of the selected gene segment (i.e., to read the complete molecular structure of a gene), and *(3)* the ability to practice genetic engineering (i.e., to perturb and control gene expression and to alter the structure of gene products by chemically modifying precise sites in the molecular structure of the genes). Within a decade the possibilities for applying this basic triad of powerful tools were dramatically revealed by two additional innovative technologies: the *polymerase chain reaction*, or PCR (by which large amounts of specific nucleic acid sequences can be produced without prior purification, cloning, or even a complete knowledge of their sequences), and the ability to create *transgenic animals* (i.e., to transfer synthetic genes into embryonic cells to make new mice, pigs, and cows to the experimentalist's specifications). Second-generation refinements of the latter two technologies have allowed PCR to serve as a means to find novel gene products within a family of genes in which some segments have been conserved (e.g., the transmembrane domains of certain receptors and transporters). The transgenic strategy can now reproducibly provide mice that are good disease models, either by creating mice lacking a specific gene (*knockout mutations*) presumed to be essential for one or another transductive pathway or by extending the original application of transgenic mice to overexpress selected genes, such as the mutated forms identified in human monogenic diseases like Huntington's disease (see Chapter 13).

All of these developments have contributed to a very rapid advance toward a truly molecular basis for the understanding of the nervous system and the way it can be altered by drug actions. In this chapter, we explore these molecular foundations.

Cellular Variation

Before we can deal effectively with the critical details of molecules that regulate the function of the nervous system and mediate the responses to drugs that act there, we must briefly consider how the cells of the brain differ from the other cells of the body and what it is that allows for differences between types of cell.

Except for erythrocytes, all mammalian cells have a nucleus that separates the basic units of genetic information from the cytoplasm. The cytoplasm and its organelles allow the cell to generate energy, which the cell uses to synthesize the structural and enzymatic molecules that give it and its enveloping plasma membrane the functional properties by which it contributes to the overall operation of the organism. Liver cells, kidney cells, skin cells, white blood cells, and cells of the nervous system all possess these basic similarities and individual properties. In an individual organism, all of these somatic cells (lymphocytes being the only exception), including the cells of the central and peripheral nervous systems, possess entirely the same basic set of genetic information. Lymphocytes can rearrange their genes extensively to refine the antibodies they must produce to meet all challenges.

The total set of potential genetic instructions of an individual, its *genome*, is composed of basic instructional units—the genes—each having a specific location on a specific chromosome. When there are multiple versions (or *alleles*) of a given gene, those expressed in a given individual are considered the *genotype* of that individual. Each type of specialized cell in an individual's organs expresses a subset of genes that encode the special structural, enzymatic, and other functional proteins that endow the cell with its size, shape, location, and other physiological characteristics. This set of characteristic features expressed by a cell is termed its *phenotype*.

Classically, the phenotype of the cell classes in the brain (i.e., neurons or glia) has been described according to its physical features (e.g., location, size, shape, connections) and transmitter-specific neurochemistry. Understanding the genetic basis for cell typing affords the opportunity to explain how neurons of the same shape and size can have different connections or transmitters, what this diversity and redundancy achieves, and why some neurons and neuronal systems may be more vulnerable to destructive insults than others. For now, we can assume that these cellular archetypes and the almost innumerable variants within each of the two large classes (neurons and glia) are the outward reflection of the corresponding specific expression of subsets of their genes. After a brief overview, additional complexities will be described. Neuropharmacology students recently exposed to molecular biology or biochemistry courses may find the next few sections "old hat." However, we include this material so that everyone reaching the end of this chapter will have the fundamentals necessary for a full appreciation of the starting material (see Fig. 3–1).

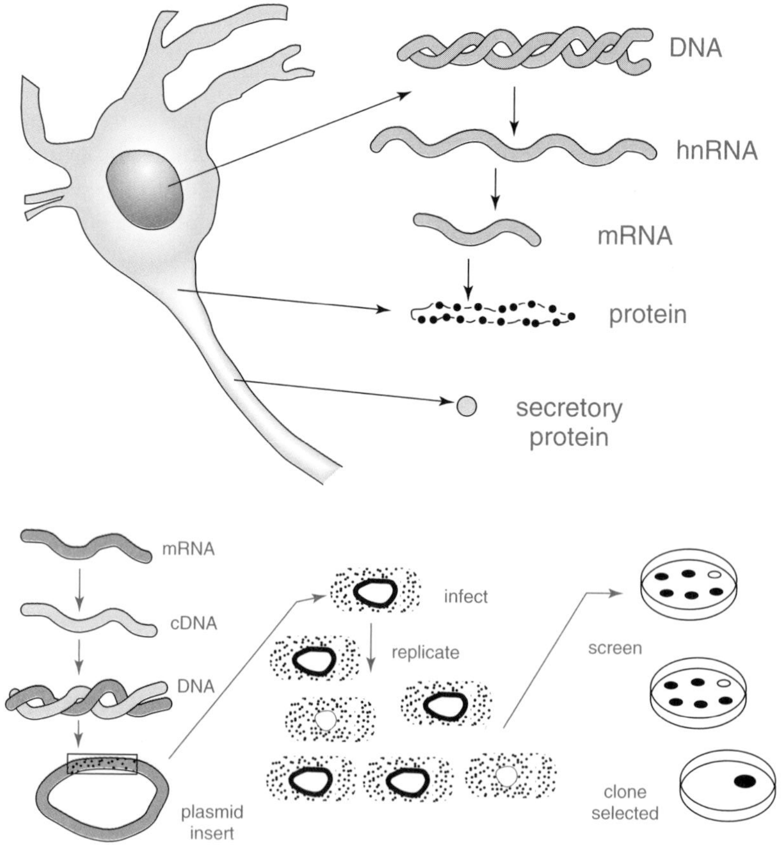

FIGURE 3–1. A schematic overview of the basic steps and cellular compartments involved in determining the specific phenotype of a neuron (*above*) and in cloning the mRNAs that allow the neuron to translate its genetic information into specific proteins (*below*). An mRNA is converted to a single-stranded DNA, which is further converted into a double-stranded segment that is then inserted into a plasmid. Bacteria are infected with individual plasmids, and individual plasmid-infected bacteria are grown into colonies. Replicates of the culture plate are screened by nucleic acid or antibody probes to select clones of interest.

1. Genetic information is stored in the form of long strands of deoxyribonucleic acid (DNA) chains and selectively expressed in specific cells. Thus, each specialized cell attains its specialized functional and structural status by expressing a subset of all of its genetic instructions. To express selective segments of the genome, the DNA-encoded information is converted, or *transcribed*, into a second similar molecular form as strands of ribonucleic acid (RNA) under the control of special proteins and RNA-synthesizing enzymes (*polymerases*) that perform the transcription steps. We will deal in slightly more detail with the actual molecular properties of DNA and RNA below, but the immense amount of detail simplified here is beyond our scope.
2. The primary transcript form (the heterogeneous nuclear RNA or hnRNA) is then edited by several rapid steps and exported from the nucleus (through nuclear pores) to the cytoplasm. The edited RNA transcript, or *messenger RNA* (mRNA), is then translated by special cytoplasmic organelles, also made of RNA and proteins, called *ribosomes*. The translation is a chemical language shift from the nucleic acid code of the RNA into the amino acid sequence of the protein that is to be expressed.
3. In some brain cells, such as the glia, which make their proteins largely for use within the cell, the ribosomes occur freely within the cytoplasm. In neurons and other dedicated secretory cells, the translated protein undergoes posttranslational processing, in which the protein's structure may be modified to attain the folded, globular, or linear structural properties that allow it to become associated with the proper intracellular compartments (e.g., within the plasma membrane or within the cytoplasm) where it is intended to function. As more and more proteins have been revealed in molecular structure and cellular location, at least three main types are now specified. There are two types of *integral membrane protein*, with the N terminus either extracellular (type I) or intracytoplasmic (type II), as well as *peripheral membrane proteins* close to the cytoplasmic surface of the membrane but not penetrating the lipid bilayers. There are also wholly cytoplasmic proteins.
4. In cells, like neurons, that transport large amounts of protein products within the cell's interior for purposes of secretion, more extensive posttranslational processing occurs. In such cells, the ribosomes are physically associated with a special set of endoplasmic reticular membranes, giving the membranes a "rough" appearance, for which they are known as the "rough endoplasmic reticulum." Within the channels of this inside-the-cell network, the newly synthesized pro-

teins are led to a set of smooth endoplasmic reticular membranes, the *Golgi apparatus*, where they are packaged into secretory organelles for transport to the secretory, or releasing, segments of the cell. Specific domains of the mRNAs, and hence the translation products of these secretory proteins, allow the internal apparatus of the cell to guide the to-be-secreted products from one compartment to another.

All of these organelle systems of the cell are essential for the selective transcription, translation, and packaging or compartmentalization of the specific proteins by which a given class of cells attains its specific phenotype. These structural elements of cell biology were well known to classic cytologists long before the molecular mechanisms underlying these events were understandable. Interested readers will now begin to comprehend the special analytic advantages that arise from the methods of molecular biology.

Fundamental Molecular Interactions

The cornerstone discovery of molecular biology was the formulation by Watson and Crick in 1953 of the double-stranded helix model of DNA structure. The insightful model they developed provided a coherent integration of the regular X-ray crystallographic structure of partially purified DNA with the previously known quantitative chemical data on the relative frequency within DNA of its four nucleotide bases, thus explaining why the purine–pyrimidine pairs adenine (A)–thymidine (T) and guanine (G)–cytosine (C) occur in precisely equal frequency. The Watson-Crick molecular model for DNA also accurately predicted the basic mechanism of DNA replication and accurate repair.

Nucleic Acid Base Pairing Complementarity

In the Watson-Crick double-helix, two right-handed helical polynucleotide chains coil around the same central axis, making a complete helical turn every 10 nucleotides (Fig. 3–2). In the interior of the helix, the purine and pyrimidine bases (A with T and G with C) are paired through hydrogen bonding of their complementary structures, placing the phosphate groups around the outside of the helix. The structural focal point for gene expression is the precise molecular complementarity between the primary sequences of nucleotide bases in one strand of the DNA helix with the antiparallel sequence of the second strand. The strand that encodes the genetic information is termed the *sense* strand. Wherever a particular base occurs in the sense strand, there will be a complementary base, and only that base in the *antisense* strand, such that A always pairs with T, and vice versa, and G always pairs with C, and vice versa.

The base pair complementarity allows for duplication of the genetic information in dividing cells. This is accomplished by enzymes known as DNA polymerases, which replicate each single strand back into double strands according to the single strand's template. The double-stranded complementarity also provides a means to repair the DNA should it be damaged since whichever single strand survives the damage can act as a template for the repair.

In a similar manner, the information-bearing, or sense, strand of the helical DNA chain is copied into a complementary single-stranded RNA during the process of transcription. RNA is thus a single-stranded complementary copy of the DNA antisense strand (so that its sequence resembles closely that of the DNA sense strand; see Fig. 3–2). RNA differs chemically with the substitutions of uridine for thymidine and ribose phosphates for deoxyribose phosphates. Transcription is accomplished by enzymes known as RNA polymerases. The affinity of the base pairs along sequences of a single DNA strand for their complementary base pair sequences in DNA or RNA are so precise that small segments can be used as probes for the detection of homologous sequences between large domains of DNA and RNA because the molecular

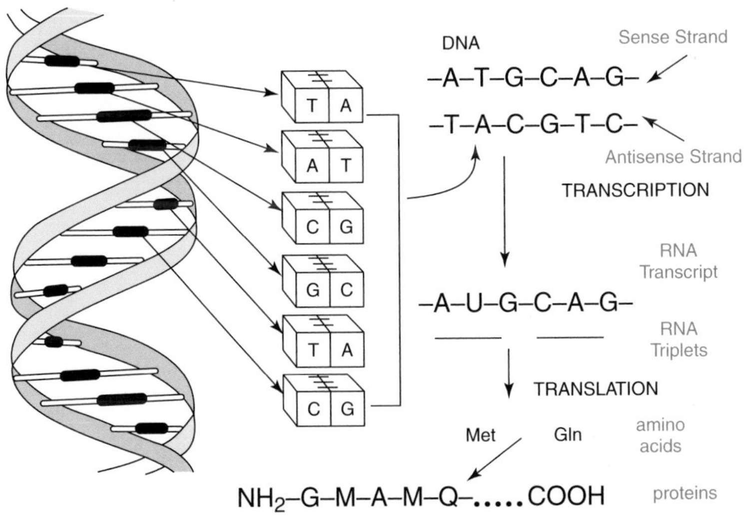

FIGURE 3–2. The arrangement of bases within the DNA double helix and the relationships between nucleic acid base pairs, RNA, and amino acids during the process of transcription and translation.

complementarity will allow the probe to bind to its complementary structure only when there is a long sequence of consistent match. The ability of a single-stranded nucleic acid to bind, or *hybridize*, to its complementary sequence is an essential component of many molecular biological techniques.

The Genetic Code

To translate genetic information from sequences of RNA into linear sequences of amino acids in proteins requires strict coding so that the 20 or so amino acids commonly found in proteins can be specified by various combinations of the four nucleotides. Through an ingenious series of important experiments, which space does not permit us to describe here, it was demonstrated that sets of three RNA bases (triplets) provide the code words that specify which amino acid will be incorporated into protein according to the order of the triplet codons in the gene and mRNA (Fig. 3–2). Other triplet sequences mark the point at which synthesis would begin, end, or be modified (by removing domains and linking the remainders together, called *splicing*).

Often, mRNAs encode far more protein sequence than is represented in the final form of the processed gene product. One feature that was detected only through recent cellular biological insight was the *signal peptide*, a 15- to 30–amino acid sequence at the N-terminal end of the encoded gene product in which the amino acids are highly hydrophobic. The signal peptide is a near-constant feature of proteins intended for secretion, such as neuropeptides; its function seems to be to guide the nascent protein chain through the endoplasmic reticular membrane for subsequent packaging by the Golgi membrane apparatus. Proteins that will not be secreted lack signal peptides. Thus, the structure of novel proteins predicts whether or not they are likely to be secretory products.

DNA Segments for Genes Are Interrupted

In addition to these remarkable explanations of gene expression and translation, another totally unanticipated wrinkle of gene regulation soon emerged. In the late 1970s, with the analysis of the genes encoding immunoglobulins and hemoglobins, it was recognized that the basic organization of genes in *eukaryotic* cells (cells with nuclei, as in all multicellular organisms) did not follow the principles that had been uncovered from the study of *prokaryotes* (cells, like bacteria, that lack a separable genetic compartment). Instead, higher organisms (and some viruses) have their gene segments split up into coding regions from which gene products are expressed (*exons*). Intervening regions of DNA (*introns*) separate the exons, but the RNA versions of introns are not found in the mRNA.

This interrupted DNA coding structure leads to two outcomes. *(1)* The primary gene transcript, formally termed the hnRNA, will contain extra RNA sequences, and the introns must be edited out before the mRNA can successfully direct protein synthesis by ribosomes (see Fig. 3–3). The editing process opens up the hnRNA, removes introns, and resplices the cut ends. *(2)* In some cells, including neurons, the composition of the transcribed mRNA can also be edited by splicing out certain exon segments. The editing–splicing process (or "alternative splicing") provides a means by which a gene containing several exons can give rise to several different gene product proteins with some shared domains and some unique domains. When gene products share similar nucleotide and protein sequences, we often speak of them as a "structural family" (see Chapter 11).

This editing and splicing may seem to be an unnecessarily complicated route to follow for a process that is intended to translate important genes with great fidelity into equally important enzymatic and structural proteins. However, the reader should be cautioned that such biological complexity almost always implies important and unanticipated regulatory control and enrichment. In the case of gene regulation and expression, the added complexities offer the means by which new life forms can evolve.

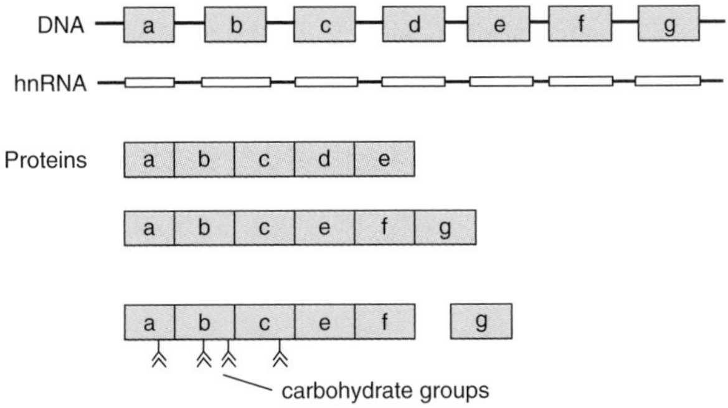

FIGURE 3–3. The relationship between DNA base sequences in introns and exons (a–g), the resulting primary RNA transcript, the subsequently edited forms of mRNA (two different forms of hypothetical editing and splicing are depicted), and the resulting proteins, which can then undergo posttranslational processing to yield small peptides or to add carbohydrates or other chemical modifications to the molecule.

Nucleic Acid Sequence Determinations

The final fundamental procedural development that accelerated discoveries made with molecular biological methods was the body of techniques to determine the sequence of DNA molecules, even if they are several thousand base pairs in length. The methods for doing this sequencing are clearly outside the province of this book. Two different approaches to sequence determination (that of Maxam and Gilbert and that of Sanger), each based on some very clever chemistry, had profound enough importance to merit Nobel recognition. Today, these methods have been automated into sequencer devices, speeding the process and reducing human error. From these DNA sequences, it is possible to deduce the nucleotide sequences of the RNA and thereby the amino acid sequence of the protein product. The structures of the DNA and of the gene product can then be analyzed in the cumulative knowledge of previously determined sequences, often via computer. In addition to structural comparison of gene or RNA sequences, important clues to the functional features of the encoded protein may be inferred from the domains of hydrophobic, hydrophilic, or other consensus structural sites for possible posttranslational modification.

Once-Over Quickly Cloning

mRNAs can also be converted into DNA by the class of virus enzymes known as *reverse transcriptases*. The copied double-stranded DNA form, or *cDNA*, can be incorporated (inserted) into specific sites within an infectious vector, or plasmid. The sites for insertion are selected by identifying DNA sequences that can be "cut" by the actions of *restriction endonucleases*, enzymes from purified bacterial sources that cleave DNA sequences at specific palindromic repeated sequence sites. Using plasmids of known DNA sequence that can be tailored to include the proper restriction cleavage sites and allow for DNA insertion, the same enzyme can then later be used to cleave out the insert. The restriction sites for insertion are typically chosen within plasmid genes that code for some discernible functional property (e.g., antibiotic resistance). Thus, interruption of coding and expression (when insertion has been successful) leads to loss of that functional property and results in a means to identify which plasmids have successful inserts. Each plasmid can generally incorporate only one cDNA insert, and, with a great excess of host bacteria, each insert-bearing plasmid will infect only a single host (almost always a specific type of *Escherichia coli*).

By growing these infected bacteria in such a way that each individual bacterium gives rise to a colony of identical bacteria bearing identical replicates of the plasmid and insert, the DNA has been *cloned*. The cDNA can then be recovered from the plasmid through another exposure to the restriction enzyme selected for the original opening of the plasmid insertion site. Thus, in a relatively few steps, one can start with a mixture of mRNAs in widely differing proportions from common to very rare and purify them individually, as well as prepare a virtually unlimited number of pure samples of the DNA insert.

Polymerase Chain Reaction

In early 1988, the final touches were put onto a startling new technology for preparing large amounts of specific and rare nucleic acid sequences through amplification in vitro without the necessity of first purifying the desired product through cloning. The changes in molecular biological research wrought by this new technology were immediate and dramatic, and the chemistry was relatively quickly rewarded with a Nobel Prize.

As originally conceived, the goal was to develop a quick diagnostic tool for the genetic disorder sickle cell hemoglobinopathy. Using previously selected restriction enzymes, human DNA was sliced into small sections, one of which contained the complete hemoglobin gene sequence. When the DNA was heated to 95°C, the double-stranded DNA "melted" and the strands separated. Two relatively short DNA sequences were synthesized (oligonucleotides) to be complementary to opposite strands of the hemoglobin gene separated by the part of the gene structure that was known to be altered in patients with the sickle cell mutation. These oligonucleotide "primers" hybridized to their complementary sequences on the single-stranded DNA. In the presence of large amounts of a purified DNA polymerase and large amounts of all the deoxynucleotides (A, G, T, and C), the primers extended to the end of the single strand. In the original experiments, this span covered only a distance of 20–30 bases; subsequent modifications now permit amplification of nucleotide sequences that are several hundred bases long.

By then separating through heat denaturation the dual helical strands of the newly synthesized material, cooling the reaction mixture, and adding fresh DNA polymerase and nucleotide substrates, the cycle of denature, anneal, and extend reactions, each lasting only a few seconds, can be rapidly repeated. As long as the polymerase and the nucleotide substrates remain in excess, the extended sequences of the first reaction each serve as a template for opposite strand synthesis in subsequent cycles, thereby providing a geometric rate of amplification. By identifying a DNA polymerase from a bacterial strain that survived in the heat of a Yellowstone geyser (*Thermus aquaticus*, or *Taq* for short), it was possible to develop a procedure that would survive the heat-denaturation step. Thus, with the *Taq* polymerase and large starting amounts of deoxynucleotides, a series of rapid cycles allows for the exponential in vitro amplification of the desired gene segment.

If the sequences selected are known to be generally constant in the genome of the species to be studied, the same primers can be used to select, amplify, and then evaluate the same intervening gene segment from many individuals. If the amplified segment is simply evaluated for its length, it is possible to determine whether a given individual has had a major structural gene mutation, such as a deletion or insertion that will alter the way the restriction enzymes fragment DNA. However, many other clever applications of the technology have pushed its advantages even more broadly than initially imagined. By first using reverse transcriptase to make double-stranded cDNA, one can also apply PCR to mRNAs and even make the assessment quantitative (so-called q-PCR). By modifying the ends of the probes to be used, it is possible to incorporate special synthetic sequences that will later make it easier to clone or sequence the amplified segments or to reinsert modified versions to assess the functional importance of a specific sequence domain. Using special synthetic nucleotides that will form complementary base pairs even if the sequences do not match precisely, it is possible to amplify homologous sequences from different but related genes (e.g., as the transmembrane domains for the seven-transmembrane domain receptors like the noradrenergic, cholinergic, and serotonergic receptors).

This highly superficial survey should indicate the facility of cloning DNA segments (taken directly from genomic digests or from mRNA copies) and then sequencing the cloned segments and deducing the structure of their product. However, we have glossed over the one potentially sour note in this rhapsody of high-tech molecular music making: how do you identify which clone carries the insert that encodes the gene product you want to identify? Suppose you do not even know what it is you want to look for and clone? As it turns out, through more cleverness and a few good breaks from Mother Nature, biologists have found multiple methods to accomplish these feats of molecular magic, although some of the screening methods have given new meaning to the phrase "searching for a needle in a haystack."

Molecular Strategies in Neuropharmacology

The immediate applications of molecular biological strategies within neuropharmacology are shortcuts in molecular isolation and sequence determination—for instance, to uncover new peptides or proteins or to provide more complete understanding of enzymes, receptors, channels, or other integral proteins of the cell. A rather likely premise holds that the phenotype of a cell within the nervous system depends on the structural, metabolic, and regulatory proteins by which it establishes its recognizable structural and functional properties. If this is valid, then complex, multifaceted neurons will probably rely on hundreds, if not thousands, of special-purpose proteins, many of which may exist in rather limited amounts. Purifying such rare proteins by the methods that existed before molecular cloning, especially in the absence of a functional assay to guide the purification process, is an overwhelming task, requiring exceptional patience, resources, and a very large supply of the proper starting tissue material. For some of the very rare hypothalamic hypophysiotropic releasing factors (see Chapter 11), hundreds of thousands of hypothalami were required, as well as the development of unique purification schemes for each subsequent factor to be pursued.

Converting the quest for the structure of specific proteins into a molecular biological quest for the mRNA or gene segment that encodes this protein greatly facilitates the experimental analysis, simply because of the powers of cloning, complementarity, and rapid sequencing. Several methods have been developed that increase the chances of finding whatever the researcher is seeking. These methods depend in part on the nature of the cDNA being pursued and how the investigator probes either for the insert or for the translation of the fusion gene product in a cell system capable of processing the primary translation product into a structural form that will resemble its natural configuration and sometimes even its natural function. Given the possibility of protein–protein interactions, synthetic genes can be translated into tools to detect unknown proteins with affinity for the newly discovered gene product, a strategy exploited shamelessly in the isolation of the vesicle proteins needed for transmitter secretion.

In addition to their capacity to accelerate the discovery of new molecules participating in the nervous system's response to disease or to self-administered drugs, molecular biological strategies can be used to determine how critical a particular gene product may be in mediating a cellular event with behavioral importance—for example, the role of a specific transmitter's receptor subtype in a cellular event (e.g., long-term potentiation), which may in turn underlie behavioral phenomena such as memory formation or recall (see Chapter 12). These functional probes can be achieved through advanced methods for transgenic animal construction (see below) in which specific mutations are engineered into any targeted gene of an experimental animal's genome, leading eventually to the production of homozygous animals lacking either that gene completely or the capacity to turn it on and off as experimentally desired.

Less demanding and less vulnerable to potential confounding roles of the targeted gene in the developing nervous system are more short-term manipulations, such as the intracerebral or intraventricular injection of special nucleotide constructs to deliver *ribozymes* (RNA constructions that can target and degrade specific mRNAs) or *antisense oligonucleotides* (which can bind to mRNAs, delay their translation, and enhance their degradation). A related strategy creates cells or whole transgenic animals with novel mutations of the proteins in question that closely resemble but can outcompete with the protein at its natural binding partner sites (a so-called dominant negative mutation). In addition to preparing novel mutant animals by transgenic technologies, it is possible to insert the special constructs into *null cells* (cells that normally do not express the transduction system under investigation) and, when they express a new receptor in their membrane, to probe these receptors for more conventional pharmacological characterization of natural or synthetic agonists and antagonists. In this way, one might not only develop novel drugs unique to such receptors but also detect previously unknown natural ligands for functionally uncharacterized "orphan" receptors (see Chapter 11).

General Strategies for Clone Screening and Selection

We now briefly explore some of the more important steps of these exciting discovery protocols.

Enrichment by Tissue Selection and Preparation

One basic starting point is to select the brain cells or regions that are presumed to express the molecule to be studied and then to enrich sources of mRNA to favor the detection of the one being pursued. Cell lines and even tumors that produce large amounts of a hormone (e.g., pheochromocytomas or Vasoactive Intestinal Polypeptide (VIP-omas) [see Chapter 11] or bear large numbers of the desired receptors or channels (e.g., electroplax or striated muscle) have proved to be excellent starting materials. Once the cell source is selected, the desired mRNAs can be further enriched (e.g., by sucrose gradient centrifugation or electrophoresis), provided that some characteristics of the mRNA being sought are known.

Recognizing the Wanted Clone

A general early strategy for detecting desired colonies of cloned bacteria, known as *colony hybridization*, illustrates a general flavor for the operations involved. Bacteria are grown on special culture plates, from which their colonies can be copied (transferred as a group by lightly pressing them to another supporting surface, called "replica plating," thereby sampling and preserving the spatial identity of all of the colonies on the plate). The bacteria on the replicate supports are treated to expose their DNA and screened with radioactive nucleic acid probes. Colonies that hybridize with the probe can then be identified by autoradiography. Alternatively, if the plasmid-carrying inserts were tailored to allow for expression of the protein encoded by the transferred genetic material, it might also be possible to identify the desired clones by immunological reactivity. When a reactive colony has been identified, its original bacterial colony is recovered from the original culture plate and the living bacteria are then grown in large amounts to provide the starting material for DNA sequence analysis. Modern methods now make these steps seem primitive.

Building Your Own DNA

If partial protein (or peptide) sequences are known, it is possible to make predictions of what the mRNA sequence should be (by back-translating the genetic code for amino acids) and including enough alternatives to overcome the ambiguous cases where a specific amino acid may be encoded by several variant triplets. From this predicted RNA structure, it is possible to design and synthesize the hypothetical cDNA. This approach has been used, for example, to create hypothetical cDNAs for the hypophysiotropic hormones whose amino acid sequences had been accurately determined "the old-fashioned way," by earning it one amino acid at a time from highly purified brain extracts. The main reason to do this with a biologically active protein or peptide whose structure is already known would be to get at the complete structure of its prohormone or to obtain its complete genomic structure and analyze its regulatory control and expression mechanisms. As will be seen in Chapter 11, when this is done, more often than not, the prohormone of the known peptide is found to encode more than one active product. However, because of the redundancy of triplet RNA codons for some amino acids (there are six different codes for leucine alone), it is generally difficult to acquire a functional full-length mRNA by predictive synthesis.

An alternative approach is to synthesize a shorter complementary single-stranded DNA (a so-called oligodeoxynucleotide probe) and use it as a probe to screen libraries of clones. Such libraries may be prepared from mRNA extracts or from special digests of the whole genome. In the former case, the starting material would be brain, while in the latter case, in theory, any somatic cells could be used to prepare the library. With the availability of automated "gene machines," it is now possible to synthesize a proper probe or two overnight and use them to screen an awaiting genomic or cDNA library, thereby determining the complete coding sequences for a partially purified protein within a few weeks.

On such a screening expedition, likely candidate colonies can be cross-screened by a second synthetic oligodeoxynucleotide probe, based on another separate domain of the full protein. Clones positive for both probes would then have to contain the gene sequence that encodes the two sequences against which the probes were made as well as the sequence between them. This strategy has been used with many neuropeptide mRNAs (see Chapter 11).

When You Haven't Got a Clue

It is also possible to penetrate the large treasure trove of cellular proteins that have not yet been identified by conventional strategies. Given the length of the mammalian genome and the relatively short list of identified specific molecules in cells of all classes, we must conclude that there is an awful lot left to be identified and few clues as to what it is we do not know. The methods of molecular biology can help here too.

However, there appears to be a rather select group of ancient conserved sequences, numbering probably less than 1000, that nature has found to be important enough to keep relatively unmodified throughout hundreds of millions of years of evolution. This core must then be significantly enriched by the other 99% of the genome. Interestingly for neuropharmacologists, most neurotransmitter receptor molecules are not among these ancient conserved sequences, although at least one 5-hydroxytryptamine (5-HT) receptor and some of the transporters are. Furthermore, with the complete sequencing of the genomes of *Caenorhabditis elegans* and *Drosophila*, the paths of evolution, conservation, and modification of ion channels and receptors are becoming traceable. The inferences drawn from the first partial inventories of gene structure and protein motifs perpetuate the conservation of critical molecular mechanisms of life.

In nonneural tissues, it has been possible to identify the unique proteins expressed in a male versus a female liver or those that are unique to thymus-derived versus bone marrow-derived lymphocytes, by exploiting the fact that a very high proportion of the proteins expressed in these pairs of tissues are, for the most part, very similar. Depending on how the pairs of tissues or cell types are defined and how similar or dissimilar they actually turn out to be, this method can be made highly sensitive and can reveal unique differences in cell-specific gene expression.

The discovery strategy can also be broadened to look for large sets of tissue-specific genes. For example, Sutcliffe, Milner, and Bloom employed molecular cloning methods to determine what proteins are generally made by the brain that are not found in other major tissues and the degree to which neurons differ in specific proteins underlying their phenotypes that at present are defined only empirically. In their original approach, a cDNA library was prepared from mRNAs extracted from whole rat brain and the individual cDNAs were characterized for their ability to hybridize to mRNAs extracted from brain, liver, or kidney. Those expressed in brain, but not detectable in extracts of liver or kidney, represented well over half of the total brain mRNA population (now estimated at 40,000 out of an anticipated total of 80,000–100,000, although some estimates emerging from the human genome have been interpreted as being perhaps half this many). Whatever the final actual inventory of brain-specific genes turns out to be, perhaps half or more of the genome contains information pertinent to the generation of neuronal function. Individual brain-specific clones are then analyzed further by determining their nucleotide sequence and deducing the amino acid sequence of its encoded protein. Proteins that are unique to the database of known sequences can then be identified further by raising antisera against synthetic peptides that mimic selected regions of the deduced protein structure. Subsequent variants of this approach sought genes expressed in one brain region and not another and even in one region of the cerebral cortex and not another. One such hypothalamus-specific mRNA cloned in this manner, hypocretin, encodes a novel neuropeptide now known to be involved in the regulation of appetite, blood pressure, and sleep and to be associated with human narcolepsy.

Yet another strategic approach involves injecting mRNAs from enriched or prepared sources into frog oocytes or mammalian cell lines whose genetic composition has been characterized. By injecting groups of mRNAs and evaluating the oocytes or cell lines for the response one seeks, it is possible through trial and error to identify the mRNA for a specific functional protein, like an ion channel or a cell surface receptor.

An All-Points Search

Would-be molecular neuropharmacologists attending a seminar involving these approaches may find initial exposure to the jargon confusing without some orientation. "Orientation" is an appropriate term, for such presentations are sprinkled with what may sound like references to compass points. When DNA is fractionated by restriction enzymes and the resulting fragments are separated by gel electrophoresis, it is possible to transfer, or *blot*, the resulting fragments, separated mainly on the basis of length, from the acrylamide gel to a nitrocellulose or nylon support and, there, to analyze them for the ability to hybridize with cDNA or RNA probes. This method is referred to as a *Southern blot* analysis, named for Dr. E. M. Southern, who started the evolution of this method. Later, a similar approach was devised in which the starting material was RNA. Here, the separated RNAs were blotted for probing with radioactive, single-stranded cDNA probes. Because the starting material is, speaking in terms of nucleic acid, the opposite of that in a Southern blot, this RNA blot is referred to as a *Northern blot*.

More recently, immunological methods have been used to probe protein extracts that were separated by acrylamide gel electrophoresis and then blotted (by an electrical transfer) for identification by peroxidase or radioisotope-labeled antibodies to specific protein antigens. The result is termed a *Western blot*. If RNA or protein samples are simply dried directly onto nitrocellulose for probing analysis without first separating them for size, the resulting blots are termed *slot blots* or *dot blots*. These blots are useful when screening a large number of clone extracts for inserts or expressed products quickly. As of this writing, no method has yet been termed an "Eastern blot."

The resourcefulness with which innovative and automatable molecular biological methods are being applied to all aspects of biology grows more and more amazing with each issue of your favorite journal. For example, the creation of gene chip arrays allows for the comparison among relatively homogeneous cell types of the order and magnitude of changes in gene expression following an experimental perturbation. Gene probes, applied to essentially microscope-scale slides, can detect changes in gene expression by array readers, and the ordering of their expression patterns and the clusters of genes following similar or opposing patterns can be extracted from the data by powerful bioinformatic computer analysis software. As the human and mouse genome inventories begin to narrow the remaining gaps, it is possible to imagine a more or less complete list of the genes we have available. One senses that we are about to experience a logarithmic increase in the number of specific molecules that will be fully characterized and from which pharmacological engineers will shape drug molecules precisely to fit the pocket of receptors or enzymes for ultimate specificity.

Beyond the Clones

Although we may rightly marvel at the advances that have been achieved through the use of molecular biological techniques, all that we have in essence discussed so far in this chapter is a set of methods that provide novel, powerful, and accurate ways to identify, isolate, and characterize the amino acid sequences of a host of intracellular proteins and their possible subsequent metabolic products and interaction partners. While this is unquestionably a major advance in the research armamentarium, it still does not really begin to deal with a wide range of other important questions that are also approachable through the molecular tools that recombinant methodologies provide.

For example, once a cDNA has been proven to represent the mRNA for a specific molecule, the deduced protein sequence can be inspected to infer potential functional properties. Thus, the acetylcholine receptor molecule and the myelin proteolipid protein exhibit several stretches of 20–24 hydrophobic amino acids in a row, which is strong presumptive evidence of membrane-crossing domains and, thus, suggestive of plasma membrane constitutive proteins. In any case, when the protein structure can be deduced, the entire molecule or selected fragments of it can be synthesized and used to raise antisera. These antisera can then be used to develop radioimmunoassays for the protein or used in Western blots. The antisera can also be used for immunocytochemical analysis of the nervous system, to determine which cell and which compartments of those cells exhibit the protein that has been identified. Synthetic fragments can be used to determine whether the protein domains may be substrates for posttranslational modification, being processed by further proteolytic cleavage or by structural modification with glycosylation, phosphorylation, sulfation, or acylation. Subcellular fractions and ultrastructural cytochemistry may suggest organelle specialization or cell surface associations.

The products of cDNA cloning can also be taken back to the genome, to probe the regions around the location of the exons to search for the molecular mechanisms that control transcription. Once the surrounding elements in the genome have been located, the cDNA probes can be used to determine the degree to which the mRNA or the underlying gene exons have been conserved across eukaryotic species and to determine the position on the chromosomes to which the gene can be mapped. The chromosomal location of many proteins has been determined in this or a similar manner. However, the human genome, and that of most mammals, is estimated to be on the order of 3×10^9 bp long. Until the race to sequence the human genome greatly accelerated the pace of sequence collection and computer-directed gene identification, fewer than 20,000 genes had been mapped to specific chromosomes. In an intermediate step of genome analysis, markers of unknown genetic functions were mapped at relatively low density across all human chromosomes. From those low-resolution maps, the powerful, increasingly automated genomic analysis routines have already yielded much more detailed mapping. While a large proportion of the genes noted so far can be related to previously discovered molecular motifs, at least 25% of the genes have no such antecedents. For students of genetic diseases of the nervous system, the situation is even more complex. Given the surprising number of mutations associated with known, but rare, inheritable disorders, from color blindness to deafness to epilepsy to mental retardation, it seems clear that there are many ways to impair brain function. Nevertheless, the genetic roots of complex illnesses such as Alzheimer's disease, depression, and schizophrenia are beginning to give way to these powerful methods (see Chapter 13).

The ability to link fragments of DNA with inheritance of genetic disorders and to specific markers within the digested fragments helped to establish approximate locations of genetic mutations, such as the localization of the gene for Huntington's disease to human chromosome 4. Given the increased detailed information on length and complexity of neuronal gene expression, efforts to link specific patterns of DNA polymorphism (the combinations of unique alleles of specific genes) may be a critical future development. Southern blot analysis can track which family members have inherited specific allele combinations by refinements of restriction fragment length polymorphism designed to detect single-nucleotide polymorphisms.

New modifications of these basic strategies are reported continuously. Although there is a steady growth of molecular information on the elements of neuronal function, most of the details that follow in this book deal with those still relatively few, but prominent, molecules (e.g., the major known neurotransmitters, their synthetic and catabolic enzymes, their receptors and response mediators), whose nature is already partly established. While this list may well be lengthened ever more rapidly by the shortcuts made possible by molecular biological methods, students should recognize that identifying new molecules per se is merely a first step toward an important pharmacological end but is nowhere near the true strength of what molecular biology may have to offer our field.

Even more promising in its implications for neuropharmacology is the microinjection of a segment of cloned DNA into the pronucleus of a single-cell *zygote* (a fertilized ovum) to create a transgenic mouse. After injecting a large number of fertilized ova (a success rate of 10% is considered good), one transfers the eggs to the uterine cavity of a foster mother that has been pseudobred with a sterile male. When the pups are born, they are evaluated through skin fibroblast cultures for incorporation of the injected DNA and, if possible, for expression of the gene product (e.g., overproduction of a circulating hormone). After puberty, their sperm can also be evaluated for integration of the foreign DNA sample into germ cells; these are best evaluated by their ability to transmit the integrated gene to the mouse's own progeny. If the integration of the foreign DNA is successful and if the resultant gene structure (at the site of integration) does not itself produce mutational consequences that may be troublesome, the founder mice give rise to lines of mice bearing the transgene. By preparing gene constructs that induce overproduction of growth hormone, it has been possible to use the technology to make not only supermice but superpigs and supergoats as well.

Transgenic mice are now being used for a variety of experimental purposes. The reverse application of transgenic technology was actually done first, namely, preparing a transgene for expression of myelin basic protein and using that to "cure" the gene mutation of the myelin-deficient mouse, which is ordinarily unable to make the protein. Other partially tested strategies include removing a natural gene (receptor or neurotransmitter transporter) to determine the effects on cellular or behavioral function—for example, receptivity to psychostimulants.

Among the more experimental applications of transgenic technology is the preparation of a gene construct in which the regulatory domains of a known gene (e.g., neuropeptide) are coupled with the expression of a novel reporter gene (e.g., an enzyme that is normally absent in the mouse); under these conditions, one can evaluate whole-animal treatments (or whole-cell treatments if cell lines are transfected instead of embryonic cells) that "turn on the reporter gene," thus providing inferential evidence of what controls (i.e., elevated cyclic adenosine monophosphate) the activation of the natural gene. Finally, applications still somewhat precarious to be reproducible include the use of a retrovirus to carry a gene segment to be transfected into cells beyond the single-cell embryo level; this is especially useful in systems where partial development can reveal the effects of the added gene. A third creative but still early application involves transfection of *embryonic stem cells* (undifferentiated blast-like cells taken from the blastocyst level of embryonic development). When these transfected stem cells are returned to blastocyst embryos, they spread throughout the developing organism; and if the researcher is fortunate, the foreign gene will appear in the germ cell lines as well. Should gene markers ever be identified that predict the development of single gene–dependent neuropsychiatric diseases (such as Huntington's, familial Alzheimer's, or scrapie), it should be possible to prepare true animal models of the diseases and their treatment.

Molecular Motifs of Transmitter Release and Response

We conclude this foray into molecular neuropharmacology by quickly considering the consistent structural patterns, or "molecular motifs," identified in selected categories of molecules that are essential for synaptic transmission and thus critical for the content of subsequent chapters. This survey will also illuminate some of the clever ways that molecular biology has moved beyond mere discovery to illumination of the molecular mechanisms of synaptic transmission.

Sodium Channels

The three basic channels (Na^+, K^+, and Ca^{2+}) underlying neuronal excitability were cloned by starting with known short segments of the proteins obtained from highly purified Na^+ channel preparations. Oligonucleotides encoding these peptide segments and antibodies raised against them were used to probe cDNA libraries of electric organ mRNA and led to the sequence determination for what turned out to be the α subunit of the Na channel (there is also a simpler β subunit). Modeled from the distribution of hydrophobicity plots for the deduced amino acid sequences, the hypothetical α-subunit structure contained four repeats of highly similar six-transmembrane α-helical domains, in which the fifth and sixth transmembrane domains are separated by extracellular loops, while the N-terminal, the C-terminal, and the cytoplasmic loops linking the sixth transmembrane domain to the first transmembrane domain of the adjacent repeat are intracellular cytoplasmic loops.

These sequence data were then used to derive the rat brain Na^+ channels, which also turned out to have a similar structure, at least in the transmembrane domains, though not the connecting loops. These modeled conformations have been refined by a variety of functional analyses, including expression in oocytes to confirm their functional properties (including sensitivity to channel-blocking toxins) and to determine pore- and voltage-sensitive domains by mutation, deletion, and transposition of sequences.

Calcium Channels

Subsequent work, again starting with highly purified protein sources and cloning from the first few short segments of amino acid sequences wrung from them, revealed the molecular characteristics of the Ca channel proteins, which had previously been categorized on their functional differences only, as voltage-sensitive Ca^{2+} channels first cloned from muscle. As more and more voltage-sensitive Ca^{2+} channel subunits and variants were recognized from muscle and neuronal sources, the complexity of the heterooligomeric Ca^{2+} channel proteins has been surprising. Present data indicate a tetrameric α_1 subunit with significant sequence similarity to the Na channel, with six presumptive transmembrane domains, a transmembrane $\alpha_2\delta$ subunit linked by disulfides that is required for complete function, as is a second adjacent transmembrane γ subunit and a peripheral intracytoplasmic β subunit. With these basic subunit characterizations in place, the challenges have been to define the basis for the functional categorizations of the previous L, N, and T types of Ca^{2+} channel recognized physiologically and their variants. Significant progress in this challenge has been obtained (see Randall and Benham, 1999).

Potassium Channels

Definition of the structures of K^+ channels began with the cloning of the *Shaker* gene of *Drosophila*, which then provided the structural tools to characterize the multiple forms of voltage-sensitive and then ligand-gated channel proteins carrying K^+ currents in mammalian neurons. The crystallization of a bacterial K^+ channel has given three-dimensional reality to the differences in structure and regulatory function. The basic K^+ channel is a presumptive six-transmembrane structure with a single ionic pore, almost identical to the conformation of one of the repeat segments in the Na^+ and Ca^{2+} α subunits. In voltage-dependent K^+ channels, there is a cytoplasmic peptide between the fifth and sixth presumptive transmembrane domains, known as "H5" senses the intracellular polarization. Similarly, the cytoplasmic loop between transmembrane domains 4 and 5 provides the means to inactivate the channel from the inside of the cell. The ligand-gated form of the K^+ channel, termed the *inward rectifier*, shows an even simpler structure, with two transmembrane domains and an intervening H5-like voltage-sensing sequence. Two other variants may account for the outward rectifier K^+ channels with dual ionic pores and either eight or four transmembrane domains.

Neurotransmitter Receptors

Here, the starting point was protein sequences from highly purified preparations of electroplax or muscle receptors (for acetylcholine), brain (for GABA and glycine), or blood cells (for norepinephrine). Once those structures were in hand, however, multiple additional members of their receptor superfamilies were snared by probing cDNA libraries with RNA probes designed to accept loose (or low-stringency) matches with similar, but not identical, nucleotide sequences.

As noted in greater detail in Chapter 4, neurotransmitter receptor cloning studies have so far split the spoils into two large groups: the *ionophore receptors*, in which the receptor is a multimeric ion channel composed of four or five subunits, each exhibiting a similar presumptive four-transmembrane domain structure, and the *metabotropic*, or *G protein–coupled, receptors*, each of which is a monomer with seven conceptual transmembrane domains, although such receptors may also dimerize. Within the ionophore receptors, the protein subunits for a given neurotransmitter (acetylcholine nicotinic receptors, GABA and glycine receptors, most of the glutamate receptors, and one of the 5-HT receptors), there is enough similarity to allow for recognition of further receptor subtyping of different assembled multimers structurally. Different combinations of GABA subunits expressed in neurons with a high degree of heterogeneity, for example, can confer cell type-specific sensitivity to ethanol or benzodiazepines (see Chapter 7). A surprising feature of one category of glutamate receptor, the AMPA (α-amino-3-hydroxy-5-methyl-4-isoxazole propionic acid) receptor (see Chapter 6), is its dynamic insertion and removal from the subsynaptic sites under conditions of activation.

The second large category of neurotransmitter receptors comprises the guanine nucleotide-binding protein-coupled receptors (for cholinergic muscarinic receptors; all catecholaminergic, histaminergic, and neuropeptide receptors; all other 5-HT receptors; and the metabotropic glutamate receptor). All of these exhibit a presumptive seven-transmembrane domain configuration, first recognized in rhodopsin and then found in the first of the cloned adrenergic receptors. Through expression of mutated mRNA constructs in null cells, it has been possible to determine quite definitively which segments of which transmembrane domains are responsible for ligand recognition and which segments of which cytoplasmic loops are responsible for interactions with the G proteins. There is good evidence that at least two G protein–coupled receptors, the β-adrenergic receptor and the substance P receptor, are internalized after activation by ligands, temporarily downregulating sensitivity.

Neurotransmitter Transporters

We now turn our attention to a family of molecules that are essential to the process of transmitter conservation-after-release. This process, which we used to call simply "reuptake," has been upgraded with the term *transporter*, following cloning. In a relatively brief burst of discoveries in the early 1990s, the transporters for GABA, glycine, all of the monoamines, as well as proline and betaine were cloned and found to express a very similar structural motif consisting of 12 transmembrane domains with substantial conservation across the transmembrane domain sequences, allowing for the information from the initial cloning of the GABA transporter to be extended to the others. In all of these molecules, concentration of the specific small molecule from the extracellular space back into the interior of the neuron, generally the presynaptic terminals, seems to be driven by cotransport of Na^+ and Cl^- ions, and the molecules bear a strong resemblance to the previously known glucose transporters and to adenylyl cyclase for as yet unclear reasons.

A second variety of transporter has also been defined that concentrates cytoplasmic neurotransmitter into the synaptic vesicles, driven by either a pH or an ionic gradient. While these transporters also exhibit apparent 12-transmembrane domain structures, the actual sequences distinguish them from those that operate in the presynaptic plasma membrane.

Synaptic Vesicle Proteins

In an era of neuropharmacology not too long ago, a minor parlor game was made of the ability to classify synaptic vesicles as to their transmitter content on the basis of size, shape, and cytochemistry. The advent of molecular biology has radically transformed our views of vesicles from mere morphological description to arguably the most completely detailed nervous system organelle in terms of a specifiable inventory of proteins and interactions during transmitter secretion. Again, the initial studies started with highly purified brain synaptic vesicles and a search for the proteins they contain and, in a marvelous display of Mother Nature's generosity, managed to reveal the molecular machinery for secretion across wide realms of the phylogenetic tree. Given the estimates for the molecular mass of the typical 500 Å and the roughly 600 kDa size of the vesicle transporter, there could be room for about a dozen more molecules of approximately 50 kDa. The inventory to date reveals the following three groups:

1. Vesicle-bound proteins: *synapsins* (which are substrates for cAMP-dependent protein kinase and can regulate transmitter release); *synaptotagmin* (possibly a Ca-sensor protein) and *synaptobrevin* (also known as VAMP for vesicle-associated membrane protein and a substrate for proteolysis by tetanus toxin and by some of the botulinum toxins), both of which seem to be essential for docking on the presynaptic plasma membrane; and *synaptophysin*, a vesicle protein of as yet unknown function (but antibodies to it have served as a useful means to quantify synapses). Each of these vesicle proteins has several isoforms such that, it is believed, one of each class will be present on every synaptic vesicle.
2. Synaptic membrane proteins: the synaptic soup thickens with two other proteins, *SNAP-25*, a synapse-associated protein of 25 kDa, and *syntaxin*, both of which are substrates for other botulinum toxin proteases and, thus, also essential players in the docking and/or fusion steps to release transmitter.
3. A series of soluble proteins called NSF (an *N*-ethylmaleimide–sensitive factor that alters the capacity to transport vesicles across the cisternae of the Golgi complex in a clever in vitro assay) and its three soluble NSF-associated proteins (called α-, β-, and γ-SNAPs, of which the β form is quite enriched in brain, while the others are more pronounced in nonneuronal secretory cells and bear a strong resemblance to secretory proteins recognized in yeasts for their role in constitutive secretion).

In the case of the vesicle-related secretion proteins, cloning provided the detailed sequences and the related isoforms, the capacity to make small mutations and deletions to define the docking and interaction domains, and the ability to reassemble the entire array in null cell assays and to establish the potential roles for each protein. There are a few more paralytic bacterial toxins whose substrates in secretion are still to be accounted for, so we are not likely to be at the end of this very interesting search. Furthermore, we still do not know exactly how the vesicles are reformed within the synaptic terminal following the docking, fusion, and release steps, presumably involving guanine exchange factor, ADP-ribosylation factor, and vesicle coat proteins similar to those budding from the Golgi membranes. A minor mystery still surrounds the nature of the signal for when it is time to ship the whole vesicle back up to the perikaryon for an overhaul.

Selected References

Bargmann, C. A. (1998). Neurobiology of the *Caenorhabditis elegans* genome. *Science* 282, 2028–2033.

Barondes, S. H. (1993). *Molecules and Mental Illness*. W. H. Freeman, New York.

Brady, S., D. R. Coleman, and P. Brophy (1998). Subcellular organization of the nervous system: organelles and their functions. In *Fundamental Neuroscience* (M. Zigmond, F. E. Bloom, S. C. Landis, J. Roberts, and L. Squire, eds.). Academic Press, New York, pp. 71–106.

Brown, D. A. (2000). The acid test for resting potassium channels. *Curr. Biol.* 10, R456–R459.

Crick, F. H. C., L. Barnett, S. Brenner, and R. J. Watts-Tobin (1961). General nature of the genetic code for proteins. *Nature* 192, 1227–1232.

Gilbert, W. (1985). Genes-in-pieces revisited. *Science* 228, 823.

Gusella, J. F., R. E. Tanzi, M. A. Anderson, W. Hobbs, K. Gibbons, R. Raschtchian, T. C. Gilliam, M. R. Wallace, N. S. Wexler, and P. M. Conneally (1984). DNA markers for nervous system diseases. *Science* 225, 1320–1325.

Hille, B., C. M. Armstrong, and R. MacKinnon (1999). Ion channels: from idea to reality. *Nat. Med.* 5, 1105–1109.

Le Novere, N. and J. P. Changeux (2001). The ligand gated ion channel database: an example of a sequence database in neuroscience. *Philos. Trans. R. Soc. Lond. B Biol. Sci.* 356, 1121–1130.

Lobe, C. G. and A. Nagy (1998). Conditional genome alteration in mice. *Bioessays* 20, 200–208.

Randall, A. and C. D. Benham (1999). Recent advances in the molecular understanding of votage-gated Ca^{2+} channels. *Mol. Cell. Neurosci.* 14, 256–272.

Sanger, F. and A. R. Coulson (1975). A rapid method for determining sequences in DNA by primed synthesis with DNA polymerase. *J. Mol. Biol.* 94, 444–448.

Sutcliffe, J. G. (2001). Open-system approaches to gene expression in the CNS. *J. Neurosci.* 21, 8306–8309.

Tjian, R. (1995). Molecular machines that control genes. *Sci. Am.* 54–61.

Watson, J. D. and F. H. C. Crick (1953). Molecular structure of nucleic acids: a structure for deoxyribose nucleic acid. *Nature* 171, 737–738.

4

Receptors

The concept that most drugs, hormones, and neurotransmitters produce their biological effects by interacting with receptor substances in cells was introduced by Langley in 1905. It was based on his observations of the extraordinary potency and specificity with which some drugs mimicked a biological response (*agonists*) while others prevented it (*antagonists*). Later, Hill, Gaddum, and Clark independently described the quantitative characteristics of competitive antagonism between agonists and antagonists in combining with specific receptors in intact preparations. This receptor concept has been substantiated in the past several years by the actual isolation of macromolecular substances that fit all of the criteria of being receptors. To date, although not all have been cloned, receptors have been identified for all of the proven neurotransmitters as well as for histamine, opioid peptides, neurotensin, vasoactive intestinal peptide, bradykinin, cholecystokinin, somatostatin, substance P, insulin, angiotensin II, gonadotropin, glucagon, prolactin, and thyroid-stimulating hormone. In addition, as noted in Table 4–1, multiple receptors have been shown to exist for all of the biogenic amines, acetylcholine (ACh), γ-aminobutyric acid (GABA), histamine, opiates, and the amino acid transmitters. If receptors for all agents (e.g., hormones, trophic factors, odorants, peptides) in addition to the neurotransmitters were counted, a total of 1000 would not be surprising. This number would also include *orphan receptors*, which are nuclear receptors that have been cloned because of their similarity to known receptors but which have no known ligands. Some of these orphan receptors, when examined, appear to be transcription factors.

TABLE 4–1. Neurotransmitter Receptors

- Adrenergic
  - α_{1A}, α_{1B}, α_{1C}, α_{1D}
  - α_{2A}, α_{2B}, α_{2C}, α_{2D}
  - β_1, β_2, β_3
- Dopaminergic
  - D_1, D_2, D_3, D_4, D_5
- GABAergic
  - $GABA_A$, $GABA_{B1a}$, $GABA_{B1\delta}$, $GABA_{B2}$, $GABA_C$
- Glutaminergic
  - NMDA, AMPA kainate, $mGluR_1$, $mGluR_2$, $mGluR_3$, $mGluR_4$, $mGluR_5$, $mGluR_6$, $mGluR_7$
- Histaminergic
  - H_1, H_2, H_3
- Cholinergic:
  - Muscarinic: M_1, M_2, M_3, M_4, M_5
  - Nicotinic: muscle, neuronal (α-bungarotoxin-insensitive), neuronal (α-bungarotoxin-sensitive)
- Opioid
  - μ, δ_1, δ_2, κ
- Serotonergic
  - $5\text{-}HT_{1A}$, $5\text{-}HT_{1B}$, $5\text{-}HT_{1D}$, $5\text{-}HT_{1E}$, $5\text{-}HT_{1F}$, $5\text{-}HT_{2A}$, $5\text{-}HT_{2B}$, $5\text{-}HT_{2C}$, $5\text{-}HT_3$, $5\text{-}HT_4$, $5\text{-}HT_5$, $5\text{-}HT_6$, $5\text{-}HT_7$
- Glycinergic
  - Glycine

GABA, γ-aminobutyric acid; NMDA, *N*-methyl-D-aspartate; AMPA, α-amino-3-hydroxy-5-methyl-4-isoxazole propionic acid; mGluR, metabotropic glutamate receptor; 5-HT, 5-hydroxytryptamine.

SOURCE: *RBI Handbook of Receptor Classification*, 1994. ed J. W. Kebabian & J. L. Neumeyer.

Multiple receptors appear to metastasize at an uncontrollable rate, but this should be viewed skeptically until a physiological response to the ligand has been shown or a specific gene has been cloned and expressed. Some of the receptor subtypes have been identified only by binding techniques (see below), which can lead to erroneous conclusions. All of the receptors for neurotransmitters and peptide hormones that have been studied, regardless of whether they have been isolated, are localized on the surface of the cell; only the receptors for steroid and thyroid hormones are intracellular.

After the discovery that the action of physostigmine (eserine) resulted from its anticholinesterase activity, it was assumed that most drugs acted by inhibiting an enzyme. However, it now appears that with few exceptions (an action on ion channels or transport proteins) the mechanism of action of neuroactive drugs usually stems from their effect on specific receptors. Predictably, the current search for receptors is among the most intense areas in the neurosciences. This interest is not purely academic. The recent identification of adrenergic, dopaminergic, muscarinic, serotonergic, and histaminergic receptor subtypes has led to the synthesis of highly selective drugs that are considerably more specific than their prototypes, which were developed after general screening for activity. With advances in gene cloning and expression, more and more receptor subtypes are being identified, each presumably having its own function. What this indicates is that future drugs can be designed to fit a single receptor subtype, thus precluding the side effects of a nonspecific drug.

Definition

In this rapidly developing field, considerable confusion has arisen as to what functional characteristics are required of an isolated, ligand-binding molecule to qualify as a receptor. This confusion, a semantic problem, developed after the successful isolation of macromolecules that exhibit selective binding properties, which made it mandatory to determine whether this material comprised both the binding element and the element that initiated a biological response or merely the former. Some investigators use the term *receptor* only when both binding and signal generation occur; they use the term *acceptor* if no biological response has been demonstrated. Others are content to ignore the bifunctional aspect and use *receptor* without specifications. In this chapter, we will define a *receptor* as the binding or recognition component and refer to the element involved in the biological response as the *effector*, without specifying whether the receptor and effector reside in the same or separate units. The criteria for receptors will be dealt with shortly; the biological response that is generated by the effector obviously has a wider range of complexity, from a simple one-step coupling to an unknown number of steps (Fig. 4–1).

Assays

Basically, there are two ways to study the interaction of neurotransmitters, hormones, or drugs with cells. The first procedure (and until relatively re-

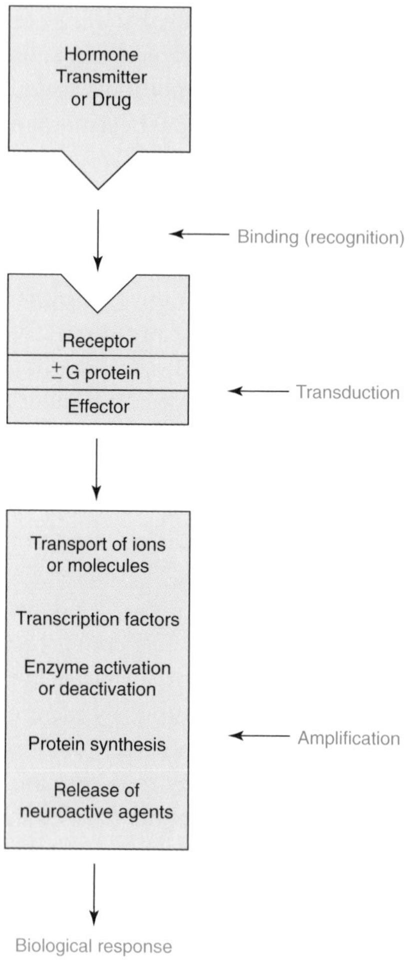

FIGURE 4–1. Schematic model of ligand–receptor interaction.

cently the only one) is to determine the biological response of an intact isolated organ, such as the guinea pig ileum, to applied agonists or antagonists. The disadvantage of this procedure is that one is obviously enmeshed in a cascade of events beginning with transport, distribution, and metabolism of the agent before it even interacts with a receptor, ranging through an unknown multiplicity of steps before the final biological response of the tissue is measured. Thus, although studies with agonists may be interpretable, it is not difficult to envision problems when antagonists are employed since these compounds may compete at a level different from receptor binding. Despite the not unusual problem of a nonlinear relationship between receptor occupancy and biological response, this approach has yielded a considerable amount of information. The second approach to studying receptors is by measuring ligand binding to a homogenate or slice preparation. This technique became feasible with the development of ligands of a high specific radioactivity and a high affinity for the receptor. Here, the direct method is to incubate labeled agonist or antagonist with the receptor preparation and then separate the receptor–ligand complex from free ligand by centrifugation, filtration, or precipitation. The indirect technique is to use equilibrium dialysis, where the receptor–ligand complex is determined by subtracting the ligand concentration in the bath from that in the dialysis sac.

Recently, a third, electrophysiological approach to identifying receptors has emerged. Intracellular stimulation and recording via microelectrodes inserted into a brain slice or neurons in culture, combined with the application of receptor agonists and antagonists, can functionally identify receptor subtypes.

Though not always appreciated, the advantage of using isolated tissues is that both the efficacy and the functional activity of an agonist are assessed, in contrast to binding procedures using broken cell preparations, where only affinity or a biochemical sequela of binding can be appraised. Ideally, both techniques should be used, but, alas, this is rarely done. It should be noted in passing that efficacy and affinity are independent. To date, it appears that the drugs that exhibit high affinity but low efficacy have more efficient coupling to the effector than the reverse situation; therefore, these are the most potent and selective agents. For a detailed discussion of the above, the review by Kenakin et al. (1992) is recommended.

Identification

In the midst of an intensive drive to isolate and characterize receptors, some zealous investigators have lost sight of the basic tenets that must be satisfied before it is certain that a receptor has indeed been isolated. Thus, on occasion, enzymes, transport proteins, and merely extraneous lipoproteins or proteolipids that exhibit binding properties have been mistakenly identified as receptors. Authentic receptors should have the following properties:

1. *Saturability.* The great majority of receptors are on the surface of a cell. Since there are a finite number of receptors per cell, it follows that a dose-response curve for the binding of a ligand should reveal saturability. In general, specific receptor binding is characterized by high affinity and low capacity, whereas nonspecific binding usually exhibits high capacity and low-affinity binding that is virtually nonsaturable.

2. *Specificity.* This is one of the most difficult and important criteria to fulfill because of the enormous mass of nonspecific binding sites compared with receptor sites in tissue. For this reason, in binding assays it is necessary to explore the displacement of the labeled ligand with a series of agonists and antagonists that represent both the same and different chemical structures and pharmacological properties as the binding ligand. One should also be aware of the avidity with which inert surfaces bind ligands. For example, substance P binds tenaciously to glass, and insulin can bind to talcum powder in the nanomolar range. With agents that exist as optical isomers, it is of obvious importance to show that the binding of the ligand is stereospecific. Even here problems arise. With opiates it is the levorotatory enantiomorph that exerts the dominant pharmacological effect. Snyder, for example, has found glass fiber filters that selectively bind the levorotatory isomer. Specificity obviously means that one should find receptors only in cells known to respond to the particular transmitter or hormone under examination. Furthermore, a correlation should be evident between the binding affinity of a series of ligands and the biological response produced by this series. This correlation, the *sine qua non* for receptor identification, is unfortunately a criterion that is not often investigated.
3. *Reversibility.* Since transmitters, hormones, and most drugs act in a reversible manner, it follows that their binding to receptors should be reversible. Also, the ligand of a reversible receptor should be not only dissociable but recoverable in its natural (i.e., nonmetabolized) form. This last dictum distinguishes receptor–agonist interactions but not receptor–antagonist binding from enzyme–substrate reactions.
4. *Restoration of function upon reconstitution.* Following the isolation and identification of the components of the receptor system, to "put Humpty Dumpty back together again" is the goal of all receptorologists.
5. *Molecular neurobiology.* The ultimate identification is to isolate the gene for a receptor, express it, and demonstrate the exact similarity of the cloned receptor to the natural one.

It is important to recognize that the quantitative and spatial measurement of receptors utilizing autoradiography is also a key problem. Where labeled ligands are employed to map receptors in brain via light microscopy, a mismatch is often encountered. Reasons offered for this problem are *(1)* except for autoreceptors, neurotransmitters and receptors are located in different neurons; *(2)* in addition to the synapse, receptors and transmitters are found throughout the neuron and in glial cells; *(3)* ligands may label only a subunit of a receptor or only one state of the receptor; and *(4)* autoradiography is subject to quenching. With immunohistochemical peptide mapping, a possible problem is the recognition by the antibody of a prohormone or, alternatively, a fragment of a peptide hormone in addition to the well-recognized problem of cross-reactivity of the antibody with a physiologically different peptide.

Finally, all drugs do not necessarily act directly on a receptor. They could bind to a site that is adjacent to a receptor and thus influence the activity of the receptor.

KINETICS AND THEORIES OF DRUG ACTION

From the law of mass action, the binding of a ligand (L) to its receptor (R) leads to the following equation:

$$\mathrm{L} + \mathrm{R} \underset{k_2}{\overset{k_i}{\rightleftharpoons}} \mathrm{LR}$$

thus
$$K_d = \frac{k_2}{k_1} = \frac{[\mathrm{L}]\,[\mathrm{R}]}{[\mathrm{LR}]}$$

where k_1 = association rate constant
k_2 = dissociation rate constant
[L] = concentration of free ligand
[R] = concentration of free receptor
[LR] = concentration of occupied binding sites
K_d = dissociation equilibrium constant

Since the total number of receptors = $[\mathrm{R_t}] = [\mathrm{R}] + [\mathrm{LR}]$,

$$[\mathrm{R}] = [\mathrm{R_t}] - [\mathrm{LR}]$$

$$K_d = \frac{[\mathrm{L}]([\mathrm{R_t}] - [\mathrm{LR}])}{[\mathrm{LR}]}$$

$$\text{and } [\mathrm{LR}] = \frac{[\mathrm{L}][\mathrm{R_t}]}{[\mathrm{L}] + K_d}$$

Since the fraction of receptors occupied (r) $= \frac{\text{bound}}{\text{total}} = \frac{[LR]}{[R_t]}$,

$$r = \frac{[L]}{[L] + K_d}$$

If experiments are performed in which the receptor concentration is kept constant and the ligand concentration is varied, then a plot of r versus [L] will produce a rectangular hyperbole, the usual Langmuir adsorption isotherm. Here r approaches the saturation value of 1. If r is plotted against log [L], a sigmoid curve will result; log [L] at half-saturation will give log K_d on the horizontal axis (Fig. 4–2).

This equation can be rearranged as follows:

$$K_d r + r[L] = [L]$$

$$\frac{r}{[L]} = \frac{-r}{K_d} + \frac{I}{K_d}$$

Now if r/[L] is plotted against r, a straight line will result (assuming only one set of binding sites) with two intercepts, the one on the x axis giving the number of binding sites per molecule and the y intercept yielding I/K_d. This type of plot is the Scatchard plot (more correctly, Rosenthal plot), widely used in studying receptor–ligand interactions (Fig. 4–3). Among the pitfalls encountered in a Scatchard analysis is the problem that the system is not in true equilibrium.

Another useful representation that can be derived from the general equation is the Hill plot (Fig. 4–4). If log $E/(E_{max} - E)$ is plotted versus log [L]

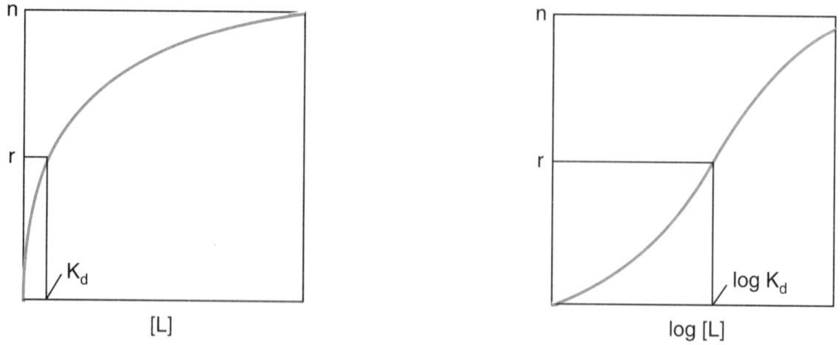

FIGURE 4–2. Ligand–receptor interactions plotted in two ways, where *[L]* is the concentration of ligand (drug, hormone, or neurotransmitter) and *r* is the biological response, proportional to the moles ligand bound per mole of protein. The number of binding sites per molecule of protein is designated by *n*.

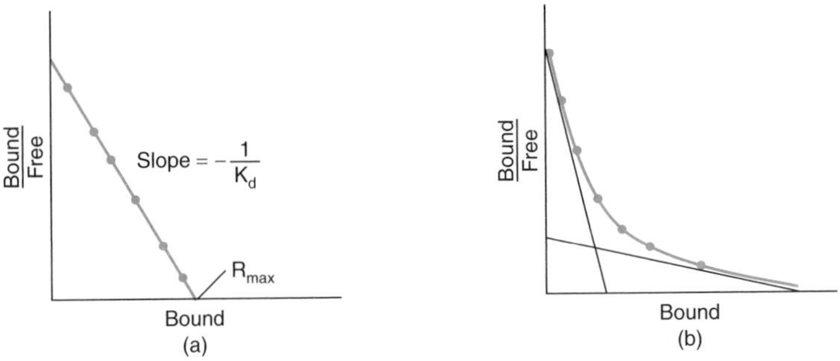

FIGURE 4–3. Scatchard plots of the binding of a ligand to a receptor. In *(a)* only one binding affinity occurs, but in *(b)* both high- and low-affinity binding sites are suggested.

when E is the effect produced and E_{max} is the maximum effect, then the slope, indicative of the nature of the binding, gives a Hill coefficient of unity in cases where E/E_{max} is proportional to the fraction of total number of sites occupied (r). In many situations, the slope turns out to be a noninteger number different from unity. This finding indicates that cooperativity may be involved in the binding of the ligand to the receptor. *Cooperativity* is the phenomenon whereby the ligand binding at one site influences, either positively or negatively, the binding of the ligand at sites on other subunits of the oligomeric protein. This idea, originally suggested in 1960 by Monod, Wyman, and Changeux to explain allosteric enzyme properties, currently offers the most attractive hypothesis for studying reactions of receptors with hormones, transmitters, or drugs. We will utilize this hypothesis later to explain drug action, including the problem of efficacy (intrinsic activity), spare receptors, and desensitization.

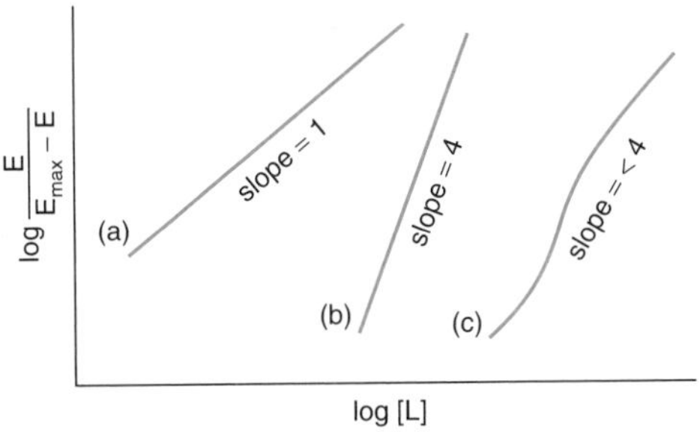

FIGURE 4–4. Hill plots for receptor–ligand binding: *(a)* noncooperative binding; *(b)* idealized plot of cooperative binding with four sites; *(c)* a typical Hill plot of multiple binding sites but less than four. (Modified from Van Holde, 1971.)

Another useful analysis when competitive inhibitors of the receptor binding ligand are studied is the equation derived by Cheng and Prusoff (1973) in their kinetic analysis of enzyme inhibitors:

$$K_i = \frac{I_{50}}{1 + [L]/K_d}$$

where K_i is the equilibrium dissociation constant of the competitive inhibitor and I_{50} is the concentration of the inhibitor producing 50% inhibition at the concentration of the labeled ligand that is used in the study.

Clark (1937) produced the first model of drug–receptor action, known as the "occupation theory," in which the response of a drug was held to be directly proportional to the fraction of receptors occupied by the drug. Here, as mentioned earlier, one should find the usual Langmuir absorption isotherm. However, instead of the expected rectangular hyperbole when drug concentration was plotted against the drug bound, in most cases a sigmoid curve resulted. A second problem with the occupancy theory is that in many instances only a small fraction of the total receptors available are occupied, and yet a maximum response is obtained. These additional receptors, which may represent as much as 95%–99% of the total, are referred to as *spare receptors*. The occupancy theory is further complicated when the activity of a series of related agents is explored and the biological response varies from maximum to 0, even though all of the agents occupy the receptor. In other words, these agents could be full agonists, partial agonists that give less than a maximal response, and antagonists whose occupancy produces no response. This phenomenon is referred to as *efficacy* or *intrinsic activity* of a drug and is obviously not directly related to the binding affinity of the drug. A fourth characteristic of drug–receptor interaction that is sometimes observed is *desensitization*, which is the lack or decline of a response to a constant stimulus.

These problems—the sigmoidal dose-response curve, some anomalous effects with antagonists that spare receptors might account for, efficacy, and desensitization—can be comfortably fitted into a two-state model of a receptor analogous to the allosteric model of Monod, Wyman, and Changeux, proposed independently by Changeux and by Karlin in 1967.

According to the two-state model, receptors exist in an active (R) and an inactive (T) state, and each is capable of combining with the drug (A):

$$\begin{array}{ccc} & K_{AR} & \\ A + R & \rightleftharpoons & AR \\ \updownarrow & & \updownarrow \\ & K_{AT} & \\ A + T & \rightleftharpoons & AT \end{array}$$

Here, an agonist prefers the R (active configuration of the receptor), and the efficacy (i.e., intrinsic action) of the drug will be determined by the ratio of its affinity for the two states. In contrast, competitive antagonists prefer the T form of the receptor and will shift the equilibrium to AT. The sigmoid relationship between the fraction of receptors activated and the drug concentration (i.e., cooperativity) can be explained by this model with its equilibrium between R and T, if one designates T as a subunit of R that binds A and thereby influences the further binding of A to R. Cooperativity can also explain anomalous effects of antagonists whenever the effect of an antagonist persists even in the presence of a high concentration of agonist. Here, it could be postulated that by tightly binding to one conformational state of the receptor, the antagonist inhibits the binding of the agonist. One might even use this two-state model to account for desensitization, where T would be a receptor that has been desensitized perhaps by a local change in the ionic environment. The ionic environment may be one of the factors that dictate the two conformational states of a receptor. Other possibilities include polymerization (clustering) of the subunits, depolymerization, or phosphorylation. It should be emphasized that this model is conjectural, subject to modifications as the need arises.

When one considers the finite number of receptors per cell and the fact that virtually all receptors are found on the plasma membrane, it is not surprising that progress in this area has been slow and laborious. It has been calculated that, assuming one binding site and an average molecular weight of 200,000 for a receptor, complete purification of a receptor protein would require about a 25,000-fold enrichment. The extraordinary density of ACh receptors in electric tissue has made this preparation so popular a choice that considerable information is now available on this cholinergic receptor. Two snake neurotoxins, *Naja siamensis* and α-bungarotoxin, which specifically bind nicotinic cholinergic receptors, have been the key agents that have helped in isolating this receptor.

Also to be considered is the relationship between receptors and effectors as they interact in the fluid mosaic membrane. As envisaged by Singer and Nicolson (1972), membranes are composed of a fluid lipid bilayer that contains globular protein. Some of these proteins extend through the membrane, and others are partially embedded in or on the surface. The hydrophilic ends of the protein protrude from the membrane, while the hydrophobic ends are localized in the lipid bilayer. Some of the proteins are immobilized, but others, floating in an oily sea of lipids, are capable of free movement. In this concept of membrane fluidity, the receptor protein would be on the surface of the membrane and the effector within the membrane. Although the ratio of receptors to effectors may in some cases be unity (thus explaining instances in which the occupancy theory is satisfied), it may also be greater than 1.

Consequently, in a situation where multiple hormones activate a response (e.g., there is only one form of adenylyl cyclase in a fat cell, but it may be activated by epinephrine, glucagon, corticotropin, or histamine), it would be concluded that an excess of receptors over effectors is present. This model would also explain spare receptors and is exemplified by the fact that only 3% of insulin receptors need to be occupied in order to catalyze glucose oxidation in adipocytes. With the possibility of easy lateral movement of effector in the membrane, it is also understandable why one receptor may activate several types of effector. Membrane fluidity will account for the observation that hormones can influence the state of aggregation of the receptor, thus giving rise to either positive or negative cooperativity as determined in kinetic studies of binding. Although direct evidence of the interaction and migration of receptors and effectors is difficult to obtain, current information is easily accommodated by the fluid mosaic membrane model.

Currently, there are four major groups of receptors known to be involved in signal transduction (Fig. 4–5), of which the first two are neurotransmitter-activated. The first group is referred to as *ligand-gated channels* or *ionotropic channels*. It is composed of multiple subunits with a central pore, which, when activated, open to permit the passage of Na^+, K^+, Ca^{2+}, or Cl^-. Thus, depending on which ion is involved, the membrane potential may be either depolarized or hyperpolarized. Since no second-messenger biochemical systems are involved, the effects of neurotransmitters on these cell surface receptors are very fast, with excitatory and inhibitory responses occurring in milliseconds. Examples of these ionotropic receptors are the nicotinic ACh receptor, the *N*-methyl-D-aspartate receptor, the $GABA_A$ receptor, and the 5-hydroxytryptamine$_3$ (5-HT_3) receptor.

The second neurotransmitter-activated receptors that are also membrane-localized are the G protein–coupled receptors, referred to as *metabotropic receptors*. These receptors mediate slower responses (seconds to minutes) that are generally modulatory, dampening or enhancing the signal. All known G protein–coupled receptors contain seven hydrophobic transmembrane domains that are linked by hydrophilic groups. Examples of these receptors are muscarinic cholinergic, adrenergic, dopaminergic, serotonergic, metabotropic glutaminergic, opiate, peptidergic, and some purinergic receptors (vida intra). The structure of these receptors, as exemplified by the β_2-adrenergic receptor, is shown in Figure 4–6.

Metabotropic receptors are a family of guanine nucleotide-binding proteins with a heterotrimeric structure consisting of α-, β-, and γ-subunits. The G protein signal-transduction cycle is shown in Figure 4–6. At last count, 22 different G protein α subunits and at least five β and 12 γ subunits had been identified. These proteins can be roughly classified into four groups: G_s, G_i,

G_q, and G_{12}. Activation of the G_s subunit family increases adenylyl cyclase activity, opens Ca^{2+} channels, and inhibits Na^+ channels. The Gi subunit family opens K^+ channels, closes Ca^{2+} channels, inhibits adenylyl cyclase, and promotes cyclic guanosine monophosphate phosphodiesterase and probably phospholipase A_2. Increased phospholipase C is the effector for G_q, while with G_{12}, which activates Rho, a guanosine triphosphate (GTP)–binding protein, there is yet no other known function (a dysfunctional family?). Also recently recognized are a family of over two dozen regulators of G protein signaling (RGS) that promote desensitization by activating GTPase. Of course, these RGS proteins, which may have other functions, are in turn regulated. To add to this bewildering complexity in neuronal signaling, in addition to the fact that a single receptor can activate multiple G proteins that may or may not interact, cloning studies have revealed multiple isoforms of adenylyl cyclase, phospholipase C, phospholipase A_2, and calcium and potassium channels.

An observation that has helped assign G proteins a role in signal transduction is that bacterial toxins catalyze the nicotinamide-adenine dinucleotide (NAD)-dependent adenosine disphosphate (ADP) ribosylation of the α subunit of many of the G proteins and inhibit their activity. Cholera toxin ri-

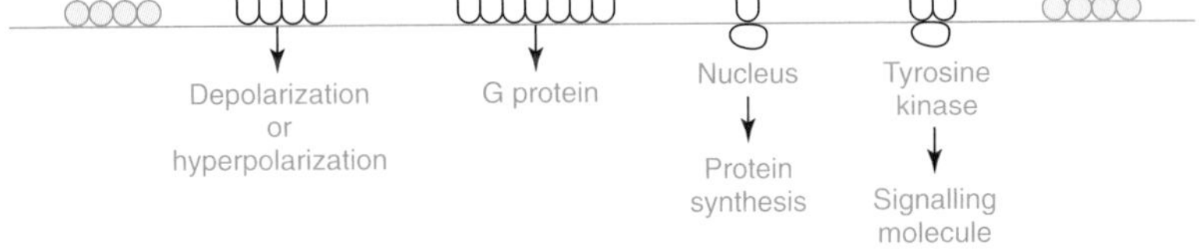

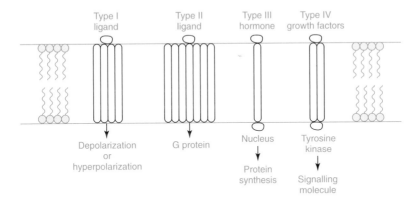

FIGURE 4–5. Type I is the ionotropic or ligand-gated receptor with a response time of milliseconds. An example is the nicotinic acetylcholine receptor. Type II is the metabotropic or G protein–linked receptor, generally modulatory, with a response time in minutes. Examples are the γ-aminobutyric $acid_B$ ($GABA_B$) and muscarinic receptors. Type III is the steroid receptor localized primarily on the nucleus rather than the cell membrane. The response time is usually minutes to hours, as exemplified by estrogen and thyroid hormones. Type IV represents the tyrosine kinase receptor, activated by insulin or growth factors, with a response time of minutes.

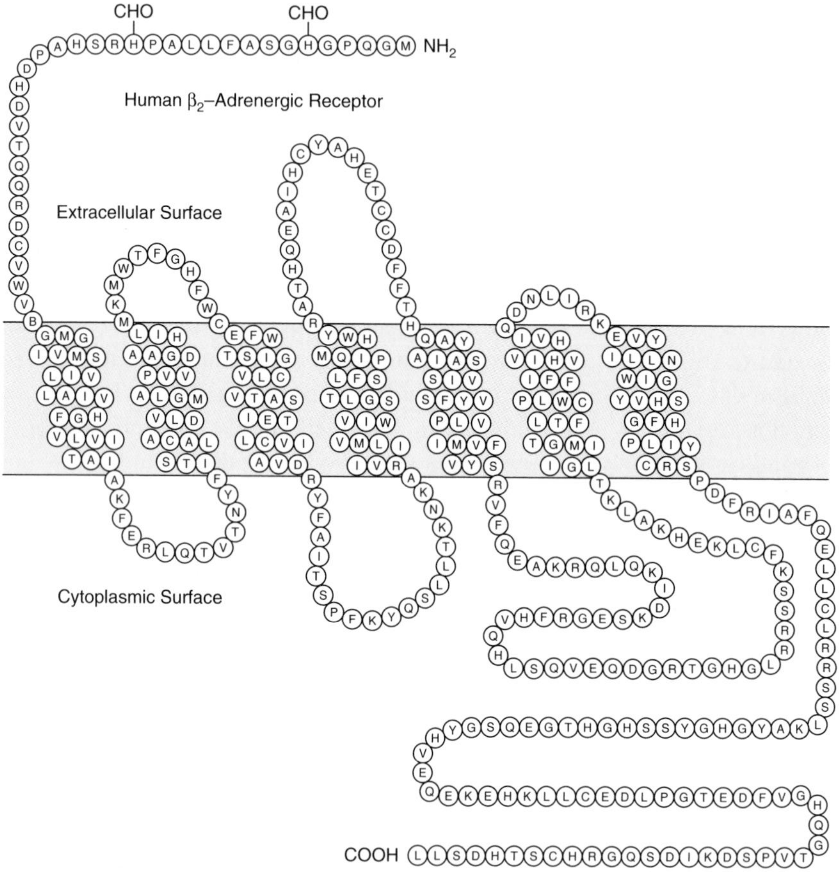

FIGURE 4–6. Topographical representation of the primary sequence of the human β_2-adrenergic receptor. The receptor protein is illustrated as possessing seven hydrophobic regions, each capable of spanning the plasma membrane, thus creating extracellular and intracellular loops as well as an extracellular terminus and a cytoplasmic C-terminal region. (From Lefkowitz et al., 1989.)

bosylates G_s and G_t, whereas G_i and G_o are substrates for pertussis toxin. Some G proteins (unclassified) are resistant to both toxins.

The third group are receptors for steroid hormones, thyroid hormones, vitamin D, and retinoic acids. These lipophilic ligands penetrate the cell, where they bind to the nuclear DNA to stimulate transcription. Response to hormones usually occurs in hours. Receptors for steroid hormones have previously been thought to be only intracellular since the hormones act on nuclear DNA to alter gene expression. This action is referred to as a *genomic* effect and usually takes hours to days to be observed. These genomic actions include the induction of neurotransmitter enzymes, receptors, and dendritic spines. Recently, however, evidence has accumulated that steroid receptors are also found on membranes. The hormone's *nongenomic* effects, occurring in seconds to minutes, act on the $GABA_A$ receptor to modulate chloride flux (see Chapter 6) and modulate a variety of other receptors (see Fig. 4–7). The major players in this activity are progesterone and its metabolites, estrogens, testosterone, and adrenal steroids. The neuroactive steroids, some of which can be synthesized in the brain, may be sedative, anxiolytic, antidepressant, or anticonvulsant. Recently, testosterone has been shown to reduce the secretion of the β-amyloid peptides that characterize Alzheimer's disease.

The fourth group are the tyrosine kinase receptors, now numbering around 100, differing from other receptors in that the kinase activity is part of the receptor. Membrane-localized, these receptors are activated by insulin and a number of growth factors, including nerve growth factor. In the presence of a ligand, the receptor dimerizes, activating the kinase to autophosphorylate. The subsequent phosphotyrosine residues provide acceptor sites for a variety of proteins.

Purinoreceptors have been classified as P_1 (G protein–linked) and P_2, with adenosine as the endogenous ligand for P_1 and adenosine triphosphate (ATP) and other purine and pyrimidine nucleotides activating P_2 receptors. Adenosine is primarily derived from the hydrolysis of ATP by ectonucleotidases. The P_1 family are G protein–coupled receptors, subdivided into A_1, A_{2a}, A_{2b}, and A_3 subtypes; all have been cloned. Interest in the P_1 receptor stems from the fact that adenosine (which is found at virtually every synapse that has been examined) exhibits an extraordinary constellation of activities. Adenosine (and its agonists) mainly inhibits the evoked release of neurotransmitters, hyperpolarizes postsynaptic membranes, decreases locomotor activity, possesses sedative and anticonvulsant activities, increases coronary and cerebral blood flow, produces bronchoconstriction, reduces neurodegeneration that occurs with stroke, and at high doses promotes catalepsy. Except for the A_3 subclass, the P_1 family of purinoreceptors is blocked by xanthines, with the classic antagonists being caffeine and theophylline.

The P_2 receptor family with ATP as the major agonist can be divided into two major classes, designated P_{2x} and P_{2y}, each with seven members to date. The former is coupled to ligand-gated ion channels and the latter is a G protein–linked family with the usual seven transmembrane domains. ATP, acting on P_{2X} ion channel receptors, is considered to be a cotransmitter in the peripheral nonadrenergic, noncholinergic nervous system, where it is re-

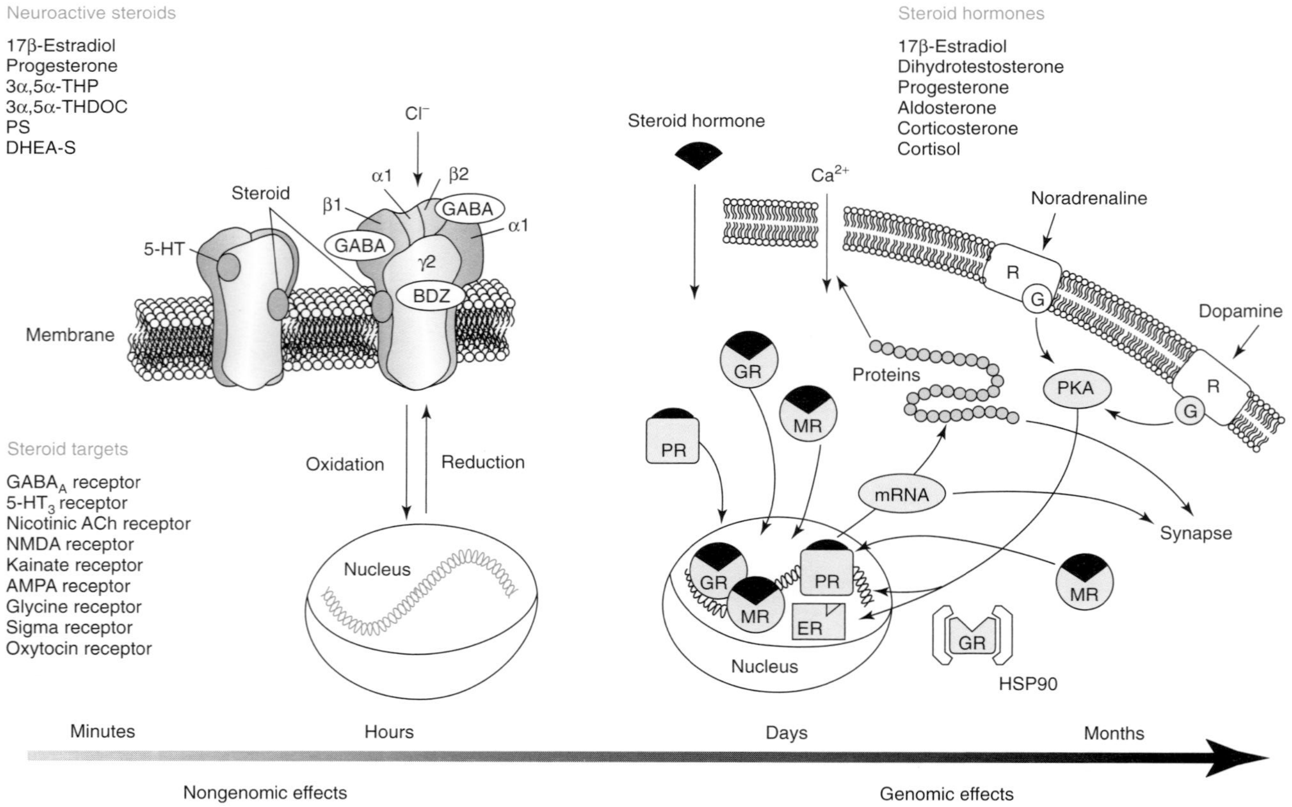
Neuroactive steroids
17β-Estradiol
Progesterone
3α,5α-THP
3α,5α-THDOC
PS
DHEA-S
Steroid
5-HT
Membrane
Cl⁻
α1
β2
β1
GABA
GABA
α1
γ2
BDZ
Steroid targets
GABAA receptor
5-HT3 receptor
Nicotinic ACh receptor
NMDA receptor
Kainate receptor
AMPA receptor
Glycine receptor
Sigma receptor
Oxytocin receptor
Oxidation
Reduction
Nucleus
Steroid hormones
17β-Estradiol
Dihydrotestosterone
Progesterone
Aldosterone
Corticosterone
Cortisol
Steroid hormone
Ca2+
Noradrenaline
R
G
Dopamine
R
G
PKA
Proteins
GR
MR
PR
mRNA
Synapse
GR
MR
PR
ER
MR
Nucleus
GR
HSP90
Minutes
Hours
Days
Months
Nongenomic effects
Genomic effects

leased with norepinephrine or ACh. ATP also has been shown to mediate fast synaptic transmission in mammalian neurons. Subclasses of the P_{2Y} receptor are activated by uridine triphosphate and ADP. The lack of selective antagonists to the P_2 receptor has hampered understanding of the function of its subclasses.

Continued administration of agonists can cause many receptors to desensitize and down regulate. Desensitization, occurring on a time scale of minutes, is reflected by a decreased response of the cell and is often related to receptor phosphorylation. Downregulation of receptors is observed on a time scale of hours after prolonged agonist exposure when receptors are internalized and degraded. Conversely, and predictably, continued administration of receptor antagonists generally causes upregulation of receptors.

Many neuroactive agents act on receptors that are coupled to the adenylyl cyclase system (see Fig. 4–8). The components of this complex are the receptor, the catalytic portion of adenylyl cyclase that converts ATP to cyclic adenosine monophosphate, and two G proteins referred to as G_s and G_i that are coupled to the catalytic unit of the enzyme. When the receptor is stimulated (e.g., a β_2-adrenergic receptor), the coupling protein is G_s. Conversely, when the receptor is inhibited (e.g., an α_2-adrenergic receptor), G_i is the coupling protein. Examples of adrenergic receptors whose activity is linked to the adenylate cyclase complex are the receptors α_2, β_1, and β_2 (but not α_1, which may be coupled to phosphatidyl inositide hydrolysis).

Finally, although not classified as receptors, another class of proteins that deserves attention is the transport proteins. With the exception of ACh, the

FIGURE 4–7. Nongenomic and genomic effects of neuroactive steroids. The term *neuroactive steroids* has been coined for steroids that interact with neurotransmitter receptors. Modulation of neuronal excitability by neuroactive steroids occurs over a very short (milliseconds to a few seconds) time period. The list in the upper left-hand corner shows steroids that fulfill the criteria for neuroactive steroids; the lower list gives neurotransmitter receptors that are targets for steroid modulation. The right-hand side of the figure describes the classical model of steroid-hormone action via the steroid-receptor cascade at the genomic level, which takes place over minutes to hours. The list on the right-hand side gives the names of typical steroid hormones. Certain steroids, such as 17β-estradiol and progesterone, have to be defined both as steroid hormones and as neuroactive steroids. BDZ, benzodiazepine; DHEA-S, dehydroepiandrosterone sulfate; ER, estrogen receptor; G, G protein; GR, glucocorticoid receptor; HSP90, heat-shock protein 90; MR, mineralocorticoid receptor; PKA, protein kinase A; PR, progesterone receptor; PS, pregnenolone sulfate; R, receptor; THDOC, tetrahydrodeoxycorticosterone; THP, tetrahydroprogesterone. (From Rupprecht and Holsboer, 1999.)

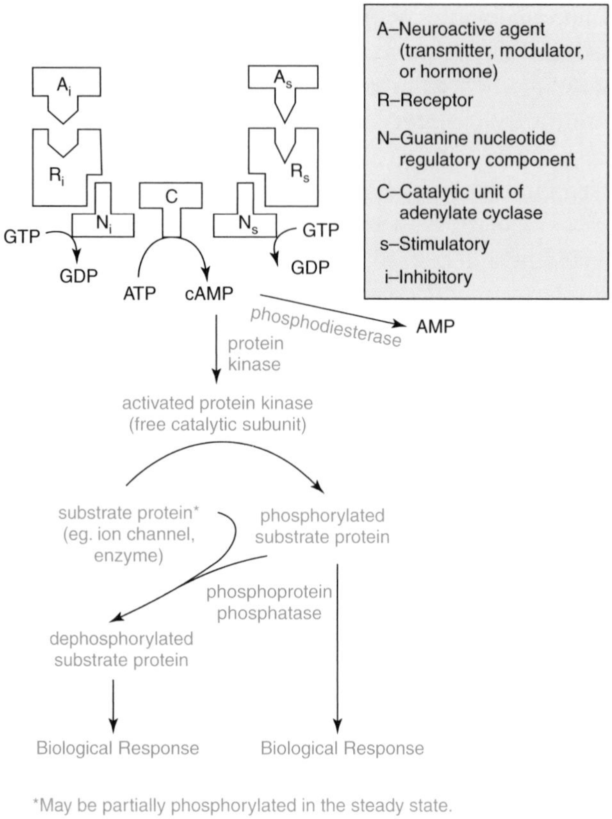

FIGURE 4–8. Components of a receptor-activated, cyclic nucleotide–linked system. (From Lefkowitz et al., 1984.)

action of all neurotransmitters that are released is terminated primarily by reuptake into their presynaptic terminals (amino acid transmitters can also be taken up by glial cells). ACh is hydrolyzed by acetylcholinesterase, and it is choline that is recaptured by the cholinergic terminal. Neurotransmitter transport proteins are relatively specific for each transmitter, are sodium-dependent, exhibit high-affinity kinetics, and are dependent on the membrane potential of the neuronal terminal. The GABA and glutamate transporters have the ability to function in reverse, transporting the transmitters out of the cell. Cloned plasma membrane neurotransmitter transporters fall into two families, one that includes transporters for monoamines and GABA, containing 12 transmembrane domains, and one for glutamate and aspartate, whose topology is still uncertain but likely to contain at least eight transmembrane domains. As discussed in the chapters on each neurotransmitter, these reuptake systems are the basis of the mechanism of action of many neuropharmacological agents, particularly the antidepressant drugs and some drugs of abuse such as cocaine.

The transporters operate at the plasma membrane of terminals and are different from intracellular vesicular transporters, which accumulate transmitters into synaptic vesicles. Vesicular transporters, some of which have been cloned, operate via a vacuolar type H^+-pumping ATPase. This proton pump generates an H^+ electrochemical gradient whereby the efflux of H^+ is coupled to the reuptake of the transmitter into the vesicles.

Because of space limitations, this chapter has not covered such topics as the factors regulating synthesis and degradation of receptors, subunit compositions of receptors, complex kinetics, isolation techniques, or the chemical compositions of these macromolecules. Students who are interested in these subjects may consult a number of recent papers and reviews listed below.

Selected References

Ariens, E. J. and A. J. Beld (1977). The receptor concept in evolution. *Biochem. Pharmacol.* 26, 913.

Berman, D. M. and A. G. Gilman (1998). Mammalian RGS proteins: barbarians at the gate. *J. Biol. Chem.* 273, 1269.

Birdsall, N. J. M., N. M. Nathanson, and R. D. Schwarz (2001). Muscarinic receptors: it's a knockout. *Trends Pharmacol. Sci.* 22, 215.

Burnstock, G. (2001). Purine-mediated signalling in pain and visceral perception. *Trends Pharmacol. Sci.* 22, 102.

Burnstock, G. (1999). Current status of purinergic signalling in the nervous system. *Prog. Brain Res.* 120, 3.

Changeux, J. P., J. Thiery, Y. Tung, and C. Kittel (1976). On the cooperativity of biological membranes. *Proc. Natl. Acad. Sci. U.S.A.* 57, 335.

Cheng, Y.-C. and W. H. Prusoff (1973). Relationship between the inhibition constant (K_i) and the concentration of inhibitor which causes 50 percent inhibition (I_{50}) of an enzymatic reaction. *Biochem. Pharmacol.* 22, 3099.

Clapham, D. E. and E. J. Neer (1997). G protein beta gamma subunits. *Annu. Rev. Pharmacol. Toxicol.* 37, 167.

Clark, A. J. (1937). General pharmacology. In *Heffter's Hanbuch d-exp. Pharmacol. Erg. Band 4*. Springer, Berlin.

Girault, J. A. and P. Greengard (1999). *Principles of Signal Transduction.* In *Neurobiology of Mental Illness* (D. S. Charney, E. J. Nestler, and B. S. Bunney, eds.). Oxford University Press, New York, p. 37.

Jonas, E. K. and L. K. Kaczmarek (1999). The inside story, subcellular mechanisms of neuromodulation. In *Beyond Neurotransmission* (P. S. Katz ed.). Oxford University Press, New York, p. 83.

Karlin, A. (1967). On the application of a plausible model of allosteric proteins to the receptor for acetylcholine. *J. Theor. Biol.* 16, 306.

Kenakin, T. P. (1997). Agonist-specific receptor confirmations. *Trends Pharmacol. Sci.* 18, 493.

Kenakin, T. P., R. A. Bond, and T. I. Bonner (1992). Definition of pharmacological receptors. *Pharmacol. Res.* 44, 351.

Laduron, P. M. (1984). Criteria for receptor sites in binding studies. *Biochem. Pharmacol.* 33, 833.

Lefkowitz, R. J., B. K. Kobilka, and M. G. Caron (1989). The new biology of drug receptors. *Biochem. Pharmacol.* 38, 2941.

Lefkowitz, R. J., M. G. Caron, and G. L. Stiles (1984). Mechanisms of membrane–receptor regulation. *N. Engl. J. Med.* 310, 1570.

Lester, H. H., S. Mager, M. W. Quick, and J. L. Corey (1994). Permeation properties of neurotransmitter transporters. *Annu. Rev. Pharmacol. Toxicol.* 34, 219.

Monod J., J. Wyman, and J. P. Changeux (1960). On the nature of allosteric transitions. *J. Biol. Chem.* 12, 88.

Morris, A. J. and C. C. Malbon (1999). Physiological regulation of G protein–linked signalling. *Physiol. Rev.* 79, 1373.

Nelson, N. (1998). The family of Na^+/Cl^- neurotransmitter transporters. *J. Neurochem.* 71, 1783.

Neubig, R. R. (1994). Membrane organization in G-protein mechanisms. *FASEB J.* 8, 939.

Rakic, P., P. S. Goldman-Rakic, and D. Gallager (1988). Quantitative autoradiography of major neurotransmitters in monkey striate and extrastriate cortex. *J. Neurosci.* 8, 3670.

Rudnick, G. (1990). Bioenergetics of neurotransmitter transport. *J. Bioenerg. Biomembr.* 30, 173.

Rupprecht, R. and A. F. Holsboer (1999). Neuroactive steroids: mechanism of action and neuropsychopharmacological perspectives. *Trends Neurosci.* 22, 410.

Seal, R.P. and S.G. Amara (1999). Excitatory amino acid transporters: a family in flux. *Annu. Rev. Pharmacol. Toxicol.* 39, 431.

Singer, S. J. and G. L. Nicolson (1972). The fluid mosaic model of the structure of cell membranes. *Science* 175, 720.

Strader, C. D., T. M. Fong, M. R. Tota, D. Underwood, and R.A.F. Dixon (1994). Structure and function of G protein–coupled receptors. *Annu. Rev. Biochem.* 63, 101.

Suenningsson, P., C. LeMoine, G. Fisone, and B. B. Fredholm (1999). Distribution, biochemistry and function of striatal adenosine A_{2A} receptors. *Prog. Neurobiol.* 59, 355.

Takai, Y. (2001). Small GTP-binding proteins. *Physiol. Rev.* 81, 153.

Van Holde, K. E. (1971). *Physical Biochemistry*. Prentice-Hall, Englewood Cliffs, NJ.

5

Modulation of Synaptic Transmission

Contrary to what one might assume, the more we learn about intercellular communication in the nervous system, the more complicated the situation appears. Up to about the mid-1970s, synaptologists smugly focused on a straight point-to-point transmission, where a presynaptically released transmitter impinged on a postsynaptic cell. Gradually, situations emerged where previously identified neurotransmitters were observed to not act in this fashion but, rather, to modulate transmission. These departures from the previous norm and the continuing discovery of peptides and small molecules such as adenosine, which exhibited neuroactivity but did not appear to be transmitters in the classic sense, supported the broader view of modulation of synaptic transmission. In retrospect, it is a concept that should have been apparent early on since it imparts to neural circuitry an extraordinary degree of flexibility, which is necessary when considering the mechanisms of behavioral changes. Also, modulation is not a neurophysiological property seen only in higher forms. In Cole Porter's words, "Birds do it, bees do it, even educated fleas do it."

Definitions

Primarily via the activation of a receptor, synaptic transmission may be modulated either presynaptically or postsynaptically. With presynaptic modulation, regardless of the mechanism, the ultimate effect is a change in the amount of transmitter released. With postsynaptic modulation, the ultimate effect is a change in the firing pattern of the postsynaptic neuron or in the activity of a postsynaptic tissue (e.g., blood vessel, gland, muscle). Because some confusion has arisen as to the correct nomenclature of pre- and postsynaptic receptors, an explanation is in order. What may be classified as a presynaptic receptor on neuron B may be a postsynaptic receptor of neuron A that is making an axoaxonic contact with neuron B (see Fig. 2–3). Thus, depending on which neuron is being investigated, the receptor will be denoted as either pre- or postsynaptic.

Procedures to localize activity at the presynaptic receptor level in a terminal include *(1)* the use of synaptosomes; *(2)* the addition of tetrodotoxin to the preparation to block action potentials in neighboring interneurons; *(3)* patch clamping presynaptic terminals; *(4)* imaging techniques using appropriate dyes; or *(5)* where feasible, either chemical destruction of terminals or lesioning of the neuron and then demonstration by ligand binding of loss of the receptor. To complete the nomenclature on presynaptic receptors, an *autoreceptor* is located on the terminal or somatic–dendritic area of a neuron that is activated by the transmitter(s) released from that neuron. A *heteroreceptor* is a presynaptic receptor activated by a modulating agent that originates from a different neuron or cell. As discussed in Chapter 1, modulators differ from transmitters in that they have no intrinsic activity but modulate ongoing neural activity. However, a transmitter may modulate at a concentration that is subthreshold for transmitter activity. As we will now detail, there exists an extraordinary number of possibilities for altering the point-to-point synaptic transmission mentioned above.

Presynaptic modulation can be affected by the following:

1. Receptor activation of a presynaptic neuron causing the following:
    a. A change in the firing frequency in the presynaptic neuron. This is probably the most common type of modulation, particularly in the central nervous system (CNS), where it can be assumed that the firing rate of virtually every neuron is governed by inputs on dendrites, soma, or axons. The firing rate determines the frequency of impulse conduction, hence the spread of action potentials into terminals or varicosities and the amount of transmitter that is liberated.
    b. A change in the transport or reuptake of a transmitter or precursor or in the synthesis, storage, release, or catabolism of a transmitter. All of these possibilities will result in a change in the concentration of a transmitter at the terminal. In practice, it has been shown that presynaptic activation of biogenic amine neu-

rons promotes the phosphorylation of both tyrosine and tryptophan hydroxylase, which increases the synthesis of norepinephrine, dopamine, and serotonin. Curiously, phosphorylation of the pyruvate dehydrogenase complex causes a decrease in enzyme activity and in theory would decrease acetylcholine levels of (ACh) and the amino acid transmitters. To date, however, modulation of this enzyme activity by presynaptic receptor activation has not been observed.

c. An effect on ion conductances at the terminal. The three ions and their respective channels that one might focus on would be K^+, Ca^{2+}, and Cl^-. Endogenous neuroactive agents or drugs could alter transmitter release by opening or blocking the channels or changing the kinetics of channel open time via the possible mediation of protein phosphorylation or other second messengers.

2. A direct effect of neuropharmacological agents on some element of the release process. This could be an effect of the modulating agent on vesicular apposition to a terminal, fusion, or fission.

Postsynaptic modulation can be affected by the following:

1. A long-term change in the number of receptors. This is not observed under normal physiological conditions. It is, however, commonly noted pharmacologically where the administration of a receptor agonist for a period of time will result in downregulation of the receptor, and, conversely, treatment with a receptor antagonist increases receptor density.
2. A change in the affinity of a ligand for a receptor. The now classic example is the salivary nerve of the cat, where both ACh and vasoactive intestinal peptide (VIP) are co-localized. When VIP is released upon electrical stimulation, it increases the affinity of ACh for the muscarinic receptor on the salivary gland up to 10,000-fold, with a consequent increase in salivation.
3. An effect on ionic conductances. As discussed in 1a above, this is frequency modulation. It is a postsynaptic effect on the first neuron in a relay, but it would be classified as a presynaptic effect on the second neuron. References to all of the neurons and modulating agents that have been investigated are given in the reviews by Starke et al. (1989), Chesselet (1994), and Levitan and Kaczmarek (2002). These reviews can be summarized by stating that virtually every neuronal pathway is modulated and virtually every endogenous neuroactive agent has been shown in one preparation or another to be capable of affecting synaptic transmission. All of this information is

descriptive. The question of the second-messenger systems that may be involved in regulating synaptic activity is addressed in the following section. Figure 5–1 depicts a major pathway for modulation of synaptic transmission. Rapidly accumulating evidence suggests that, in most cases of receptor-activated inhibitory presynaptic modulation, the ultimate effect is to open K channels. This hyperpolarizes terminals; less Ca^{2+} enters; and, as a consequence, less transmitter is released. Another major presynaptic mechanism is the

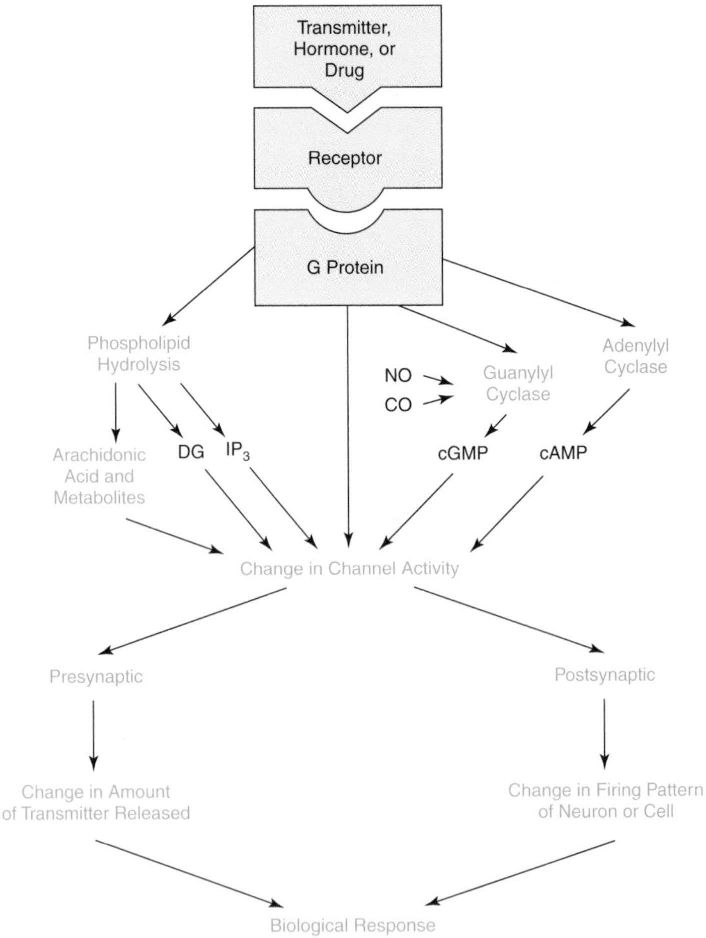

FIGURE 5–1. Major pathways for modulation of synaptic transmission. DG, diacylglycerol; IP_3, inositol trisphosphate; NO, nitric oxide; CO, carbon monoxide; cGMP, cyclic guanosine monophosphate; cAMP, cyclic adenosine monophosphate.

inhibition of a calcium channel that would also decrease the evoked release of a transmitter. With the less common receptor-activated excitatory modulation, closing of a K channel has been implicated. A well-documented postsynaptic modulatory mechanism is that of a transmitter that inhibits a voltage-dependent K current, causing a subsequent depolarizing stimulus to produce an enhanced response.

Second Messengers

Three major biochemical cascades and two new gaseous messengers, nitric oxide (NO) and carbon monoxide, have been described:

1. *Protein phosphorylation.* Following the identification of cyclic nucleotides by Sutherland and associates and the implication that they act as a second-messenger system preceded by an initial nerve impulse or hormonal signal, Krebs and colleagues demonstrated a multistep sequence of events that linked cyclic adenosine monophosphate (cAMP) generation in muscle to the regulation of carbohydrate metabolism. Since that time, the cyclic nucleotides have been shown to regulate an enormous diversity of processes ranging from axoplasmic transport to neuronal differentiation and including transmitter synthesis and release and the generation of postsynaptic potentials. All of these effects are attributable to the cAMP- or cyclic guanosine monophosphate (cGMP)–activating protein kinases and thus protein phosphorylation. Second-messenger activity, achieved via protein phosphorylation, can be mediated by cAMP- and cGMP-dependent protein kinases as well as by calcium–calmodulin-dependent protein kinase and calcium–phosphatidylserine- or calcium–diacyglycerol-dependent protein kinase (protein kinase C). The phosphoryl acceptor in these proteins is the hydroxyl group of serine, threonine, or tyrosine, with the first two being the most prominent. All protein kinases can themselves be autophosphorylated, a process that usually increases kinase activity. Although in most instances biological activity results from kinase-activated phosphorylation of a substrate protein, in some cases it is phosphatase-activated dephosphorylation of a phosphorylated protein that produces the biological response. Currently, eight phosphoprotein phosphatases have been identified (protein phosphatase-2B is also referred to as calcineurin and is regulated by proteins referred to as immunophilins). Phosphatases fall into two broad classes, phosphoserine/phosphothreonine phosphatases and phosphotyrosine phosphatases.

Yet another level of regulation, noted above, is suggested by the finding that protein phosphatase activity can be inhibited by other proteins, the most interesting of which is DARPP-32, found in D_1 dopaminoceptive neurons. DARPP-32 (dopamine- and cAMP-regulated phosphoprotein of 32 kDa), by acting as a protein phosphatase inhibitor when it is phosphorylated, can regulate the postsynaptic effects of dopamine in dopaminoceptive cells. Phosphorylated DARPP-32 is inactivated by protein phosphatase-2B. In some instances, Ca^{2+} acts as a second messenger without the participation of protein phosphorylation. The diversity of signals that are coupled to protein phosphorylation is depicted in Figure 5–2. (For a molecular illustration of receptor coupling, see Fig. 4–6.)

Despite the vast number of systems in which protein phosphorylation is implicated, around 500, there are currently only about a dozen cases where direct evidence links this process to modulation of synaptic transmission, either pre- or postsynaptically. It is not yet known if the proteins that make up the channels are phosphorylated or if the phosphorylated proteins are morphologically associated with the channels. At any rate, with over four dozen proteins in the brain that are known to be phosphorylated (Table 5–1), the story is far from complete. Questions that remain to be answered include the regulation of the enormous number of steps in the cascade and the substrate specificity of the phosphodiesterases, protein kinases, and phosphoprotein phosphatases. What cannot be overemphasized is the involvement of phosphorylation in virtually every aspect of neuronal function, including transmitter release, synthesis, and reuptake; ion channel activity; regulation of receptors; cytoskeleton proteins; gene expression; and possibly short-term memory (Nestler & Greengard, 1999).

2. *Phosphoinositide hydrolysis.* In 1953, Hokin and Hokin showed that the incorporation of inorganic phosphate (P_i) into phosphatidylinositol (PI) and phosphatidic acid (PA) in pancreatic slices was stimulated by ACh and ultimately resulted in the release of amylase. This receptor-activated hydrolysis of phosphoinositides is referred to as the *phosphatidylinositol effect.* For nearly 30 years after this report, the literature on this effect was replete with the traditional scientific jargon "it is tempting to speculate that . . . ," with no one having solid evidence as to whether the phosphoinositides or the inositol phosphates were the message and, if so, exactly what was the medium for the exchange. That this situation has now dramatically improved is shown in Figure 5–3.

The signals that initiate this transduction process in neuronal systems include ACh, norepinephrine, serotonin, histamine, glutamate bradykinins, substance P, vasopressin, thyrotropin-releasing hormone, neurotensin, VIP, nerve growth factor, and angiotensin acting on brain, sympathetic ganglia, salivary glands, iris smooth mus-

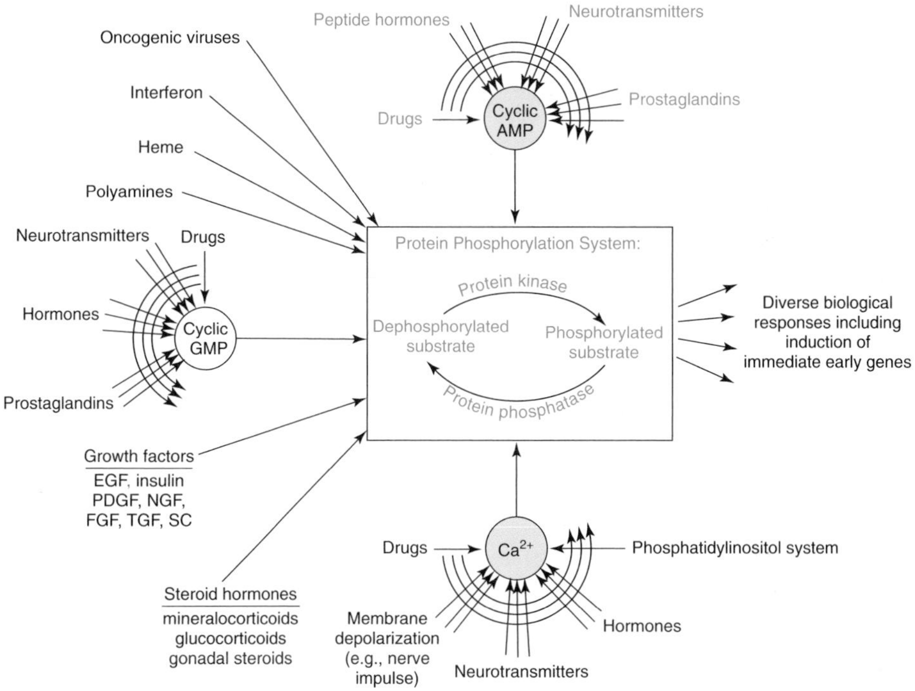

Figure 5–2. Schematic diagram of the role played by protein phosphorylation in mediating some of the biological effects of a variety of regulatory agents. Many of these agents regulate protein phosphorylation by altering intracellular levels of a second messenger, cyclic adenosine monophosphate (AMP), cyclic guanosine monophosphate (GMP), or Ca^{2+}. Other agents appear to regulate protein phosphorylation through mechanisms that do not involve these second messengers. Most drugs regulate protein phosphorylation by affecting the ability of first messengers to alter second-messenger levels (*curved arrows*). A small number of drugs (e.g., phosphodiesterase inhibitors, Ca^{2+} channel blockers) regulate protein phosphorylation by directly altering second-messenger levels (*straight arrows*). EGF, epidermal growth factor; PDGF, platelet-derived growth factor; NGF, nerve growth factor; FGF, fibroblast growth factor; TGF, transforming growth factor; SC, somatomedin C. (Modified from Nestler and Greengard, 1989.)

TABLE 5–1. Classes of Neuronal Proteins Regulated by Phosphorylation

Enzymes involved in neurotransmitter biosynthesis and degradation
Tyrosine hydroxylase
Tryptophan hydroxylase
Neurotransmitter receptors
β-Adrenergic receptors
α_2-Adrenergic receptors
Muscarinic cholinergic receptors
Opioid receptors
$GABA_A$ receptor subunits
NMDA glutamate receptor subunits
Non-NMDA glutamate receptor subunits
Nicotinic acetylcholine receptor subunits
Neurotransmitter transporters
Monoamine-reuptake transporters
Monoamine vesicle transporters (VAT)
Ion channels
Voltage-dependent Na^+, K^+, Ca^{2+} channels
Ligand-gated channels
Ca^{2+}-dependent potassium channels
Enzymes and other proteins involved in the regulation of second messenger
G proteins
Phospholipases
Adenylyl cyclases
Guanylyl cyclases
Phosphodiesterases
IP_3 receptor
Protein kinases
Autophosphorylated protein kinases (protein kinases phosphorylating themselves)
Protein kinases phosphorylated by other protein kinases (many examples)
Protein phosphatase inhibitors
DARPP-32
Inhibitors 1 and 2
Cytoskeletal proteins involved in neuronal growth, shape and motility
Actin
Tubulin
Neurofilaments (and other intermediate filament proteins)

(*continued*)

TABLE 5–1. *(Continued)*

- Myosin
- Microtubule-associated proteins
- Actin-binding proteins

Synaptic vesicle proteins involved in neurotransmitter release

- Synapsins I and II
- Clathrin
- Synaptophysin
- Synaptobrevin

Transcription factors

- CREB family members
- Fos and Jun family members
- STATs
- Steroid and thyroid hormone receptors
- NFκB-IKB family

Other proteins involved in DNA transcription or mRNA translation

- RNA polymerase
- Topoisomerase
- Histones and nonhistone nuclear proteins
- Ribosomal protein S6
- eIF
- eEF
- Other ribosomal proteins

Miscellaneous

- Myelin basic protein
- Rhodopsin
- Neural cell adhesion proteins
- MARCKS
- GAP-43

This list is not intended to be comprehensive but, instead, to indicate the wide array of neuronal proteins regulated by phosphorylation. Some of the proteins are specific to neurons, but most are present in many cell types in addition to neurons and are listed here because their multiple functions in the nervous system include the regulation of neuron-specific phenomena. Not included are the many phosphoproteins present in diverse tissues, including brain, that play a role in generalized cellular processes, such as intermediary metabolism, and that do not appear to play a role in neuron-specific phenomena. NMDA, *N*-methyl-D-aspartate; CREB, cAMP response element-binding protein; STAT, signal-transducing activator of transcription; GAP-43, growth-associated protein of 43 kDa; MARCKS, myristoylated alanine-rich C, kinase substrate; IP_3, inositol trisphosphate; DARPP-32, dopamine and cAMP-regulated phosphoprotein of 32-kDa; GABA, γ-aminobutyric acid; eIF, eukaryotic initiation factor; eEF, eukaryotic elongation factor.

SOURCE: Modified from Hyman and Nestler (1993).

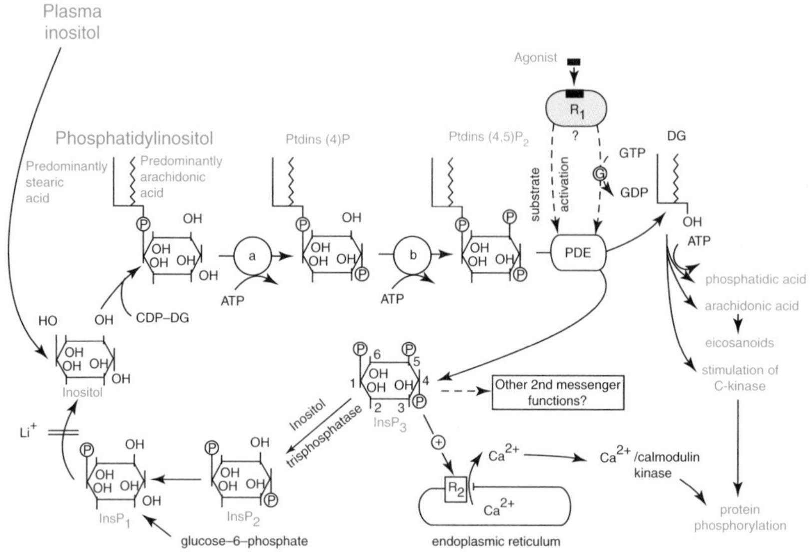

FIGURE 5–3. Receptor-activated phosphoinositide metabolism. The binding of an agonist to a receptor on the plasma membrane stimulates the hydrolysis of phosphatidylinositol 4,5-bis-phosphate [Ptdins (4,5)P_2] by a phosphodiesterase (PDE, phosphoinositidase, phospholipase C), a specific phospholipase whose activity is controlled by a guanine nucleotide regulatory protein to form inositol 1,4,5-trisphosphate ($InsP_3$) and diacylglycerol (DG). The former binds to a receptor (R_2) on the endoplasmic reticulum to release calcium, which can directly produce a biological response or can activate calmodulin kinase to promote protein phosphorylation and a subsequent biological response. In some cells (e.g., mouse atria, neuroblastoma, and glioma hybrid NG108-15), receptor-activated production of $InsP_3$ requires extracellular Ca^{2+}.

The latter parallel arm of the cycle, diacylglycerol, can also promote a biological response via the production of prostaglandins, thromboxanes, and leukotrienes from released arachidonic acid (arachidonic acid has also been reported to stimulate guanylate cyclase to generate cyclic guanosine monophosphate) or via stimulation of protein kinase C (C-kinase) and subsequent protein phosphorylation. Diacylglycerol has also been reported to promote fusion of synaptic vesicles to terminal membrane. Diacylglycerol can also be generated from phosphatidylcholine via phospholipase D followed by phosphatidic acid phosphatase.

The phosphoinositides are synthesized from inositol with cytidine diphosphate:diacylglycerol (CDP-DG) as intermediary and the stepwise phosphorylation by kinases (a and b). As shown in the figure, lithium blocks the cycle by inhibiting inositol-1-phosphatase. Although the antimanic activity of lithium has been ascribed to this inhibitory effect, the evidence is not compelling. ATP, adenosine triphosphate; GTP, guanosine triphosphate; GDP, guanosine diphosphate (Modified from Berridge and Irvine, 1984.)

cle, adrenal cortex, and neuronal tumor cells. Specific receptors that have been implicated are muscarinic cholinergic receptors, α_1-adrenergic receptors, the H1 histaminergic receptor, substance P, and the V1 vasopressin receptor. In each case, Ca^{2+} appears to be the intracellular second messenger that activates phosphoinositide hydrolysis. Like the specific guanosine triphosphate (GTP)–binding proteins that link receptors to the adenylate cyclase system discussed earlier, a specific GTP-binding protein is also coupled to the phosphodiesterase that catalyzes the hydrolysis of phosphatidylinositol 4,5-bisphosphate. In addition to the G protein–linked receptor, it is now known that a tyrosine kinase–linked receptor is coupled to a specific phosphodiesterase (phospholipase C), which yields inositol 1,4,5-trisphosphate ($InsP_3$) and diacylglycerol.

Thus, as noted in Figure 5–3, the key participants in the transduction process are a G protein–linked receptor, a tyrosine kinase–linked receptor, and phosphodiesterases, yielding two separate second messengers, the water-soluble $InsP_3$ and the lipid-soluble diacylglycerol. The former mobilizes calcium (released in a wave form, i.e., *oscillatory*), which can act through calmodulin to phosphorylate specific proteins, and the latter, by activating protein kinase C, a calcium–phosphatidylserine-dependent family of protein kinases, also phosphorylates specific proteins. The $InsP_3$ receptor, associated with the smooth endoplasmic recticulum, is a tetramer of identical subunits that is regulated by adenosine triphosphate (ATP) and cAMP. Since diacylglycerol can activate guanylate cyclase to produce cGMP, cGMP-dependent activity (protein kinase or otherwise) must be considered. With an assumed ambidexterity of neuronal cells, these two arms could function singly, cooperatively, or antagonistically, depending on the situation, thus providing subtle variations on the modulatory mechanism. In addition, as noted in Figure 5–3, calcium may produce a physiological response directly without invoking activation of calmodulin, so yet another control is indicated. On the subject of control, the activities of the various kinases, esterases, and phospholipases in the PI cycle would be expected to be vital control points. For example, five isoforms of phospholipase C have now been identified, of which some are enriched in specific brain areas. Free inositol in the brain must be derived from glycolysis since plasma inositol cannot pass the blood–brain barrier to any significant degree. Glycolysis, therefore, would be another regulatory factor in the response mechanism.

Finally, although the origin of the mobilized calcium is now clear (it is released from endoplasmic reticulum and mitochondria), some controversy remains as to whether phosphoinositide hydrolysis releases only internal calcium or whether external calcium influx is also invoked. Current evidence suggests that for neuronal modulation both or either may be involved, depending on the preparation. The same answer can also be given as to whether the PI effect is presynaptic or postsynaptic. A complication in the PI cycle that has recently surfaced is the finding that the inositol trisphosphate that is produced is not always or only Ins(1,4,5)P_3 but may be Ins(1,3,4)P_3 as well as $InsP_4$, $InsP_5$, and $InsP_6$. The physiological role of these polyphospate compounds is unknown, but they may function as phosphate donors.

3. *Eicosanoids (arachidonic acid metabolites).* Arachidonic acid, synthesized from dietary linoleic acid, derived on demand by either a G protein–regulated phospholipase A_2 or diglyceride lipase activation (Fig. 5–4), yields a bewildering array of bioactive metabolites, as shown in Fig. 5–5. The three major groups are prostaglandins, thromboxanes, and leukotrienes.

It is well known that the eicosanoids, particularly the prostaglandin series, play an important modulatory role in nervous tissue, but it is difficult to write a lucid account of specifically how and where they act. This is primarily due to the fact that they are not stored in tissue but synthesized on demand, particularly in pathophysiological conditions; they act briefly (some with a half-life of seconds) and at extremely low concentrations (10^{-10} M). Although indomethacin is a good inhibitor of cyclooxygenase, blocking the conversion of arachidonic acid to prostaglandins, there are few specific inhibitors available to block lipoxygenase and epoxygenase. Thus, although it had been postulated that the E series of prostaglandins modulates noradrenergic release; blocks the convulsant activity of pentylenetetrazol, strychnine, and picrotoxin (possibly by increasing the level of γ-aminobutyric acid in the brain); and increases the level of cAMP in cortical and hypothalamic slices, these effects were noted in vitro with the addition of substantial amounts of the prostaglandins. There was very little evidence in intact animals to support these neuronal findings, and since we all believe in "in vivo veritas," the physiological relevance of the effect was in doubt.

Recently, however, direct evidence has implicated arachidonic acid and lipoxygenases as second messengers. The cascade begins with the binding of a neuroactive agent to its receptor. Then, according to findings from the Axelrod laboratory, the receptor is coupled to G proteins, which may either activate or inhibit phospholipase A_2, although this has not been conclusively established for all neural tissues. The activated enzyme promotes the release of arachidonic acid, which will then act intracellularly as a second messenger. Arachidonic acid and its metabolites can also leave the cell to act extracellularly as first messengers on neighboring cells. Eicosanoids have been shown to mediate the somatostatin-induced opening of an M channel in hippocampal pyramidal cells and the release of VIP in mouse cerebral cortical slices. It is thus becoming clear, despite enormous technological difficulties in assaying eicosanoids, that these agents are major messengers. Before we

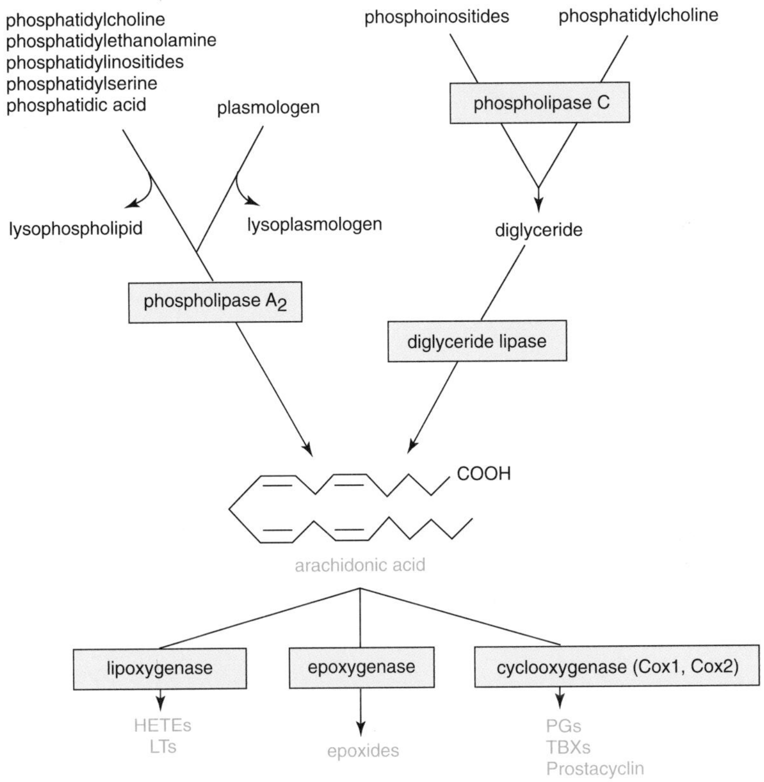

FIGURE 5–4. Pathways for the generation and metabolism of arachidonic acid. Arachidonate can arise directly from phospholipids through the action of phospholipase A_2 or prior action of phospholipase C, followed by the action of diglyceride lipase. Alternatively, the diglyceride may be phosphorylated to phosphatidic acid by the action of diglyceride kinase, and arachidonate then can be released through the action of phospholipase A_2. The released arachidonate may then be metabolized by lipoxygenase, cyclooxygenase, or epoxygenase enzymes to form leukotrienes (LTs), hydroxyeicosatetraenoic acids (HETEs), prostaglandins (PGs), thromboxanes (TBXs), and epoxides. (Modified from Axelrod et al., 1988.)

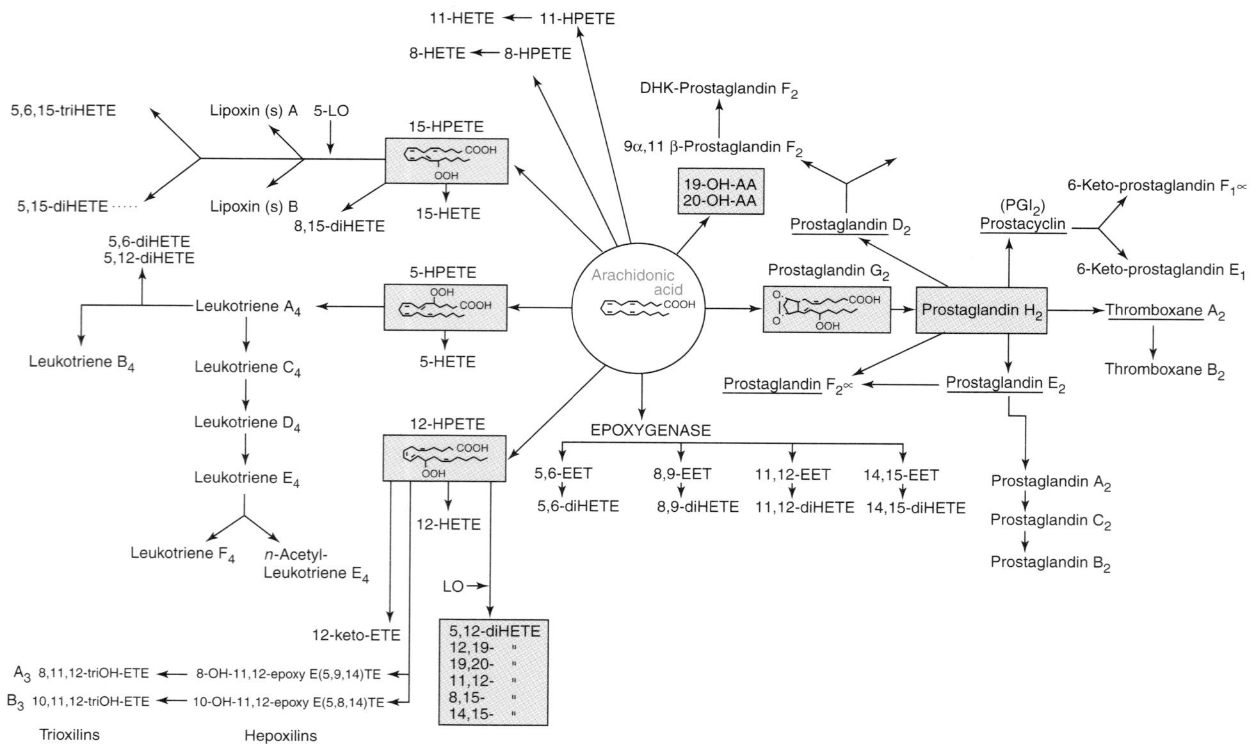

FIGURE 5–5. Metabolism of arachidonic acid (AA) by cyclooxygenase, lipoxygenases (LO), epoxygenase, and corresponding end products (prostanoids). PGI_2, prostaglandin I_2; HETE, hydroxyeicosatetraenoic acid. (From Schaad et al., 1991.)

leave the arachidonic acid story, an exciting new finding should be mentioned. Cannabinoids are a group of psychoactive compounds found in marijuana, of which the principal component is Δ^9-tetrahydrocannabinol. In binding studies, two cannabinoid receptors were isolated and cloned, referred to as CB1 and CB2 receptors. The CB1 receptor is found in the brain and peripheral nervous system, while the CB2 receptor is expressed mainly in the immune system. Both receptors are G protein–coupled. Subsequently, several cannabinoid ligands were identified, with the major ones in brain being arachidonylethanolamide (a.k.a., anandamide) and 2-arachidonylycerol. Recent evidence suggests that anandamide may react with G protein–coupled receptors that are not CB1 or CB2. Anandamide is synthesized by the transfer of an arachidonic acid–containing phospholipid to the amine group of phosphatidylethanolamine to form *n*-arachidonoyl-phosphatidylethanolamine, followed by phospholipase and hydrolysis to yield arachidonylethanolamide (anandamide) and phosphatidic acid. Cannabinoids are hydrolyzed by fatty acid amide hydrolase. Cannabinoid agonists appear to be involved in neuroprotection, as appetite stimulants, antiemetics, and analgesics. The naturally occurring canabinoid 2-arachidonylglycerol appears to have hypotensive activity.

Nitric Oxide

In 1980, Furchgott and Zawadzki observed that stimulation of the endothelium released a factor that relaxed blood vessels. This factor, referred to as the endothelium-derived relaxing factor, was subsequently identified by the Moncada laboratory as NO. Acting as a second messenger, NO is now known to be involved in an incredible number of systems. Aside from its role in the nervous system, NO is a mediator in the cardiovascular, renal, pulmonary, endocrine, and immune systems.

NO is synthesized from arginine via the enzyme NO synthase, a flavin adenine dinucleotide and flavin mononucleotide enzyme, requiring molecular O_2 and with reduced nicotinamide-adenine dinucleotide phosphate as coenzyme and tetrahydrobiopterin as cofactor. The neuronal and endothelial enzyme that is constitutively expressed is activated by Ca^{2+} and calmodulin, whereas the macrophage enzyme that is inducible by cytokines is not. The synthetic reaction sequence in the brain is shown in Figure 5–6.

NO synthase is inhibited by a variety of arginine analogs, and this finding has proved to be invaluable in delineating the functions of NO in vivo. The constituitive neuronal NO synthase is regulated by phosphorylation. A number of protein kinases can phosphorylate the enzyme, and this process decreases enzyme activity. NO itself is destroyed by reacting with hemoglobin

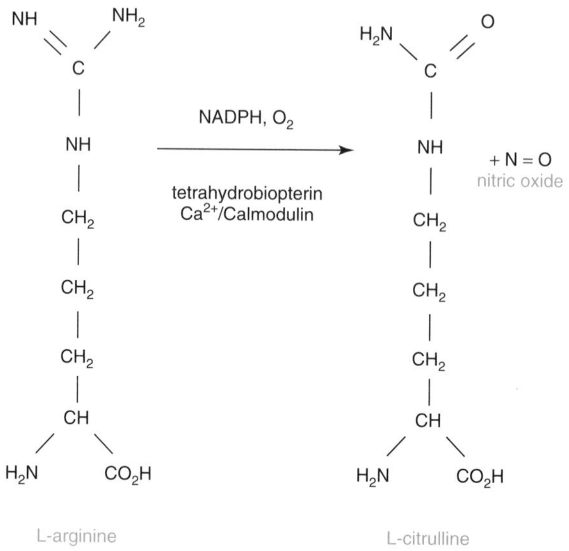

FIGURE 5–6. Synthesis of nitric oxide.

and other iron-containing compounds. Interestingly, vasodilators such as nitroprusside and nitroglycerin produce NO, and this is considered to be their mechanism of action.

NADPH diaphorase immunoreactivity is co-localized in most, but not all, cells with NO synthase activity; and this has been used to map the distribution of neurons that release NO. Currently, three isoforms of NO synthase have been reported that arise from distinct genes and share approximately 50% amino acid identity. An unusual feature of this enzyme is that it is a member of the family of cytochrome P-450 proteins, which catalyze the hydroxylation of a variety of metabolites as well as drugs.

So much for the background. Now to the neuronal effects of NO. First, it should be noted that NO is not present in synaptic vesicles, is not released from terminals by exocytosis, and is not stored in any reservoir. Rather, it is released upon stimulation and diffuses from a neuronal (and glial) population to act on enzymes and other elements. The first and primary action of this gaseous molecule is to complex the iron of soluble guanylyl cyclase to stimulate the enzyme and increase the concentration of cGMP. Another target, aside from guanylyl cylase, is cytosolic adenosine diphosphate (ADP)-ribosyl transferase. The consequence of this latter activity is unknown. The rise in cGMP will activate cGMP-dependent protein kinases, which catalyze the phosphorylation of substrate proteins, largely unidentified, which then give rise to myriad effects. The first indication of a role of NO in neuronal systems came from the Garthwaite laboratory, when investigators found that in cerebellar slices excitatory amino acids led to NO release, as did stimulation of nonadrenergic, noncholinergic peripheral neurons. Subsequent research has provided evidence that NO synthase inhibitors block *N*-methyl-D-aspartate (NMDA) receptor activation, preventing the well-documented NMDA-induced neurotoxicity. However, the situation is confusing because prolonged glutamate stimulation produces high NO levels that can kill neuronal cells. In addition to this activity and its relaxation effect in stimulating peripheral nonadrenergic noncholinergic neurons, NO is thought to play a role in both long-term potentiation, where it acts as a retrograde messenger, and long-term depression. These functions concerning synaptic plasticity are complex and remain to be resolved. Clearly, the role of NO in both physiological and pathophysiological processes is still evolving.

Carbon Monoxide

Another gaseous molecule has recently surfaced as a neuromodulating second messenger that also activates guanylyl cyclase. This is carbon monoxide (CO). CO is generated via heme oxygenase, which catalyzes the conversion of heme to biliverdin with the liberation of CO. Two heme oxygenases, expressed by separate genes, have been described. Heme oxygenase-1 is inducible and localized mainly in the spleen and liver, with a small (but inducible) concentration in glial cells and in a few neurons. In contrast, heme oxygenase-2 is constitutive, not inducible, and found in high concentrations in the brain, particularly cerebellum, olfactory bulb, and hippocampus. It is not present in glial cells under normal conditions but can be induced there. It is activated by protein kinase C.

As noted earlier, CO, like NO, raises the level of cGMP. While the localization of NO synthase does not always correlate with the localization of guanylyl cyclase, the localization of heme oxygenase and guanylyl cyclase is virtually identical and coincides with the presence of cytochrome P-450 reductase, a necessary electron donor for heme oxygenase as well as NO synthase.

Two other possible second messengers are protein carboxyl methylation and phospholipid methylation. Both processes involve *S*-adenosylmethionine as the methyl donor. Although it has been shown that protein carboxyl methylation inhibits calmodulin-linked enzymes and that phospholipid methylation alters terminal membrane viscosity, these activities currently are not seen as playing an important role in modulation.

Among endogenous modulators, the neuroactive peptides are by far the most mysterious with respect to their widespread activity. These agents can be released presynaptically, postsynaptically in response to presynaptic receptor activation, from glial cells, or from blood vessels. Another mysterious modulator is adenosine, which is found at virtually every synapse that has been examined. Electrophysiologically, it tends to inhibit the evoked release of transmitters; but it also acts postsynaptically, exhibiting a variety of behavioral effects ranging from evoking premature arousal in hibernating ground squirrels to anticonvulsant activity, increasing cerebral blood flow, and interacting with the benzodiazepine receptor. Finally, recent evidence implicates D-serine, synthesized in astrocytes, as a neuromodulator of NMDA receptors (see Chapter 6).

Reflecting on the mechanisms available to nervous tissue to modulate synaptic transmission, one cannot fail to be overwhelmed by the almost infinite possibilities that provide the fine tuning for behavioral changes. It may well turn out that the key player in this scenario is calcium. Since Ca^{2+} is released as a wave, its concentration as well as its translocation at discrete sites might dictate the modulatory effect of the attachment of a ligand to its receptor. It may also be the key to nonsynaptic cellular effects, such as cell movement or proliferation, gene expression, and metabolism, that involve second-messenger systems. At any rate, one should be aware that the ultimate effect of cascades of second-messenger systems, transcription factors, neurotrophic factors, transport factors, gene expression, or the endless numbers of signaling factors (irritatingly referred to by acronyms) is to cause a neuron to fire at a certain frequency or not to fire. It is that simple except that a neuron may have 10,000 inputs dictating its ultimate effect on other neurons to which it is coupled.

Somewhere over the rainbow, one hopes that one day somebody will reveal the grand design in neuronal communication that leads to behavioral changes ranging from the subtle to the dramatic.

Selected References

Arancio, O., M. Kiebler, C. J. Lee, V. Lev-Ram, R. Y. Tsien, E. R. Kandel, and R. D. Hawkins (1996). Nitric oxide acts directly in the presynaptic neuron to provide long-term potentiation in cultured hippocampal neurons. *Cell* 87, 1025.

Axelrod, J., R. M. Burch, and C. L. Jelsema (1988). Receptor-mediated activation of phospholipase A_2 via GTP-binding proteins: arachidonic acid and its metabolites as second messengers. *Trends Neurosci.* 11, 117.

Baranano, D. E., C. D. Ferris, and S. H. Snyder (2001). Atypical neural messengers. *Trends Neurosci.* 24, 99.

Berridge, M. J. (1993). Inositol trisphosphate and calcium signaling. *Nature* 361, 315.

Berridge, M. J., P. Lipp, and M. D. Bootman (2000). The versatility and unversatility of calcium signalling. *Nat. Rev. Mol. Cell. Biol.* vol 1, p. 11.

Buttner, N., S. A. Siegelbaum, and A. Volterra (1989). Direct modulation of *Aplysia* S-K^+ channels by a 12-lipoxygenase metabolite of arachidonic acid. *Nature* 342, 553.

Chesselet, M.-F. (1994). Presynaptic regulation of neurotransmitter release in the brain. *Neuroscience* 12, 347.

Clapham, D. E. and E. J. Neer (1993). New roles for G-protein $\beta\chi$ dimers in transmembrane signaling. *Nature* 365, 403.

deMendonca, A., A. M. Sebastiao, and J. A. Ribeiro (2000). Adenosine: does it have a neuroprotective role after all. *Brain Res. Rev.* 33, 258.

Dennis, E. A., S. G. Rhee, M. M. Billah, and Y. A. Hannun (1991). Role of phospholipases in generating lipid second messengers in signal transduction. *FASEB J.* 5, 2068.

Dunwiddie, T.V. and S. A. Masino (2001). The role and regulation of adenosine in the central nervous system. *Annu. Rev. Neurosci.* 24, 31.

Fisher, S. K., A. M. Heacock, and B. W. Agranoff (1992). Inositol lipids and signal transduction in the nervous system: an update. *J. Neurochem.* 58, 18.

Furchgott, R. F. and J. V. Zawadzki (1980). The obligatory role of endothelial cells in the relaxation of arterial smooth muscle by acetylcholine. *Nature* 288, 373.

Garthwaite, J. (1991). Glutamate, nitric oxide, and cell–cell signaling in the nervous system. *Trends Neurosci.* 14, 60.

Halushka, P. V., D. E. Mais, P. R. Mayeux, and T. A. Morinelli (1989). Thromboxane, prostaglandin, and leukotriene receptors. *Annu. Rev. Pharmacol. Toxicol.* 29, 213.

Hokin, L. E. (1985). Receptors and phosphoinositide-generated second messengers. *Annu. Rev. Biochem.* 54, 205.

Kerwin, J. R., Jr. and M. Heller (1994). The arginine–nitric oxide pathway: a target for new drugs. *Med. Res. Rev.* 14, 23.

Kruszka, K. K. and R. W. Gross (1994). The ATP- and CoA-independent synthesis of arachidonoylethanolamide. *J. Biol. Chem.* 269, 14345.

Levitan, I. B. (1994). Modulation of ion channels by protein phosphorylation and dephosphorylation. *Annu. Rev. Physiol.* 56, 193.

Levitan, I. B. and L. K. Kaczmarek (2002). *The Neuron: Cell and Molecular Biology*, 2nd ed. Oxford University Press, New York.

Maines, M. D., J. A. Mark, and J. F. Ewing (1993). Heme oxygenase, a likely regulator of cGMP production in the brain: induction in vivo of HO-1 compensates for depression in NO synthase activity. *Mol. Cell. Neurosci.* 4, 398.

Mechoulam, R., L. Hanus, and B. R. Martin (1994). Search for endogenous ligands of the cannabinoid receptor. *Biochem. Pharmacol.* 48, 1537.

Moncada, S. and A. Higgs (1993). The L-arginine–nitric oxide pathway. *N. Engl. J. Med.* 329, 2002.

Nestler, E. J. and P. Greengard (1999). Serine and threonine phosphorylation. In *Basic Neurochemistry: Molecular, Cellular and Medical Aspects*, 6th ed. (G. J. Siegel, et al., eds.). Raven Press, New York. Chapter 24, p. 471.

Nestler, E. J., S. E. Hyman and R. C. Malenka (2001). *Molecular Neuropharmacology*. McGraw-Hill, New York.

Nicoll, R. A. (1988). The coupling of neurotransmitter receptors to ion channels in the brain. *Science* 241, 545.

Piomelli, D. (1994). Eicosanoids in synaptic transmission. *Crit. Rev. Neurobiol.* 8, 65.

Piomelli, D., E. Shapiro, R. Zipkin, J. H. Schwartz, and S. J. Feinmark (1989). Formation and action of 8-hydroxy-ii, 12-epoxy-5,9,14-icosatrienoic acid in *Aplysia:* a possible second messenger. *Proc. Natl. Acad. Sci. U.S.A.* 86, 1721.
Popoli, M., N. Brunello, J. Perez, and G. Racagni (2000). Second messenger–regulated protein kinases in the brain: their functional role and the action of antidepressant drugs. *J. Neurochem.* 74, 21.
Rottingen, J. A. and J. G. Iversen (2000). Ruled by waves? Intracellular and intercellular calcium signalling. *Acta Physiol. Scand.* 169, 203.
Schaad, N. C., P. J. Magistretti, and M. Schorderet (1991). Prostanoids and their role in cell–cell interactions in the central nervous system. *Neurochem. Int.* 18, 303.
Schimizu, T. and L. S. Wolfe (1990). Arachidonic cascade and signal transduction. *J. Neurochem.* 55, 1.
Schuman, E. M. and D. V. Madison (1994). Nitric oxide and synaptic function. *Annu. Rev. Neurosci.* 17, 153.
Sessa, W. C. (1994). The nitric oxide synthase family of proteins. *J. Vasc. Res.* 31, 131.
Snider, S. H., S. R. Jaffrey, and R. Zakhary (1998). Nitric oxide and carbon monoxide: parallel roles as neural messengers. *Brain Res. Rev.* 28, 167.
Starke, K., M. Gothert, and H. Kilbinger (1989). Modulation of neurotransmitter release by presynaptic autoreceptors. *Physiol. Rev.* 69, 864.
Vogel, Z., J. Barg, R. Levy, D. Saya, E. Heldman, and R. Mechoulam (1993). Anandamide, a brain endogenous compound, interacts specifically with cannabinoid receptors and inhibits adenylate cyclase. *J. Neurochem.* 61, 352.
Weng, G., V. S. Bhalla, and R. Inengar (1999). Complexity in biological signaling systems. *Science* 784, 92.
Williams, J. H. and T. V. Bliss (1989). An invitro study of the effect of lipoxygenase and cyclooxygenase inhibitors of arachidonic acid on the induction and maintenance of long-term potentiation in the hippocampus. *Neurosci. Lett.* 107, 301.
Zhang, J. and S. H. Snyder (1995). Nitric oxide in the nervous system. *Annu. Rev. Pharmacol. Toxicol.* 35, 213.

6

Amino Acid Transmitters

On the basis of their functional actions, amino acid transmitters have been divided into two general categories: excitatory amino acid transmitters (glutamate [Glu], aspartate [Asp], cysteate, and homocysteate), which depolarize neurons in the mammalian central nervous system (CNS), and inhibitory amino acid transmitters (γ-aminobutyric acid [GABA], glycine [Gly], taurine, and β-alanine), which hyperpolarize mammalian neurons. A few amino acids have been demonstrated to fulfill most of the criteria for neurotransmitter candidates in the mammalian CNS. Among them are GABA, the major inhibitory transmitter in the brain; Glu, the major excitatory transmitter in the brain; and Gly, another important inhibitory transmitter in the brain stem and spinal cord. This has not been an easy task since many amino acids are also involved in intermediary metabolism and obviously in protein synthesis, which makes it difficult to separate their biochemical role from their transmitter role. Agmatine, the decarboxylation product of arginine, has fulfilled many of these transmitter criteria and is the most recent addition to this family of transmitters. From a quantitative standpoint, the amino acids are probably the major transmitters in the mammalian CNS, while the better-known transmitters discussed in other chapters (acetylcholine, norepinephrine, dopamine, histamine, and 5-hydroxytryptamine) probably account for transmission at only a small percentage of central synaptic sites.

GABA

Neurotransmitter Role in the Mammalian Central Nervous System

GABA was identified as the neurotransmitter for an inhibitory neuromuscular junction in the walking leg of the lobster. Since its discovery over 50 years ago, numerous biochemical and neurophysiological observations have been made about brain GABA and GABA systems that make a strong case for its neurotransmitter role in mammalian brain.

Like other neurotransmitters or neurotransmitter candidates, GABA and its biosynthetic enzyme glutamic acid decarboxylase (GAD) have a discrete, nonuniform distribution in the brain. The brain contains a high-affinity, sodium-dependent transport system that serves to terminate GABA action. Storage of GABA can be demonstrated in selected synaptosomal populations, and a vesicular GABA transporter has been cloned and sequenced. Release of endogenous or radioactively labeled, exogenously accumulated GABA can be evoked by the appropriate experimental conditions. The presence of GABA-containing neurons has been verified, and the anatomical distribution of GABAergic neurons has been mapped out, using in situ hybridization for GAD mRNA and immunocytochemical detection of GAD. However, the most compelling evidence that GABA plays a neurotransmitter role in mammalian brain has emerged from intracellular recording studies, which show that GABA causes a hyperpolarization of neurons similar to that evoked by the naturally occurring transmitter substance and that these inhibitory actions of synthetic GABA and of GABA-containing pathways can be antagonized by drugs selective for the GABA receptor, such as bicuculline.

Distribution

Synthesized in 1883, GABA was known for many years as a product of microbial and plant metabolism. Not until 1950, however, did investigators identify GABA as a normal constituent of the mammalian CNS. Moreover, no other mammalian tissue, with the exception of the retina, contains more than a mere trace of this material. Obviously, a substance with such an unusual enrichment in the brain must have some specific physiological effects that would make it important for the function of the CNS. Much evidence has now accumulated in support of the hypothesis that the major share of GABA found in the brain functions as an inhibitory transmitter. The probability that GABA functions as an inhibitory transmitter in the brain has spurred a prodigious research effort to implicate GABA in the etiology of a host of neurological and psychiatric disorders. Although the present evidence is not overwhelming, GABA has been most convincingly implicated, both directly and indirectly, in the pathogenesis of epilepsy.

In mammals, GABA is found in high concentrations in the brain and spinal cord but is absent or present only in trace amounts in peripheral nerve tissue, such as sciatic nerve, splenic nerve, and sympathetic ganglia, or in any other peripheral tissue, such as liver, spleen, and heart. These findings give some idea of the uniqueness of the occurrence of GABA in the mammalian CNS. Like the monoamines, GABA appears to have a discrete distribution within the CNS. However, unlike the monoamines, the concentration of GABA found in the CNS is on the order of millimoles per gram rather than nanomoles per gram. The brain also contains large amounts of glutamic acid (8–13 mmole/g), which is the main precursor of GABA and itself a neurotransmitter candidate (see Glutamic Acid).

Since GABA does not easily penetrate the blood–brain barrier, brain concentrations of GABA cannot be increased by systemic administration unless one opens the blood–brain barrier. GABA-lactam (2-pyrrolidinone), a less polar and more lipid-soluble compound, can reach the CNS but is not significantly hydrolyzed to GABA. A more successful approach has been use of the drug progabide, which not only penetrates the blood–brain barrier but is subsequently metabolized into GABA.

Metabolism

Three primary enzymes are involved in the catabolism of GABA before its final metabolite, succinate, enters the Krebs cycle. The relative activity of enzymes involved in the degradation of GABA suggests that, as with monoamines, they play only a minor role in the termination of the action of any neurally released GABA.

Figure 6–1 outlines the metabolism of GABA and its relationship to the Krebs cycle and carbohydrate metabolism. As mentioned previously, GABA is formed by the α-decarboxylation of L-glutamic acid, a reaction catalyzed by GAD, an enzyme that occurs uniquely in the mammalian CNS and retinal tissue. The precursor of GABA, L-glutamic acid, can be formed from α-oxoglutarate by transamination or reaction with ammonia. GABA is intimately related to the oxidative metabolism of carbohydrates in the CNS by means of a "shunt," involving its production from glutamate, its transamination with α-oxoglutarate by GABA-α-oxoglutarate transaminase (GABA-T) yielding succinic semialdehyde and regenerating Glu, and finally its entry into the Krebs cycle as succinic acid via the oxidation of succinic semialdehyde by succinic semialdehyde dehydrogenase. In essence, then, the

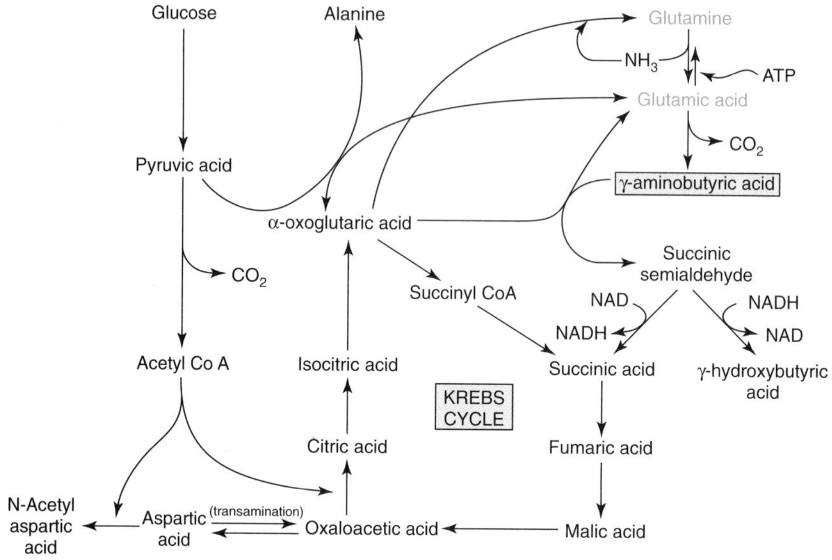

FIGURE 6–1. Interrelationship between γ-aminobutyric acid and carbohydrate metabolism.

shunt bypasses the normal oxidative metabolism involving the enzymes α-oxoglutarate dehydrogenase and succinyl thiokinase.

From a metabolic standpoint, the significance of the shunt is unknown; energetically, at least, it is less efficient than direct oxidation through the Krebs cycle (3 adenosine triphosphate [ATP] equivalents versus 3 ATP + 1 guanosine triphosphate [GTP] for the Krebs cycle). The most comon precursor for GABA formation is glucose, although pyruvate can also act as a precursor.

Glutamic Acid Decarboxylase

GAD is the only synthetic enzyme responsible for the conversion of L-glutamic acid to GABA, and the reaction is irreversible. In mammals, this relatively specific decarboxylase is found primarily in the CNS, where it occurs in higher concentrations in the gray matter. In general, the localization of this enzyme in mammalian brain correlates quite well with the GABA content. The brain enzyme has been purified to homogeneity and its properties studied in detail. It has a pH optimum of about 6.5 and requires pyridoxal phosphate, a form of vitamin B_6 as a coenzyme. This purified enzyme is inhibited by structural analogs of Glu, carbonyl (pyridoxal phosphate

[PLP])–trapping agents, sulfhydryl reagents, thiol compounds, and anions such as chloride. Two isoforms of GAD have been identified, which are encoded by two distinct genes. These two isoforms, designated GAD_{65} and GAD_{67} in accordance with their molecular weights, differ in animo acid sequence, antigenicity, cellular and subcellular location, and interaction with the GAD cofactor PLP. Their different intracellular distributions suggest that the two GAD forms may be regulated in different ways. GAD_{65} and GAD_{67} differ significantly in their affinity for the pyridoxal cofactor: GAD_{65} shows a relatively high affinity for the cofactor, whereas the larger GAD isoform does not. The affinity of GAD_{65} for the cofactor results in the ability of GAD_{65} enzyme activity to be efficiently and quickly regulated. In contrast, the activity of GAD_{67} is determined through induction of new enzyme protein rather than through posttranslational mechanisms.

GABA-Transaminase

GABA-T, unlike the decarboxylase, has a wide tissue distribution. Therefore, although GABA cannot be formed to any extent outside the CNS, exogenous GABA can be rapidly metabolized by both central and peripheral tissue. However, since endogenous GABA is present only in nanomolar amounts in cerebrospinal fluid, it is unlikely that a significant amount of endogenous GABA leaves the brain intact. The brain transaminase has a pH optimum of 8.2 and requires PLP. It appears that the coenzyme is more tightly bound to this enzyme than it is to GAD. The brain ratio of GABA-T/GAD activity is almost always greater than 1. Sulfhydryl reagents tend to decrease GABA-T activity, suggesting that this enzyme requires the integrity of one or more sulfhydryl groups for optimal activity. Transamination of GABA catalyzed by GABA-T is a reversible reaction, so if a metabolic source of succinic semialdehyde were made available, it would be theoretically possible to form GABA by the reversal of this reaction. However, as indicated below, this does not appear to be the case in vivo under normal or experimental conditions, at least those investigated so far.

Recent studies with more sophisticated cell fractionation techniques and electron microscopic monitoring of the fractions obtained have borne out the original claims that both GAD and GABA-T are particulate to some extent. GAD was associated with the synaptosome fraction, whereas GABA-T was largely associated with mitochondria. Further studies on the mitochondrial distribution of GABA-T have suggested that the mitochondria released from synaptosomes have less activity than the crude, unpurified mitochondrial fraction, and it has been inferred that the mitochondria within nerve endings have little GABA-T activity. This has led to the speculation that

GABA is metabolized largely at extraneuronal intercellular sites or in the postsynaptic neurons. Gabaculine is the most potent GABA-T inhibitor available. Similar to γ-acetylenic and γ-vinyl GABA, this agent is a catalytic inhibitor of GABA-T and, unfortunately, will inhibit GAD. However, gabaculine has a fair degree of specificity since it is about 1000-fold less effective as a GAD inhibitor than as a GABA-T inhibitor.

Succinic Semialdehyde Dehydrogenase

Brain succinic semialdehyde dehydrogenase (SSADH) has a high substrate specificity and can be distinguished from the nonspecific aldehyde dehydrogenase found in the brain. The enzyme purified from human brain has a pH optimum of about 9.2 and a K_m for succinic semialdehyde of 5.3×10^{-6} and a K_m for NAD of 3×10^{-5}. SSADH from rat brain has a similarly low K_m for succinic semialdehyde of 7.8×10^{-5} and for NAD of 5×10^{-5}. The high activity of this enzyme and the low Michaelis constant, which allow the enzyme to function effectively at low substrate concentrations, probably account for the fact that succinic semialdehyde has not even been detected as an endogenous metabolite in neural tissue, despite the active metabolism of GABA in vivo.

Since GABA's rise to popularity, the literature has been inundated with claims that many pharmacological and physiological effects can be ascribed to and correlated well with changes in GABA levels in the brain. Since both GAD and GABA-T are dependent on the coenzyme PLP, it is not surprising that pharmacological agents or pathological conditions affecting this coenzyme can cause alterations in the GABA content of the brain. Epileptiform seizure can be produced by a lack of this coenzyme or by its inactivation. Conditions of this sort also lead to a reduction in GABA levels, since GAD appears to be preferentially inhibited over the transaminase, presumably due to the fact that GAD has a lower affinity for the coenzyme than does GABA-T. A diet deficient in vitamin B_6 in infants can lead to seizures that respond successfully to treatment consisting of addition of pyridoxine to the diet. However, many other enzymes, including some of those involved in the biosynthesis of other bioactive substances, are also pyridoxal-dependent.

A number of observations indicate that there is no simple correlation between GABA content and convulsive activity. Administration to experimental animals of a variety of hydrazides, such as thiosemicarbozide, has uniformly resulted in the production of repetitive seizures following a rather prolonged latent period. The finding that hydrazide-induced seizures could be prevented by parenteral administration of various forms of vitamin B_6 led to the suggestion that some enzyme system requiring PLP as a coenzyme was being inhibited and that the decrease in the activity of this enzyme was somehow related to the production of the seizures observed. At this time, attention focused on GABA and GAD because of their unique occurrence in the CNS and because GAD had been shown to be inhibited by carbonyl-trapping agents in vitro. The hydrazide-induced seizures were accompanied by substantial decreases in GABA levels and reductions in GAD activity in various areas of the brain. Subsequently, it was recognized that these decreases were underestimated by the agonal increases in GABA that accompany terminal ischemia. However, even preferential inhibition of GABA-T with carbonyl reagents such as hydroxylamine (NH_2OH) or amino-oxyacetic acid, which increased GABA levels (up to 500% of control) in the CNS, gave no protective effect against the hydrazide-induced convulsions, even though they prevented the depletion of GABA. Administration of very high doses of only amino-oxyacetic acid, instead of producing the normally observed sedation, caused some seizure activity in spite of the extremely high brain levels of GABA.

Administration of amino-oxyacetic acid to a strain of genetically spastic mice in a single dose of 5–15 mg/kg resulted in marked improvement of their symptomatology for 12–24 hr. Although this improvement was associated with an increase in GABA levels, the GABA level increased with a similar time course and to the same extent as in normal control mice, making the improved spasticity hard to attribute to GABA. All studies to date indicate that the principal genetic defect in these mice is not in the operation of the GABA system. Perhaps, the drug-induced increase in GABA quells an excess of excitatory input in some unknown area of the CNS.

Alternate Metabolic Pathways

In addition to undergoing transamination and subsequently entering the Krebs cycle, GABA can apparently undergo various other transformations in the CNS, forming a number of compounds whose importance, and in some cases natural occurrence, has not been conclusively established. Figure 6–2 depicts a variety of derivatives for which GABA may serve as a precursor. Perhaps the simplest of these metabolic conversions is the reduction of succinic semialdehyde (a product of GABA transamination) to γ-hydroxybutyrate (GHB). The transformation of GABA to GHB has been demonstrated in rat brain both in vivo and in vitro. Recent studies have demonstrated that GHB administered in physiologically relevant concentrations can also be converted to GABA by transamination. GHB aciduria, a rare inborn error in the metabolism of GABA, has been reported in children and appears to be the

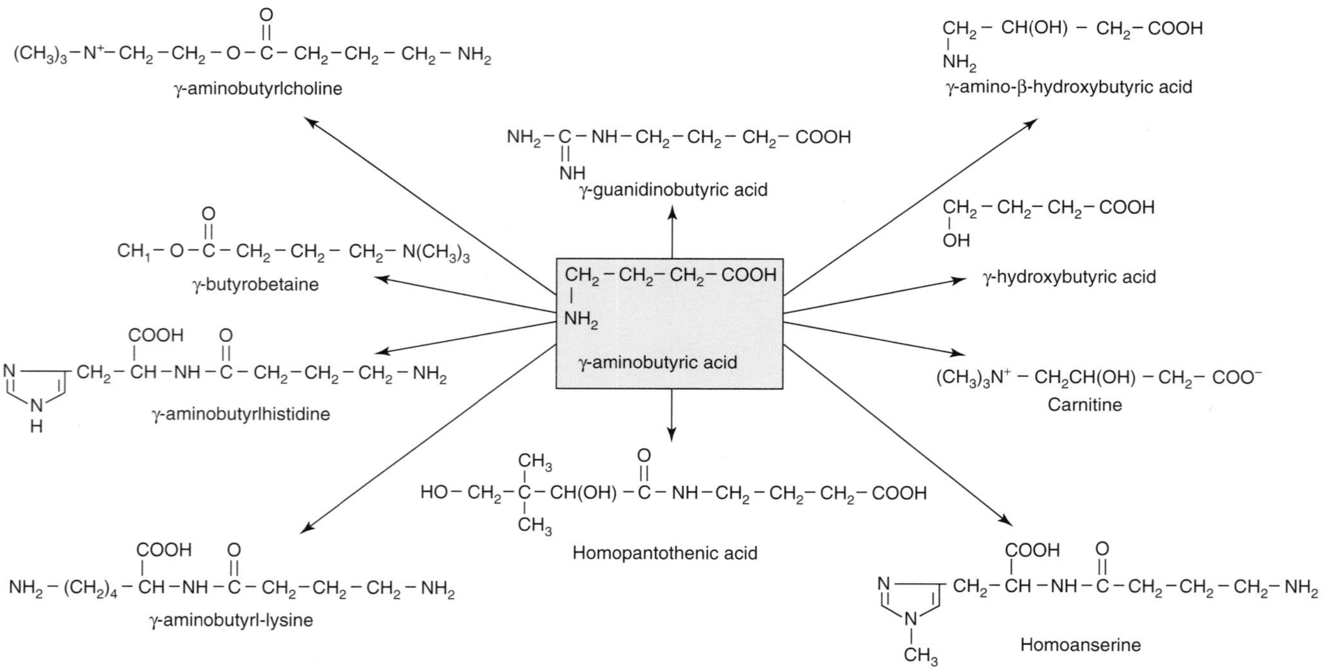

Figure 6–2. Possible alternate metabolic pathways for γ-aminobutyric acid.

result of a deficiency of SSADH, the enzyme that oxidizes succinic semialdehyde to succinic acid (see Fig. 6–1). GHB levels are increased in urine, plasma, and cerebrospinal fluid; but it is unclear whether the main clinical symptoms of motor and mental retardation, muscular hypotonia, and ataxia are related to the elevated levels of GHB.

GHB has recently achieved notoriety as a "date rape drug." Ingested in doses as low as 10 mg/kg, GHB or its lactone precursor γ-butyrolactone produces euphoria, impairment of judgment, anxiolysis, and amnestic deficits in short-term memory. Higher doses lead to unconsciousness, seizures, respiratory depression, and coma; several deaths have been attributed to GHB.

Storage

Like most classical neurotransmitters, GABA is packaged and stored in vesicles in the presynaptic terminals, from which it is released into the synaptic cleft. A vesicular transporter that accumulates GABA has been identified in GABAergic cells. This transporter was cloned on the basis of homology to *unc-47* in *Caenorhabditis elegans*, a strategy of moving from the worm to the mammalian species that has proven to be very useful for identifying a variety of mammalian transmitter-related genes. The vesicular GABA transporter (GAT) differs from the two vesicular monoamine transporters (VMATs, see Chapters 8–10) by having 10 rather than 12 presumptive transmembrane domains and a very large cytoplasmic N terminus of approximately 130 amino acids. Like the vesicular Glu transporter, GAT is highly dependent on the electrical potential across the vesicular membrane and differs from VMATs in terms of this bioenergetic dependence. Specific inhibitors of the vesicular inhibitory GAT have not been identified.

However, the vesicular GAT shares with the VMATs a lack of substrate specificity and will transport the inhibitory transmitter Gly as well as GABA. Consistent with this pharmacology, the vesicular GAT has been found in Gly- as well as GABA-containing neurons. Accordingly, some have suggested that it can be more accurately designated a vesicular inhibitory amino acid transporter. Interestingly, a limited number of GABAergic neurons appear to lack this transporter, raising the suspicion that another (related) transporter may exist in these neurons or that this small population has an alternate mechanism of storage.

Release and Reuptake

The arrival of an action potential or other depolarizing stimulus in the presynaptic GABAergic terminal initiates a sequence of events that ultimately results in vesicular fusion and release of GABA into the synaptic cleft, as is presumed to be the case for all other neurotransmitters. After release, the action of GABA is terminated largely by removal from the synaptic cleft by the actions of several types of plasma membrane transporter (see below). The uptake of several transmitters by both glia and neurons has been reported. This dual glial–neuronal reuptake is a common property in neurons using amino acid transmitters, probably because amino acids can play dual roles as both transmitters and metabolic intermediates.

The ability of glia to avidly accumulate GABA and other amino acids distinguishes amino acid transmitters from other classical transmitters. Reuptake is the primary mode of inactivation of GABA that is released from neurons. Molecular cloning techniques have suggested greater heterogeneity in the GATs than previously suspected, with the genes for four distinct GATs being detected. At least three specific GAT proteins are expressed in the CNS. In addition, a betaine transporter that accumulates GABA has been cloned. All known GATs are expressed in both neurons and glia. There is as yet no obvious answer to the question of why are there are multiple transporters for GABA?. GATs are expressed in both GABAergic neurons and non-GABAergic cells (presumably cells that receive GABA innervation) as well as glia. The presence of multiple transporter proteins for the same transmitter, localized in neurons as well as glia, differs from the situation for catecholamine transmitters, in which a single membrane-associated transporter protein with relatively selective substrate specificity is found in a neuron defining its chemical identity. One possibility is that the cloned GATs may be cotransporters for other amino acids. For example, no transporters for β-alanine and taurine have yet been cloned, but these amino acids are accumulated by GATs. Finally, it is possible that one or more of these transporters may have the ability to function in the outward direction, serving as a paradoxical mechanism for the release, rather than removal, of GABA.

The fact that the GATs transporters are not uniquely concentrated in the plasma membrane of the presynaptic GABA terminals has important functional ramifications. The GABA that is taken up in glia or non-GABAergic postsynptic cells will not be available to recycle for another round of exocytotic release. This lost GABA will have to be replaced by de novo synthesis, placing enhanced demands on the synthetic capacity of the GABAergic neuron.

GABA Receptors

In vertebrate species, GABA receptors are found primarily in nerve cell membranes and are sufficiently widespread that most neurons in the CNS possess them. However, GABA receptors are not exclusively associated with neurons. They are also expressed by astrocytes, where they appear to be involved in the regulation of chloride channels. Interestingly, GABA receptors are also found outside the CNS on neurons of the autonomic nervous system.

In vertebrates, there are two major types of GABA receptor: the inotropic $GABA_A$ receptor and the metabotropic $GABA_B$ receptor. GABA receptors were initially subdivided into these two groups based on pharmacological evidence. However, the functional separation also extends to second-messenger mechanisms, differences in the location of these receptor subtypes in the mammalian CNS, and their molecular composition. In addition, both receptor subtypes have pre- and postsynaptic locations and are thought to participate independently in synaptic transmission.

Autoreceptor Regulation of GABA Release

Pharmacological studies indicate that autoreceptor-mediated regulation of GABA neurons takes place predominantly through $GABA_B$ receptors located on GABAergic nerve terminals (see $GABA_B$ Receptor). Immunohistochemical studies have revealed that both $GABA_B$ and $GABA_A$ receptors are present on postsynaptic non-GABAergic neurons. Presumably, these $GABA_A$ postsynaptic receptors respond to GABA released from a presynaptic GABAergic neuron. An anatomical arrangement of one GABA neuron terminating on another GABA cell would have the same functional consequence as an autoreceptor (decreasing subsequent transmitter release), making it difficult to distinguish between true autoreceptor and heteroceptor regulation of GABA release.

$GABA_A$ Receptor

$GABA_A$ receptors are the major inhibitory neurotransmitter receptors in the brain and the site of action of many clinically important drugs (Fig. 6–3). These receptors are believed to be involved in mediating anxiolytic, sedative, anticonvulsant, muscle-relaxant, and amnesic activity.

The inotropic $GABA_A$ receptor is by far the most prevalent of the two known GABA receptor types in mammalian CNS and has been extensively studied and characterized. Like the nicotinic acetylcholine receptor (nAChR), the $GABA_A$ receptor is composed of four subunits comprising an integral transmembrane ion channel that is gated by the binding of two agonist molecules. However, unlike the nAChR, the receptor-associated GABA channel predominantly conducts chloride ions. Since the Cl^- equilibrium potential is near the resting potential in most neurons, increasing chloride permeabil-

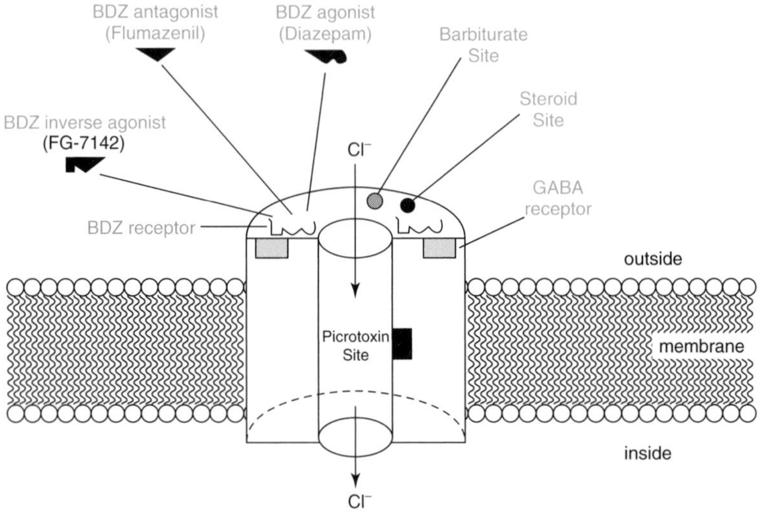

Figure 6–3. Schematic illustration of the $GABA_A$ receptor complex and the sites of action of different agents on the receptor. BDZ, benzodiazepine.

ity hyperpolarizes the neuron and thereby decreases the depolarizing effects of an excitatory input, thus depressing excitability.

The $GABA_A$ receptor, a multisubunit receptor–channel complex, can be allosterically modulated by two important classes of drugs: the benzodiazepines and the barbiturates. The primary structure of the $GABA_A$ receptor, described in 1987, revealed that it is a member of a large superfamily of ligand-gated ion channels that includes the nicotinic–cholinergic, inotropic Glu, Gly, and 5-hydroxytryptamine$_3$ ($5HT_3$) receptors. The $GABA_A$ receptor–ion channel complex is believed to be a heteropentameric glycoprotein of approximately 275 kDa composed of a combination of multiple polypeptide subunits (cf. Fig. 6–4). Seven distinct classes of polypeptide subunits (α, β, γ, δ, ϵ, θ, and ρ) have been cloned, and multiple isoforms of each have been shown to exist so that the total number of identified subunits now stands at 18. The existence of a large family of genes coding for diverse subunits (α_{1-6}, β_{1-4}, γ_{1-3}, δ, ϵ, θ, ρ_{1-2}) provides the basis for the extraordinary structural diversity of $GABA_A$ receptors.

The subunit composition of the $GABA_A$ receptors appears to vary from one brain region to another and even between neurons in a given region, but the exact composition of most native $GABA_A$ receptors is unknown. The distribution of mRNA in the CNS determined by in situ hybridization is very different for each subunit subtype. A recently cloned α subunit (α_6), which confers a unique pharmacology (binding of the partial inverse agonist

RO-15-4513, a putative ethanol antagonist) to a recombinantly expressed $GABA_A$ receptor, is expressed in only a single type of neuron, the cerebellar granule neuron. Thus, it is now becoming clear that the heterogeneity of the $GABA_A$ receptor subunit isoforms confers a diversity of pharmacological and perhaps physiological response characteristics upon the $GABA_A$ receptor. For example, coexpression of an additional γ subunit is necessary for the potentiation of GABA responses by benzodiazepines. In addition, coexpression of individual γ-subunit variants (γ_1 γ_2, or γ_3) with α and β subunits results in varying degrees of modulation by benzodiazepine receptor ligands (agonist, antagonist, inverse agonist). Photoaffinity-labeling studies have further suggested that the benzodiazepine-binding site resides on the α subunit, while the GABA-binding site itself resides on the β subunit. Finally, it appears that the α-subunit heterogeneity determines the diversity of physiological and pharmacological response characteristics of native $GABA_A$ receptors, even though expression of the γ subunits is essential for conferring the modulatory actions of benzodiazepines on $GABA_A$ receptors. Thus, when coexpressed with β_1, the α_1 subunit yields a receptor with a relatively high affinity for GABA. By contrast, coexpression of the α_2 or α_3 subunit with the β_1

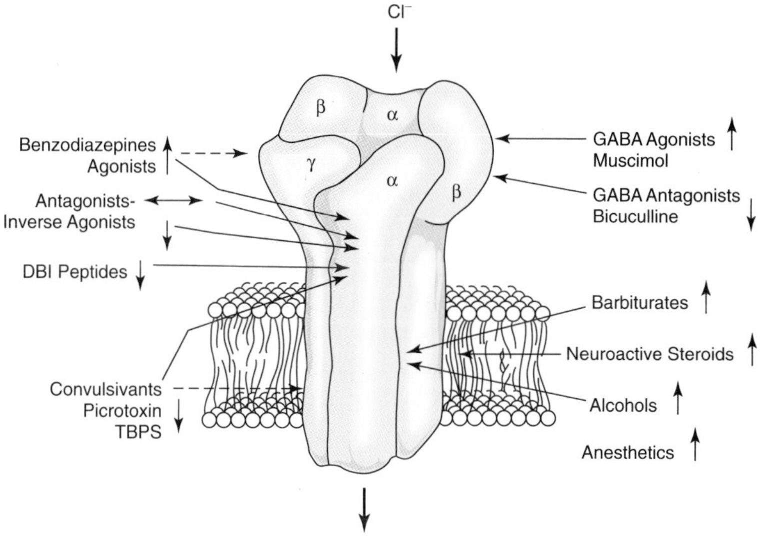

FIGURE 6–4. Schematic illustration of the $GABA_A$ receptor structure containing two α and β subunits and a single γ subunit to form an intrinsic Cl^- ion channel. Putative ligands and drugs known to interact at one of the major sites associated with the $GABA_A$ receptors and to either positively or negatively modulate GABA-gated Cl^- ion conductance are also illustrated. TBPS, *t*-butylbicyclophosphothianate; DBI, diazepam-binding inhibitor. (From Paul, 1995.)

subunit results in $GABA_A$ receptors with far lower affinity for GABA. Thus, the subunit composition of a given receptor may actually determine the local response of the $GABA_A$ receptor to synaptically released GABA. These subtle differences in subunit organization may result in subpopulations of $GABA_A$ receptors that have different regional and cellular locations, each with differential sensitivity to GABA and allosteric modulators.

This extraordinary heterogeneity of $GABA_A$ receptors clearly provides a hitherto unexplored diversity in the function of receptor subtypes affecting their sensitivity to GABA, modulation by allosteric effectors, adaptation to stimulus conditions, distribution within a neuron and between neurons, ontogenetic development, and alterations in pathological states. An exciting new pharmacology is emerging from the recognition of the functional relevance of $GABA_A$ receptor subtypes, providing a rational basis for the development of subtype-specific ligands Several studies have suggested that phosphorylation of the $GABA_A$ receptor channels may also be of importance for both short-term and long-term regulation of $GABA_A$ receptor function and expression. At present, however, the physiological significance and specific consequences of the phosphorylation of $GABA_A$ receptor channels are unknown.

An increased understanding of the benzodiazepine GABA receptor chloride channel complex has led to the development of selective anxiolytic and anticonvulsant agents that lack significant sedative and muscle-relaxant action, properties that often limit the usefulness of traditional agents such as benzodiazepines and barbiturates. A better understanding of the molecular characteristics and regulation of the multiple allosteric sites of the supramolecular complex and the endogenous substances that may physiologically subserve these sites should not only contribute to our understanding of the possible etiology of anxiety and seizure disorders but also aid in the development of more effective and specific therapeutic agents. Once the functional properties of the $GABA_A$ subunits and their subtypes are more clearly defined, it should be possible to use this knowledge in the rational screening and/or design of new, clinically useful subtype-specific agents. The generation of animal models in which particular $GABA_A$ receptor subunits are either inactivated (knockout strategy) or selectively point-mutated (knockin strategy) should help define the functional properties of $GABA_A$ subunits and their subtypes. These animals will also accelerate the recognition of the role of these receptor subtypes as potential drug targets.

$GABA_B$ Receptor

$GABA_B$ receptors belong to the superfamily of G protein–coupled receptors and are classified as metabotropic receptors. Their ligand-binding domain is not directly associated with their ion channel effector. The $GABA_B$ receptor, as mentioned above, is present at lower levels in the CNS than the $GABA_A$ receptor and is not linked to a chloride channel.

Since its pharmacological discovery in 1980, much progress has been made. Selective agonists and antagonists have been developed, and a functional role for this receptor as a mediator of slow inhibitory postsynaptic potentials in many brain regions has emerged. $GABA_B$ receptor activation also plays a role in attenuating the release of biogenic amines, acetylcholine, excitatory amino acids, neuropeptides, hormones, as well as GABA via an interaction with autoreceptors. Whereas $GABA_A$ receptors are directly associated with a Cl^- channel, $GABA_B$ receptors seem to be coupled to Ca^{2+} or K^+ channels via second-messenger systems. The inhibitory hyperpolarizing action of $GABA_B$ receptor activation appears to be mediated through either increases in potassium conductance or decreases in calcium conductance.

Molecular cloning studies have revealed that $GABA_B$ receptors, like the metabotropic glutamate receptors (mGluRs, see EAA Receptors), are members of the G protein–coupled receptor superfamily and contain seven presumptive transmembrane domains. Two major $GABA_B$ subunits have been cloned ($GABA_BR1a$ and -R1b) and a novel $GABA_B$ receptor subunit has been identified ($GABA_BR2$). These G_i-coupled $GABA_B$ receptors are larger than most G protein–coupled receptors, being comprised of 850–960 amino acids. The $GABA_B$ and mGluRs receptors can be distinguished from most other G protein–coupled receptors by their large exracellular N-terminal domains.

$GABA_B$ receptors are expressed on both pre- and postsynaptic membranes, where, as mentioned earlier, they decrease Ca^{2+} conductance, open K^+ channels, and inhibit adenylyl cyclase. In contrast to $GABA_A$ receptors, postsynaptic $GABA_B$ receptors elicit a slower, longer-lasting form of inhibition, an effect that is attributed to the opening of inwardly rectifying K+ channels. The $GABA_B$ receptor can be distinguished pharmacologically from the $GABA_A$ receptor by its selective affinity for the agonist baclofen and its lack of affinity for muscimol and bicuculline (see Table 6–1). The $GABA_B$ receptor is believed to be linked through GTP-sensitive proteins to a calcium channel. Activation of the $GABA_B$ presynaptic receptors by baclofen decreases calcium conductance and transmitter release. Postsynaptic $GABA_B$ receptors are indirectly coupled to K^+ channels via G proteins, and they mediate late inhibitory postsynaptic potentials. Unlike the $GABA_A$ receptor, the $GABA_B$ receptor is not modulated by the benzodiazepines or barbiturates. Pharmacological studies have demonstrated that blockade of $GABA_B$ receptors produces none of the profound behavioral sequelae observed following administration of $GABA_A$ antagonist (e.g., seizures). These observations suggest that, unlike $GABA_A$ receptors, which are believed to be in a continuous

TABLE 6–1. Subdivision of GABA Receptors

Receptor Class	*Pharmacology*			*Channels*	*Second Messengers*
	Agonists	*Antagonists*	*Modulators*		
		COMPETITIVE			
$GABA_A$	GABA	Bicuculline	Benzodiazepines	Cl^-	None
		GABAzine			
		NONCOMPETITIVE			
	Muscimol	Picrotoxin	Barbiturates		
	Isoguvacine	TBPS[a]	Steroids		
			DBI peptides		
$GABA_B$	GABA	Phaclofen	None	↑ K^+	Adenyl cyclase
	R(+) Baclofen	CGP-36742[b]		↓ Ca^{2+}	Phosphatidyl inositol turnover
		3-APPA[a]			

[a]TBPS, *t*-butylbicyclophosphothianate; 3-APPA, 3-aminopropylphosphinic acid; DBI, diazepam-binding inhibitor.

[b]See Figure 6–5.

tonically activated state, $GABA_B$ receptors may be activated only under certain physiological conditions.

With regard to the functions of the $GABA_B$ receptor in the brain, it seems premature to assign a physiological or pathological role. However, the discovery of selective $GABA_B$ antagonists that cross the blood–brain barrier has aided in evaluating the functions of this receptor (see Fig. 6–5). With the development of a potent, orally effective $GABA_B$ antagonist, CGP-54626, it became possible to evaluate better the physiological role of this receptor. In vivo and in vitro studies clearly demonstrated that blockade of $GABA_B$ re-

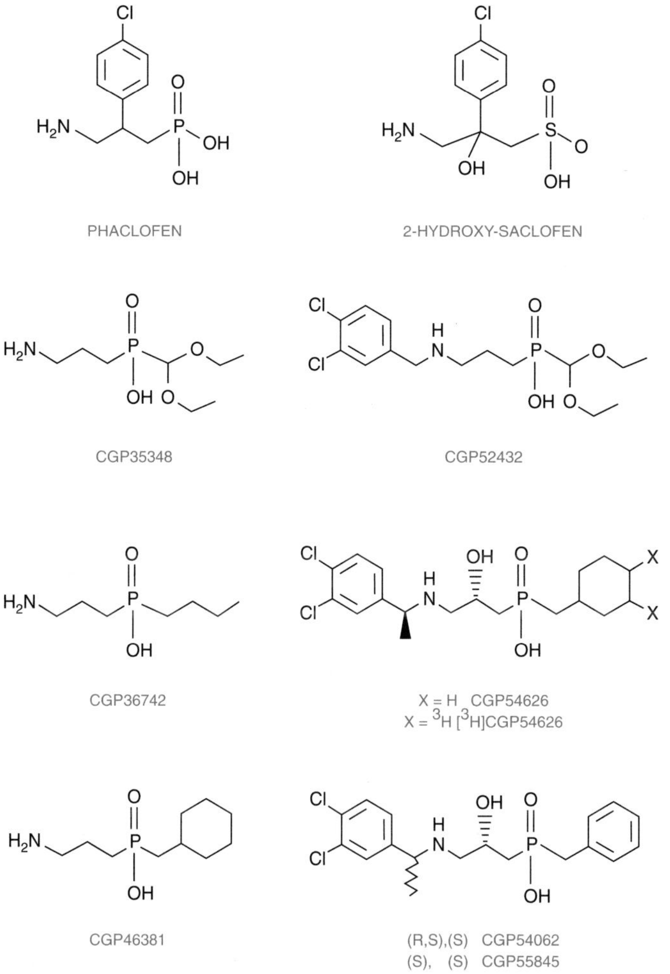

FIGURE 6–5. Chemical structure of $GABA_B$ receptor antagonists.

ceptors with this agent increased neurotransmitter (GABA and Glu) release, reduced late inhibitory postsynaptic potentials of CA1 hippocampal pyramidal neurons, and led to an increase in neuronal excitability. Behavioral studies in several species have suggested that $GABA_B$ receptor blockade can improve cognition in rats (social learning), mice (passive avoidance), and rhesus monkeys (conditional spatial color test). However, baclofen is the only drug in clinical use that interacts with $GABA_B$ receptors. This drug is used as a muscle-relaxant to decrease spasticity in a diversity of neurological disorders.

Pharmacology of GABAergic Neurons

Drugs can influence GABAergic function by interacting at many different sites, both pre- and postsynaptic (Fig. 6–6). Drugs can influence presynaptic events and modify the amount of GABA that ultimately reaches and interacts with postsynaptic GABA receptors. In most cases, presynaptic drug effects do not involve an interaction with GABA receptors. The most extensively studied presynaptic drug actions involved inhibitory effects exerted on enzymes involved in GABA synthesis (GAD) and degradation (GABA-T) and the neuronal reuptake of GABA. The major exception is the interaction of drugs with GABAergic autoreceptors to modulate both the physiological activity of GABA neurons and the release and synthesis of GABA in a manner analogous to the role played by dopamine autoreceptors in the regulation of dopaminergic function.

A great deal of emphasis has been directed recently to the study of drug interactions with GABA receptors. Drugs interacting at the level of GABA receptors can be classified into two general categories: GABA antagonists and GABA agonists. Figure 6–6 depicts possible sites of drug interaction in a hypothetical GABAergic synapse. (The structures of compounds that act at GABAergic synapses are depicted in Figs. 6–7 and 6–8.) Picrotoxinin is the active component of picrotoxin.

GABA Antagonists

The action of GABA at the receptor–ionophore complex may be antagonized by GABA antagonists either directly, by competition with GABA for its receptor, or indirectly, by modification of the receptor or inhibition of the GABA-activated ionophore. The two classic GABA antagonists (Fig. 6–6) bicuculline and picrotoxin appear to act by different means. Bicuculline acts as a direct competitive antagonist of GABA at the receptor level, while picrotoxin acts as a noncompetitive antagonist, presumably due to its ability to block GABA-activated ionophores. Although early studies raised some doubts

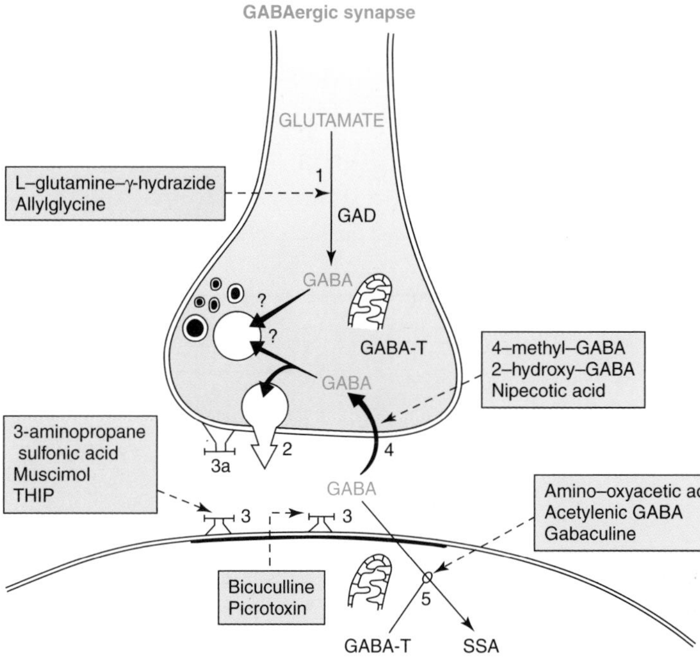

FIGURE 6–6. Schematic illustration of a GABAergic neuron indicating possible sites of drug action.

Site 1: *Enzymatic synthesis.* Glutamic acid decarboxylase (GAD-1) is inhibited by a number of various hydrazines. These agents appear to act primarily as pyridoxal antagonists and are therefore very nonspecific inhibitors. L-glutamate-γ-hydrazide and allylglycine are more selective inhibitors of GAD-1, but these agents are also not entirely specific in their effects.

Site 2: *Release.* GABA release appears to be calcium-dependent. At present, no selective inhibitors of GABA release have been found.

Site 3: *Interaction with postsynaptic receptor.* Bicuculline and picrotoxin block the action of GABA at postsynaptic receptors. 3-Aminopropane sulfonic acid and the hallucinogenic isoxazole derivative muscimol appear to be effective GABA agonists at postsynaptic receptors and autoreceptors. THIP, tetrahydroisoxazolopyridinol.

Site 3a: *Presynaptic autoreceptors.* Possible involvement in the control of GABA release.

Site 4: *Reuptake.* In the brain, GABA appears to be actively taken up into presynaptic endings by a sodium-dependent mechanism. A number of compounds will inhibit this uptake mechanism, such as 4-methyl-GABA and 2-hydroxy-GABA, but these agents are not completely specific in their inhibitory effects.

Site 5: *Metabolism.* GABA is metabolized primarily by transamination by GABA-transaminase (GABA-T), which appears to be localized primarily in mitochondria. Amino-oxyacetic acid, gabaculline, and acetylenic GABA are effective inhibitors of GABA-T.

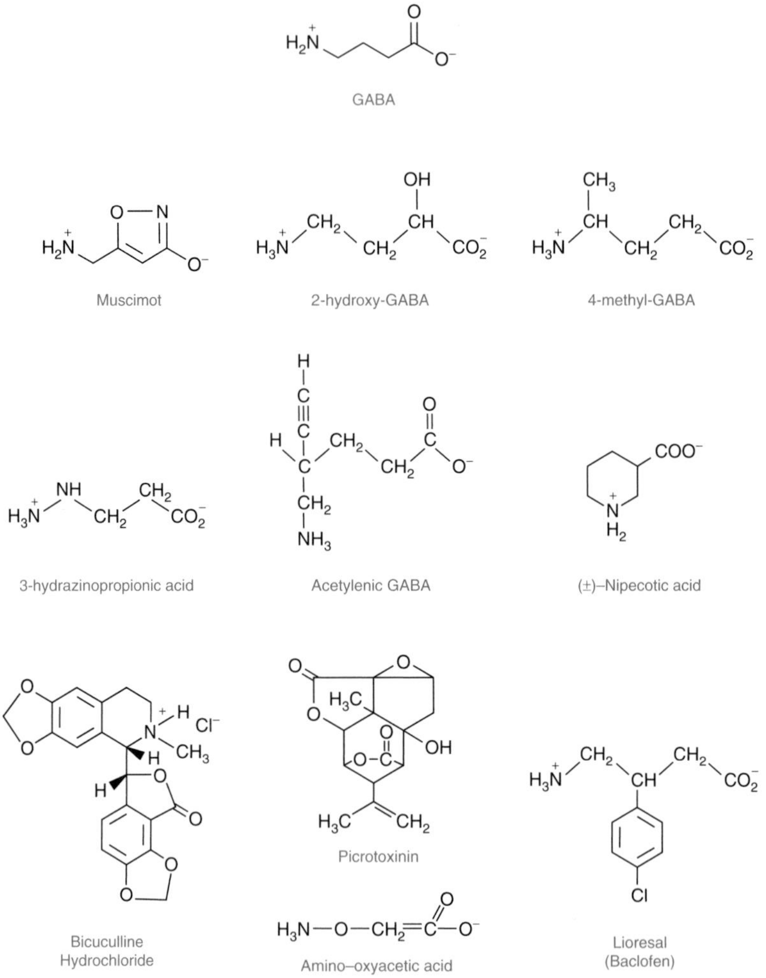

FIGURE 6–7. Structures of compounds that act at GABAergic synapses.

about the usefulness of bicuculline as a selective GABA antagonist, this skepticism has been largely resolved and appears to be primarily related to the instability of bicuculline at 37°C and physiological pH. Under normal physiological conditions, bicuculline is hydrolyzed to bicucine, a relatively inactive GABA antagonist with a short half-life of several minutes. The quaternary salts now used for most electrophysiological experiments (bicuculline methiodide and bicuculline methochloride) are much more water-soluble and stable over a broad pH range of 2–8. However, these quaternary salts are not suitable for systemic administration because of their poor penetration into the CNS.

GABA Agonists

Electrophysiological studies have demonstrated a wide variety of compounds that are capable of directly activating bicuculline-sensitive GABA receptors. These agonists can be readily subdivided into two groups based on their ability to penetrate the blood–brain barrier, dictating whether they will be active or inactive following systemic administration. Agents such as 3-aminopropane-sulfonic acid, β-guanidinoproprionic acid, 4-aminotetrolic acid, *trans*-4-aminocrotonic acid, and *trans*-3-aminocyclopentane-1-carboxylic acid are effective direct-acting GABA agonists. However, entry of these agents into the brain following systemic administration is minimal. In addition, compounds such as *trans*-4-aminotetrolic acid and 4-aminocrotonic acid also inhibit GABA-T and GABA uptake; therefore, their action is not totally attributable to their direct agonist properties.

In contrast to this class of direct-acting GABA agonists, other GABA agonists readily pass the blood–brain barrier and are active following systemic administration. Muscimol (3-hydroxy-5-aminomethylisoxazole) is the agent in this group that has been most extensively studied. Some other agents in this group include (5)-(2)-5-(1-aminoethyl)-3-isoxazole, tetrahydroisoxa-

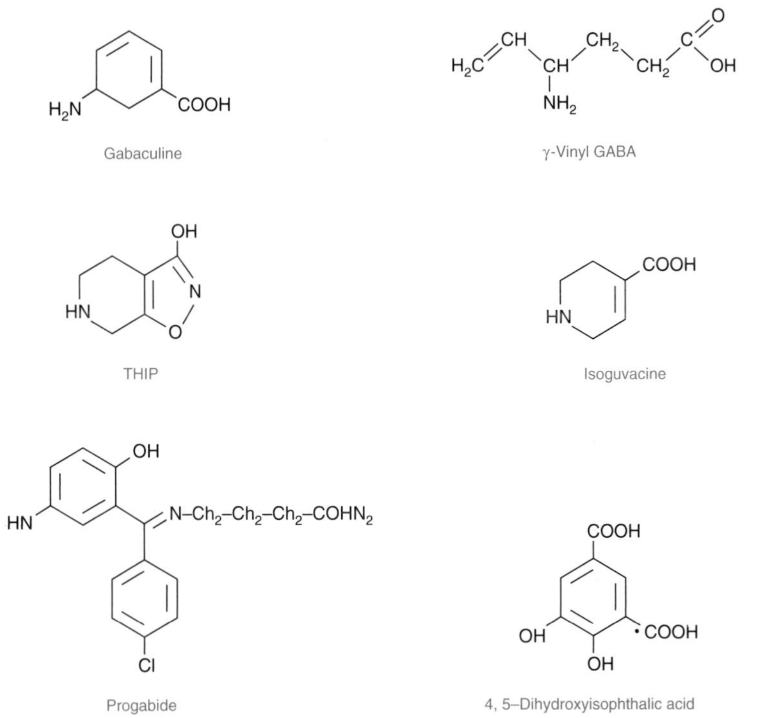

FIGURE 6–8. Structures of compounds that act at GABAergic synapses.

zolopyridinol (THIP, a bicyclic muscimol analog), SL-76002 (α[chloro-4-phenyl]fluro-5-hydroxy-2 benzilide-neamino-4H butyramide), and kojic amine (2-aminomethyl-3-hydroxy-4H-pyran-4-one). In addition, GABAergic substances may be further categorized as direct or indirect GABA receptor activators. For example, muscimol, isoguvacine, and THIP are GABA mimetic agents that interact directly with GABA receptors. Indirectly acting GABA mimetics facilitate GABAergic transmission by increasing the amount of endogenous GABA that reaches the receptor or by altering in some manner the coupling of the GABA receptor–mediated change in chloride permeability. Thus, many drugs often classified as indirect GABA agonists act presynaptically to modify GABA release and metabolism rather than by interacting directly with GABA receptors. For this reason, drugs like gabaculine (a GABA-T inhibitor), nipecotic acid (a GABA uptake inhibitor), and baclofen (an agent that, in addition to many other actions, causes release of GABA from intracellular stores) are often classified incorrectly as GABA agonists. The benzodiazepines mentioned earlier also appear to potentiate the action of tonically released GABA at the receptor by displacement of an endogenous inhibitor of GABA receptor binding, allowing more endogenous GABA to reach and bind receptors. Thus, benzodiazepines are sometimes also classified as GABA agonists. A GABA-like action can also be elicited by agents that bypass GABA receptors and influence GABA ionophores. It has been suggested that pentobarbital acts at the level of the GABA ionophore, but it is unclear whether its CNS-depressant effects are explained by this action.

The structures of some of the more potent and widely used direct-acting GABA agonists are illustrated in Figures 6–7 and 6–8. Included are muscimol, isoguvacine, THIP, and (1)-*trans*-3-aminocyclopentane carboxylic acid. Useful therapeutic effects have not yet been obtained by use of agents of this sort, which have direct GABA mimetic effects (e.g., muscimol), inhibit the active reuptake of GABA (e.g., nipecotic acid), or alter the rate of synthesis or degradation of GABA (e.g., amino-oxyacetic acid and gabaculine). However, useful therapeutic effects are achieved with the anxiolytic benzodiazepines (e.g., diazepam [Valium] and chlordiazepoxide [Librium]), which may exert their actions by facilitating GABAergic transmission.

If the anatomical distribution and functional properties of $GABA_A$ receptor subtypes and subunits become clearly defined, this knowledge may enable the development of therapeutically useful subtype-specific agonists that can be directed to modify GABAergic function in selective brain areas.

Endogenous Modulators

The large number of recognition sites associated with $GABA_A$ receptors has led to speculation that a host of endogenous regulatory factors exists. A number of candidates have been identified, but with the exception of the neurosteroids and the endogenous diazepam-binding inhibitor (DBI), there is little compelling evidence that any play an important role in modulating $GABA_A$ receptor function in vivo. DBI is an endogenous peptide that has been purified to homogeneity from rat and human brain and its structure determined by recombinant DNA technology. Several lines of evidence suggest that DBI functions as a precursor of a family of allosteric modulatory peptides of the $GABA_A$ receptor, causing negative modulation of the GABA-operated Cl^- ion fluxes (see Chapter 12). Consistent with this action, DBI appears to have anxiogenic properties similar to those associated with inverse benzodiazepine agonists in experimental animals. A similar endogenous peptide has been found in human brain and cerebrospinal fluid. The amount of human DBI immunoreactivity is elevated in the cerebrospinal fluid of severely depressed patients.

The other postulated endogenous ligands for the $GABA_A$ receptor include two naturally occuring reduced steroid metabolites of deoxycorticosterone and progesterone (allotetrahydro-deoxycorticosterone (DOC) and allopregnanolone, respectively). These neurosteroids are formed in the brain and bind with high affinity to $GABA_A$ receptors, eliciting a barbiturate-like action that potentiates GABA-elicited Cl^-conductance. These are among the most potent known endogenous ligands of $GABA_A$ receptors found in the CNS. Their plasma and brain (cortex and hypothalamus) levels are increased dramatically in rats following exposure to stress. Plasma levels of allopregnanolone are also elevated during the third trimester of pregnancy and decrease dramatically following parturition. The observations that brain and plasma levels of allopregnanolone and allotetrahydro-DOC increase rapidly in the brain after stress (4- to 20-fold in less than 5 minutes) suggest that these neurosteroids may have a physiological role in stress and anxiety. In addition, conditions that may lead to large increases in neuroactive steroid levels, such as puberty, pregnancy, or the menstrual cycle, could also alter neurochemical and behavioral adaptations to stress. To date, however, none of these putative natural ligands (neurosteroids or DBIs) have been unequivocally demonstrated to subserve a physiological action or to modulate a pathological state.

Glycine

As an Inhibitory Transmitter

Structurally, Gly is the simplest amino acid. It is found in all mammalian body fluids and tissue proteins in substantial amounts. Although Gly is not an essential amino acid, it is an essential intermediate in the metabolism of protein, peptides, one-carbon fragments, nucleic acids, porphyrins, and bile salts. It is also considered to be an established inhibitory neurotransmitter, enriched in the medulla, spinal cord, and retina. Thus, Gly appears to have a more circumscribed function in the CNS than the more ubiquitously distributed GABA. As with the other major amino acid transmitters, numerous neurochemical studies have attempted to separate and distinguish between the general metabolic and transmitter functions of Gly within the CNS. Gly also appears to be an exclusively vertebrate transmitter, making it unique among the transmitter substances.

Glycinergic neurons appear to respond to activation as other chemically defined neurons do. Arrival of an action potential in the presynaptic nerve terminal initiates a calcium-dependent cascade of events, which ultimately involves fusion of the presynaptic membrane and release of Gly into the synaptic cleft. Gly is removed from the synaptic cleft by uptake transporters located on glial cells and on the presynaptic terminals of the glycinergic neurons. However, in the last several years, very little progress has been achieved in developing pharmacological tools that act selectively on Gly systems or in generating more information concerning Gly metabolism in neuronal tissue.

In the spinal cord and brain stem, specific uptake of Gly has been demonstrated in regions exhibiting high densities of inhibitory Gly receptors. Two Gly transporter proteins have been cloned and shown to be expressed in brain as well as in peripheral tissues. Both are members of the large family of Na^+/Cl^--dependent neurotransmitter transporters (see Chapter 10) and share approximately 50% sequence identity with the GABA transporter. The Gly transporters have been named GLYT-1 and GLYT-2, in the order in which they were reported. These transporters have very similar kinetics and pharmacological properties but differ in the distribution of their transcripts, measured by in situ hybridization. The distribution of GLYT-1 mRNA closely parallels the distribution of the Gly receptor, suggesting that GLYT-1 is primarily a glial transporter and GLYT-2 is associated primarily with neurons. GLYT-1 exist in three isoforms, which are probably generated by alternate splicing. These isoforms do not exhibit any known variation in their uptake properties but do possess distinct patterns of expression in the CNS. GLYT-1 is expressed in both astrocytes and neurons, whereas GLYT-2 is localized on axons and the terminal boutons of neurons that contain vesicular Gly. The GLYT-1 isoforms can be distinguished pharmacologically from GLYT-2 since they are sensitive to the effects of sarcosine (*N*-methylglycine).

Both GLYT-1 and GLYT-2 are expressed in the brain stem and spinal cord, a location consistent with their role in terminating glycinergic transmission. However, GLYT-1 is also expressed in several regions of the forebrain that are devoid of glycinergic neurotransmission. Thus, GLYT-1 may regulate *N*-methyl-D-aspartate (NMDA) Glu receptor function in these areas by controlling the levels of extracellular Gly available to allosterically modulate the activity of these receptors (see Gly as modulator or NMDA receptor). If that is the case, then GLYT-1 inhibitors may prove to be useful clinically to augment NMDA receptor function.

The strychnine-sensitive Gly receptor has also been cloned and expressed and appears to be structurally quite homologous to the multimeric subunits of other ligand-gated ion channels, including the $GABA_A$ receptor. The native Gly receptor appears to be a pentameric structure, and photoaffinity labeling reveals that both Gly- and strychnine-binding sites are located on the α subunit. Several Gly receptor α-subunit variants have been identified ($\alpha_{1\text{–}4}$) and shown to differ in their pharmacological properties and levels of expression. Expression of α_1 and α_2 subunits is developmentally regulated, with a switch from the neonatal α_2 subunit (strychnine-insensitive) to the adult α_1 form (strychnine-sensitive) at about 2 weeks postnatally in the mouse. It is interesting that the timing of this switch corresponds with the development of spasticity in the mutant spastic mouse, prompting the speculation that insufficient expression of the adult strychnine-sensitive isoform may underlie some forms of spasticity.

To date, very little is known about the factors controlling the release of Gly from the spinal cord. Again, as with GABA, the efficient uptake process may explain why it is difficult to detect Gly release from the CNS. The main problem (as with GABA, Glu, etc.) is that there is no distinct neuronal pathway that may be isolated and stimulated; thus, all of the induced activity is very generalized, making the significance of any demonstrable release (metabolite or excess transmitter) very difficult to interpret.

Despite these well-characterized functional properties, our understanding of some aspects of the metabolism of Gly in nervous tissue remains rudimentary. For example, we still do not know whether biosynthesis is important for the maintenance of Gly levels in the spinal cord or whether the neurons depend on the uptake and accumulation of preformed Gly. As indicated in Figure 6–9, Gly can be formed from serine by a reversible folate-dependent reaction catalyzed by the enzyme serine *trans*-hydroxymethylase. Serine itself can also be formed in nerve tissue from glucose via the intermediates 3-phosphoglycerate and 3-phosphoserine. It is also conceivable that Gly might be formed from glyoxylate via a transaminase reaction with Glu. Although not established definitively, it appears likely that serine serves as the major precursor of Gly in the CNS and that serine hydroxymethyltransferase and D-glycerate dehydrogenase are the best candidates for the rate-limiting enzymes involved in the biosynthesis of Gly. Not only is our knowledge of

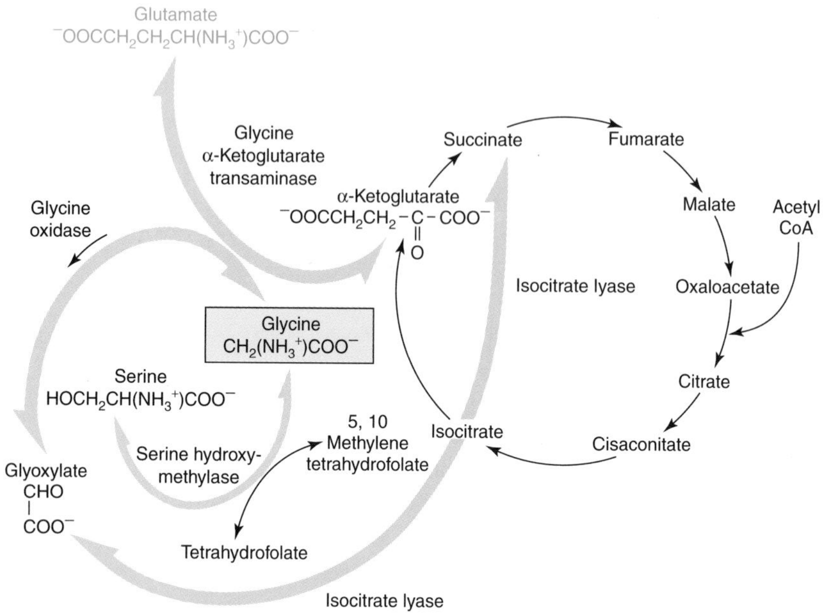

FIGURE 6–9. Possible metabolic routes for the formation and degradation of glycine by nervous tissue. (Modified from Roberts and Hammerschlag, 1972.)

the metabolism of Gly in nervous tissue minimal, but at the present time only scanty information is available on the factors regulating the concentration of Gly in the CNS. Gly in the spinal cord is labeled only slowly from radioactive glucose via one of the glycolytic intermediates indicated in Figure 6–9. It is also possible to label Gly by administration of labeled glyoxylate, which is readily transmitted to Gly by nervous tissue.

In summary, probably the most critical missing piece of evidence to establish the inhibitory role of Gly in the spinal cord is the demonstration that it is the main substance contained in the terminals of the interneurons that synapse on the motoneurons and that it is released from these terminals when direct inhibition is produced. The fact that these effects are blocked by strychnine, an antagonist selective for the Gly receptor, makes a strong presumptive case that Gly is responsible for the inhibitory actions. Some partial support for the localization of Gly has been provided by autoradiographic localization of Gly-uptake sites as visualized by electron microscopy. However, demonstration of discretely evoked release of Gly from spinal cord interneurons has been more difficult to obtain.

Gly as a Modulator of NMDA Receptors

A new role has been proposed for Gly that is distinct from its established role as an inhibitory transmitter in lower brain stem areas and in spinal cord mediated by a strychnine-sensitive chloride conductance. Several groups have shown that nanomolar concentrations of Gly increase the frequency of opening of one of the subsets of Glu receptors, namely, the NMDA receptor channel. This effect of Gly is strychnine-insensitive, suggesting a mechanism involving allosteric regulation of the NMDA receptor complex through a distinct Gly-binding site (see Fig. 6–10). This action can be mimicked with Gly agonists and blocked by other Gly antagonists, although not strychnine. This allosteric concept is supported by the existence of strychnine-insensitive Gly-binding sites that have an anatomical distribution identical to that of the NMDA receptor. When compared with the effects of benzodiazepines

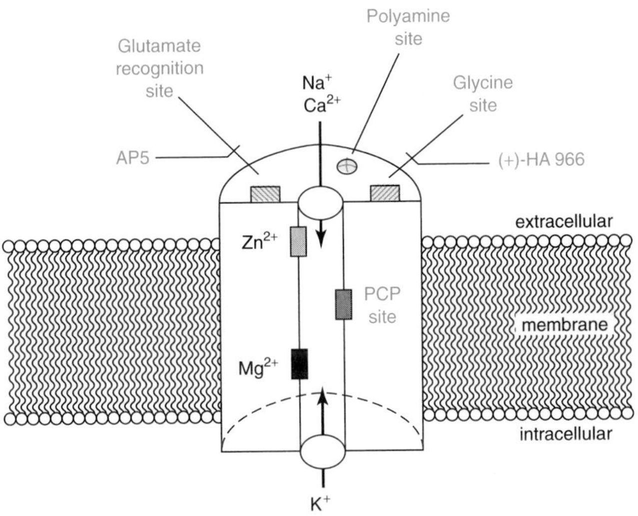

FIGURE 6–10. Schematic illustration of the *N*-methyl-D-aspartate (NMDA) receptor and the sites of action of different agents on the receptor. The NMDA receptor gates a cation channel that is permeable to Ca^{2+} and Na^{+} and gated by Mg^{2+} in a voltage-dependent fashion; K^{+} is the counterion. The NMDA receptor channel is blocked by phencyclidine (PCP) and MK801, and the complex is regulated at two modulatory sites by glycine and polyamines; 2-amino-5-phosphonopentanoic acid (AP_5) and 3-(2-carboxypiperazin-4-yl)-propyl-1-phosphonic acid (CPP) are competitive antagonists at the NMDA site.

on the GABA receptor, the enhancement of NMDA responses observed with Gly is much greater. This suggests that the main effect of Gly is to prevent desensitization of the NMDA receptor during prolonged exposure to agonists. Gly appears to accomplish this by accelerating the recovery of the receptor from its desensitized state rather than by blocking the onset of desensitization. Gly facilitation of synaptic responses mediated by NMDA receptors may be a common regulatory mechanism for excitatory synapses, which raises the question of the processes that regulate extracellular Gly concentration in brain regions where NMDA receptors play a critical role in excitatory transmission. An important question is whether endogenous Gly antagonists (e.g., kynurenate) may also play a role in regulating neuronal function where NMDA receptors are involved. However, recent clinical trials of a selective Gly antagonist for the NMDA receptor effect, previously reported to be neuroprotective in animal models of stroke, have shown no such efficacy when evaluated in the first 6 hours after a stroke.

Glutamic Acid

Long before a role for Glu in neurotransmission was established, it was recognized that certain amino acids, such as Glu and Asp, occur in uniquely high concentrations in the brain and that they can exert very powerful stimulatory effects on neuronal activity. Thus, if any amino acid is involved in the regulation of nerve cell activity, as an excitatory transmitter or otherwise, it seems unnecessary to look beyond these two candidates. The excitatory potency of Glu was first demonstrated in crustacean muscle and later by direct topical application to mammalian brain. However, except for the invertebrate model, where substantial evidence has accumulated to support a role for Glu as an excitatory neuromuscular transmitter, its status as a neurotransmitter in mammalian brain was uncertain for many years. This is probably in part explainable by the fact that Glu (and Asp) is a compound that is also intimately involved in intermediary metabolism in neural tissue. For example, it has an important function in the detoxification of ammonia in the brain, is an important building block in the synthesis of proteins and peptides including glutathione, and plays a role as a precursor for the inhibitory neurotransmitter GABA. Thus, it has been extremely difficult to dissociate the role this amino acid plays in neuronal metabolism and as a precursor for GABA from its possible role as a transmitter substance. Transport of circulating Glu to the brain normally plays only a very minor role in regulating the levels of brain Glu. In fact, the influx of Glu from the blood across the blood–brain barrier is much lower than the efflux of Glu from the brain.

Synthesis and Metabolism

In brain L-Glu is synthesized in the nerve terminals from two sources: from glucose via the Krebs cycle and transamination of α-oxoglutatrate (Fig. 6–1) and from glutamine that is synthesized in glial cells, transported into nerve terminals, and locally converted by glutaminase into Glu (see Fig. 6–11). In the Glu-containing nerve terminals, Glu is stored in synaptic vesicles and, upon depolarization of the nerve terminal, it is released by a calcium-dependent exocytotic process. The action of synaptic Glu is terminated by a high-affinity uptake process via the plasma membrane Glu transporter on the presynaptic nerve terminal and/or on glial cells. The Glu taken up into glial cells is converted by glutamine synthetase into glutamine, which is then transported via a low-affinity process into the neighboring nerve terminals, where it serves as a precursor for Glu. In astrocytes, glutamine can also be oxidized (via the Krebs cycle) into α-ketoglutarate, which can be actively transported into the neuron to replace the α-ketoglutarate lost during the synthesis of neuronal Glu. As noted earlier in this chapter, glutamine can replenish the transmitter pool of GABA via this pathway as well.

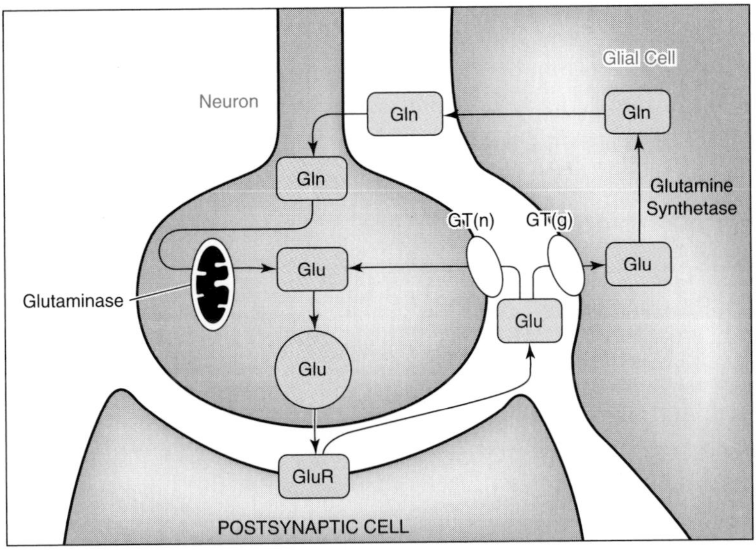

FIGURE 6–11. Pathways for glutamate utilization and metabolism. Glutamate (Glu) released into the synaptic cleft is recaptured by neuronal-type ($GT_{[n]}$) and glial-type ($GT_{[g]}$) Na^+-coupled glutamate transporters. Glial glutamate is converted to glutamine (Gln) by the enzyme glutamine synthetase. Gln is present at high concentrations in the cerebrospinal fluid (about 0.5 mM) and can enter the neuron to help replenish glutamate after hydrolysis by mitochondrial glutaminase. (Modified from Nicholls, 1994.)

Even though neuronal systems believed to utilize Glu or Asp as transmitter substances have been described in the CNS, because of their role in intermediary metabolism, it seems quite unlikely that it will be possible to map these systems accurately by simply tracking the presence of Glu or Asp or their synthesizing enzymes, as has been done in the past for the monoamines and GABA. The development of antibodies against excitatory amino acids (EAAs), especially antisera that can distinguish between Glu and Asp with a high degree of selectivity as well as antibodies against their receptors and transporters, has facilitated the anatomical mapping of EAA pathways. The development of reliable methods for combining anterograde labeling of primary afferent terminals with immunocytochemistry has helped to identify afferent nerve terminals enriched in Glu or Asp and to dissociate the role played by these amino acids in neurotransmission from their general role in metabolism. Nerve terminal enrichment in a specific EAA provides the most direct anatomical evidence that a pathway uses a particular amino acid as a neurotransmitter. The elegant studies of Rustioni and co-workers employing immunocytochemistry at the electron microscopic level have identified primary afferent fibers terminating in spinal laminae of the lumbar spinal cord with nerve terminals enriched in Glu and/or Asp, providing direct anatomical evidence that these primary afferents use these EAAs as neurotransmitters. Use of these techniques in conjunction with other less direct approaches, including mapping the EAA receptors by ligand binding autoradiography, has provided strong support for a neurotransmitter role of EAAs in the mammalian CNS.

Release

Although the status of Glu and Asp as neurotransmitters has suffered through many cycles of acceptability and nonacceptability in the past several decades, the rush to clone the Glu receptors and the extensive research on long-term potentiation demonstrating a function for Glu have stopped the doubters from speaking. In spite of considerable evidence that Glu may be an excitatory transmitter in the CNS, little is known about the biosynthesis and release of the pool of releasable transmitter Glu. Utilizing the molecular layer of the dentate gyrus of the hippocampal formation to provide a definitive system in which the major input appears to be glutaminergic, Cotman and co-workers addressed these questions. Glu was shown to be released by depolarization from slices of the dentate gyrus in a Ca^{2+}-dependent manner, and lesions of the major input to the dentate gyrus originating from the entorhinal cortex diminished this release and the high-affinity uptake of Glu. Glu biosynthesis in the releasable pools was rapidly regulated by the activity of glutaminase and by uptake of glutamine. These properties are consistent with the properties expected of a neurotransmitter, and the observations strengthened the premise that Glu may be an important neurotransmitter in the molecular layer of the dentate gyrus. Furthermore, these studies demonstrated that the regulation of Glu synthesis and release share many properties with other transmitters. For example, similar to acetylcholine synthesis, Glu synthesis is regulated in part via the accumulation of its major precursor, glutamine, and newly synthesized Glu, like acetylcholine, is preferentially released. In addition, Glu synthesis is regulated by end-product inhibition. This is similar to the mechanism by which the rate-limiting enzyme in catecholamine synthesis, tyrosine hydroxylase, is regulated in catecholaminergic neurons by dopamine and norepinephrine. It is interesting that these similarities are demonstrable despite the involvement of Glu in general brain metabolism.

Storage

The vesicle hypothesis describing the quantal release of acetylcholine at the neuromuscular junction was introduced in the mid-1950s. Since then, the concept of vesicular storage and release of acetylcholine has become firmly established and has been extended to a number of other synapses and neurotransmitters. However, there was no direct experimental evidence for the participation of synaptic vesicles in the storage and release of EAA neurotransmitters until recently. The concept has received strong support from the studies of Jahn and Ueda and co-workers on isolated synaptic vesicles from mammalian brain. These studies have shown that vesicles are capable of Glu uptake and storage and that they have a specific carrier for L-Glu. This concept was finally validated with the cloning of the vesicular transporter for Glu. The difficulty in cloning this vesicular transporter is most likely due to the fact that it is unrelated to other known transmitter transporters, although it shares significant sequence homology with EAT-4, a worm protein implicated in glutamatergic transmission. The vesicular Glu transporter was identified as a protein that was previously suggested to mediate the sodium-dependent transport of inorganic phosphate across the membrane. It is found predominantly in axon terminals, particularly those that form asymmetric (excitatory) synapses. In contrast to the GABA transporter, for which Gly is also a substrate (see Release and Reuptake), the Glu vesicular transporter has a high specificity for Glu. This transporter has been extensively characterized biochemically and shown to play a key role in exocytosis.

Plasma Membrane Glu Transporter

Most of the molecular biological research on EAA transmission has focused on receptors rather than the process of transmitter inactivation. Inactivation is an especially important process in EAA neurotransmission. The rapid removal of Glu from the synapse by high-affinity uptake not only serves to terminate the excitatory signal and recycle the Glu but also plays an important role in the maintenance of extracellular levels of Glu below those that could induce excitotoxic damage. To date, five distinct subtypes of EAA transporter (EAAT) have been identified, which together with the neutral amino acid transporters appear to be part of a novel gene family. These high-affinity, sodium-dependent EAATs exhibit distinct anatomical and cellular distributions and appear to have marked differences in pharmacological specificity (Table 6–2).

These Glu transporter genes share 50% sequence homology and exhibit minimal homology with other eukaryotic proteins, including the superfamily of neurotransmitter transporters that mediate the uptake of GABA, Gly,

TABLE 6–2. Distribution and Pharmacology of the Excitatory Amino Acid Transporters (EAATs)

Subtype	*Primary Localization*	*Pharmacological Properties*
EAATI (GLAST1, rat)	Cerebellar glia	4-MG and L-SOS as substrates
EAAT2	Forebrain glia	DHK, MPDC, L-*trans*-2,3-PDC and 3-TMG as nontransportable inhibitors
EAAT3 (EAAC1, rabbit)	Cortical neurons	L-aspartate-β-hydroxamate as an inhibitor
EAAT4	Cerebellar Purkinje neurons	L-α-AA as a substrate
EAAT5	Retina	THA and L-*trans*-2,4-PDC as nontransportable inhibitors

4-MG, (2*S*,4*R*)-4-methylglutamate; L-SOS, L-serine-*O*-sulfate; DHK, dihydrokainic acid; 3-TMG, (±)-*threo*-3-methylglutamic acid; L-α-AA, L-α-aminoadipate; THA, β-*threo*-hydroxyaspartate; MPDC, L-anti-endo-3,4-methanopyrrolidine-3,4-dicarboxylate; PDC, pyrrolidine dicarboxylate; GLT1, glutamate transporter-1; FAAC1, excitatory amino acid carrier-1; GLAST1, glutamate-aspartate transporter.

SOURCE: Bridges (2001).

choline, and the biogenic amines. These Glu transporters have distinct brain distributions, and even the glial transporters exhibit regional and intracellular differences in expression, underscoring the heterogeneity of glia as well as neurons. All of the Glu transporters demonstrate a strong Na^+ dependence and are enantioselective (i.e., D-Asp, L-Asp, and L-Glu are substrates, whereas D-Glu is not). These transporters are also inhibited by well-characterized uptake blockers including β-threo-hydroxy-Asp, dihydrokainate, and L-*trans*-2,4-pyrrolidine decarboxylate. Preferential inhibition of several subtypes of EAAT can be achieved with several of the newer nonsubstrate antagonists. Neurons and glial cells appear to possess a similar plasma membrane glutamate uptake carrier that serves to terminate the postsynaptic action of neurotransmitter Glu and to maintain the extracellular Glu concentrations below levels that may damage neurons. The presence of certain Glu transporters on glial cells is consistent with the intricate interplay of glial and neuronal elements in the synthesis and fate of Glu. Because the Glu that is released from neurons is accumulated by glia and then metabolized to glutamine, there is an ultimate recycling of the released amino acid. The fate of Glu released from a neuron containing the neuronal Glu transporter is unclear. It has not been established if Glu released from a given neuron is taken up in part by a Glu transporter on that particular neuron or primarily by Glu transporters on other neurons or glia.

The normal mechanism for terminating the synaptic action of Glu may involve all three processes to varying degrees. Radiotracer studies with labeled Glu applied to the brain indicate that most of the label is taken up into glial cells. However, this may be an artifact of the method employed since synaptically released, compared to exogenous, Glu may be better positioned with regard to the uptake carrier in the presynaptic nerve membrane than to the glial transporter. Regardless, uptake appears to be the major process for terminating the action of released Glu. There does not appear to be any significant role for enzymatic inactivation of Glu similar to that observed with GABA and the other classical neurotransmitters.

It is troublesome to some analysts that Glu, Asp, and synthetic derivatives of these dicarboxylic acids result in almost universal excitation of neuronal discharge, perhaps reflected in their far more ubiquitous distribution. Glu does not bring the cell membrane potential to the same level as the natural excitatory transmitter. In addition, both the D and the naturally occurring L isomers of EAA are active, although in the case of Glu the D isomer is often reported to be somewhat less active. These findings have led some highly skeptical investigators to suggest that the response to amino acids represents a nonspecific receptivity of the neuron to these agents and is therefore not necessarily indicative of a transmitter function.

Based on the abundant current evidence, however, Glu appears to have satisfied four of the main criteria for classification as an excitatory neurotransmitter in the mammalian CNS: *(1)* it is localized presynaptically in specific neurons, where it is stored in and released from synaptic vesicles; *(2)* it is released by a Ca-dependent mechanism by physiologically relevant stimuli in amounts sufficient to elicit postsynaptic responses; *(3)* a mechanism (reuptake and specific transporters) exists that will rapidly terminate its transmitter action; and *(4)* it demonstrates pharmacological identity with the naturally occurring transmitter.

The most clear-cut evidence that EAAs can act physiologically as excitatory neurotransmitters at a given synapse comes from experiments in which intracellular recordings of pre- and postsynaptic events have been made. In studies of this nature, the criterion of identical and parallel change induced by antagonists on synaptic events elicited by stimulation and those elicited by the action of the exogenously administered putative transmitter substance is of critical importance in identifying the excitatory amino transmitter involved. In many situations, however, such studies are not feasible, so more indirect studies like those summarized above have been utilized.

Quite often at excitatory synapses the actual molecule acting at the postsynaptic receptor has not been definitively identified, even though pharmacological analysis indicates that the synaptic response is mediated by a particular EAA receptor. Thus, a critical link in establishing a neurotransmitter role for an EAA within a specific pathway is the demonstration and characterization of the EAA receptor that mediates synaptic transmission at that synapse.

EAA Receptors

Our understanding of EAA transmitters and their function and regulation has been greatly enhanced by studies directed at the identification, characterization, localization, and isolation of receptors for these amino acids. In fact, progress on the definition of receptor subtypes and the availability of more selective agonists and antagonists has produced a quantum leap in knowledge about EAAs at synaptic sites throughout the vertebrate CNS. Furthermore, many aspects of Glu receptor dynamics have suggested an extraordinary degree of functional regulation.

Until the mid-1980s, neuropharmacologists were content with two major classes of EAA receptors, the NMDA receptors and the non-NMDA receptors, the latter composed at that time of kainate and quisqualate receptors. With the development of more selective agonists and antagonists, however, the classes of EAA receptors have expanded to at least five different types

(NMDA, kainate, α-amino-3-hydroxy-5-methylisoxazole-4-propionic acid [AMPA], 1,2-amino-4-phosphonobutyrate [AP4], and 1-aminocyclopentane-1,3-dicarboxylic acid [ACPD]; see below) in the CNS, each displaying distinct physiological characteristics. Three of these receptors have been defined by the depolarizing excitatory actions of select synthetic agonists (NMDA, kainate, and AMPA) and the blockade of the effects of these agonists by selective antagonists. A fourth, the AP-4 receptor, appears to represent an inhibitory autoreceptor. The fifth receptor, activated by ACPD, modifies inositol phosphate metabolism and has been called a metabotropic Glu receptor (mGluR). A summary of representative agonists and antagonists for each of these EAA receptor classes is given in Table 6–3.

Synaptic transmission in synapses using EAAs does not appear to follow the simple model of fast-acting synaptic transmission mediated by a single receptor class. In fact, individual synapses that use EAAs may not be restricted to distinct receptors but rather may have a combination of receptors and thereby exhibit different input/output properties and second-messenger responses.

One specific subtype of EAA receptor, the NMDA receptor, has become a major focus of attention because of evidence that it may be involved in a wide range of both neurophysiological and pathological processes as important and diverse as memory acquisition (see Chapter 13), developmental plasticity, epilepsy, and the neurotoxic effects of brain ischemia.

NMDA Receptor–Ionophore Complex

The NMDA receptor is a ligand-gated ion channel composed of two different protein subunits, NMDAR1 and NMDAR2. NMDAR1 can exist in seven splice variants, and there are four different genes encoding variants of NMDAR2 (NMDAR2A, -B, -C, and -D). At present, it is not clear how many NMDAR1 and NMDAR2 subunits are present in each functional NMDA receptor. This receptor complex has been extensively characterized physiologically and pharmacologically and is widely distributed in mammalian brain and spinal cord, with particularly high receptor densities found in hippocampus and cerebral cortex. NMDA receptors appear to have a pivotal role in long-term depression, long-term potentiation (LTP), and developmental plasticity. However, overactivation or prolonged stimulation of NMDA receptors can damage and eventually kill target neurons via a process referred to as excitotoxicity. Like the $GABA_A$ receptor, the NMDA receptor is a complex molecular entity endowed with a number of distinct recognition sites for endogenous and exogenous ligands, each with discrete binding domains. At present, there appear to be at least six pharmacologi-

TABLE 6–3. Classification of Excitatory Amino Acid Receptors in the Mammalian Central Nervous System

Currently Accepted Name	NMDA				
	Glutamate Site	Glycine Site	AMPA	Kainate	Metabotropic
Subtype-selective agonists	NMDA	Glycine D-serine R(+)HA-966 (partial) L-687,414 (partial)	AMPA S(−)-5-Flu Quisqualic acid	Kainic acid Domoic acid	L-AP_4 ACPD L-CCG-I
Subtype-selective antagonists	D(−)-AP-5 D(−)-AP-7 CGS-19755 CGP-37849 CGP-40116 CPP, (±)-D-D-CPPene	7-Chlorokynurenic 5,7-Dichlorokynurenic MNQX L-689,560	NBQX GYK1 52466		MCPG MPEP
Channel blockers	MK-801 Phencyclidine (PCP)				
Receptor-selective agonists	NMDA		AMPA	Kainic acid	L-AP-4
Receptor-selective antagonists	D(−)-AP-5		CNQX DNQX	CNQX DNQX	MCPG
Effector pathways	$Na^+/K^+/Ca^{2+}$		$Na^+/K^+/Ca^{2+}$	$Na^+/K^+/Ca^{2+}$	IP_3DAG

ACPD, 1-aminocyclopentane-1,3-dicarboxylic acid; AMPA, α-amino-3-hydroxy-5-methylisoxazole-4-propionic acid; AP-5, 2-amino-5-phosphonopentanoic acid; AP-7, amino-7 phosphonoheptanoic acid; CNQX, 6-cyano-7-nitroquinoxaline-2,3-dione; CPP, 3-(2-carboxypiperazin-4-yl)-propyl-1-phosphonic acid; DAG, diacylglycerol; D-CPPene, D-3-(2-carboxypiperazin-4-yl)-propyl-1-phosphonene; DNQX, 6,7-dinitroquinoxaline-2,3-dione; HA-966, 3-amino-1-hydroxypyrrolidone-2; IP_3, inositol trisphosphate; L-AP4, L-2-amino-4-phosphonobutanoic acid; L-687,414, R(+)-*cis*-β-methyl-3-amino-1-hydroxypyrrolid-2-one; L-CCG-I, (2S,3S,4S)-α-(carboxycyclopropyl)glycine; MCPG, α-methyl-4-carboxyphenylglycine; MNQX, 5,7-dinitroquinoxaline-2,3-dione; MPEP, 2-methyl-6-(phenylethynyl)-pyridine; NBQX, 2,3-dihydro-6-nitro-7-sulfamoylbenzo(f)quinoxaline; NMDA, *N*-methyl-D-aspartic acid; S(−)-5-Flu, S(−)-5-fluorowillardine.

cally distinct sites through which compounds can alter the activity of this receptor (see Fig. 6–10): *(1)* a transmitter binding site that binds L-glutamate and related agonists, promoting the opening of a high-conductance channel that permits entry of Na and Ca into target cells, L-Glu being virtually ineffective unless the site that binds Gly, the strychnine-insensitive Gly-modulatory site *(2)*, is also occupied (this Gly site is distinct from the Gly-binding site on the strychnine-sensitive Gly-inhibitory receptor, see discussion of Gly receptors above Glycine as an Inhibitory Transmitter, which is present in high density in the brain stem and spinal cord); *(3)* a site within the channel that binds phencyclidine (PCP site) and related noncompetitive antagonists (MK-801, ketamine), which act most effectively when the receptor is activated (i.e., open channel block); *(4)* a voltage-dependent Mg^{2+}-binding site; *(5)* an inhibitory divalent cation site near the mouth of the channel that binds Zn^{2+} to produce a voltage-independent block; and *(6)* a polyamine-regulatory site whose activation by spermine and spermidine facilitates NMDA receptor–mediated transmission.

In addition, two distinct binding sites are apparently associated with the transmitter recognition site, one that preferentially binds agonists and one that preferentially binds antagonists. It is of interest that quinolinic acid, a metabolite of tryptophan and thus a natural brain constituent, may be a specific antagonist for a particular subtype of NMDA receptor since it shows regional variation in potency. A number of Gly agonists and antagonists have already been identified for the Gly-regulatory site, even though the Gly-binding component of the NMDA receptor was discovered only recently.

The Gly-modulatory site has attracted a great deal of interest as a potential site for the action of new antiepileptic drugs or agents that might be useful in preventing ischemic brain damage. D-Serine is a potent agonist at this site, and (+)HA-966 is a selective antagonist. Gly in submicromolar concentrations increases the frequency of the NMDA receptor channel opening in a strychnine-insensitive manner, and, even though brain and cerebrospinal fluid concentrations of Gly are in the millimolar range, Gly is effective in vivo, indicating that the Gly site is not saturated. Thus, conditions that alter the extracellular concentration of Gly or compete with its binding site can dramatically alter NMDA receptor–mediated responses.

D-Serine is an endogenous ligand for the Gly site of the NMDA receptor. Strikingly, D-serine occurs exclusively in glia and not in neurons. It appears to be selectively localized in astrocytes containing its biosynthetic enzyme, serine racemase, which ensheathe nerve terminals in brain areas enriched in NMDA receptors. Thus, D-serine has been suggested to play a role in the modulation of synaptic transmission at NMDA receptors, providing a mechanism by which astrocytes may play an active role in supporting synaptic transmission via the activation of NMDA receptors.

Polyamines such as spermine and spermidine function as allosteric modulators of NMDA receptors and potentiate NMDA currents in the presence of saturating concentrations of Glu and Gly. However, in contrast to Gly or D-serine, their presence is not a requirement for NMDA receptor activation. Under pathological conditions, such as brain ischemia or trauma, in which the production of polyamines is dramatically increased, polyamines may mediate or potentiate the excitotoxic mechanisms responsible for the neuronal damage produced. This idea is supported by findings that phenylethanolamines, for instance, ifenprodil and eliprodil (SL820715), which are potent antagonists of the polyamine modulatory site of the NMDA receptor complex in a number of biochemical models, exhibit effective neuroprotective action in ischemia and trauma.

Kynurenate is a tryptophan metabolite that blocks the high-affinity Gly-binding site. It can be thought of as an endogenous neuroprotective agent that is released from glial cells following the transamination of L-kynurenine. Manipulation of endogenous kynurenate formation may provide further insight into the role it plays in modulating the action of Gly on the NMDA receptor.

Quinolinic acid is an endogenous excitatory neurotoxin, synthesized from L-tryptophan via the kynurenine pathway, that has the potential of mediating NMDA-induced neurotoxicity and dysfunction. Its endogenous occurrence in normal brain is relatively low (approximately 50–100 pmoles/g). Toxicity induced by exogenous quinolinic acid can be prevented or reversed by noncompetitive or competitive NMDA antagonists, suggesting that the neurotoxicity produced by locally administered quinolinic acid is mediated through the NMDA subtype of EAA receptors. Although quinolinic acid was first identified in the human brain in 1983, knowledge of the possible role played by this toxin in neuropathology was uncertain while speculation abounded. The observation by Heyes and co-workers in the late 1980s that quinolinate levels were dramatically elevated in acquired immunodeficiency syndrome (AIDS) was a major advance toward understanding the pathobiology of quinolinate. Further studies have demonstrated that the motor signs of AIDS dementia correlate strikingly with the levels of quinolinate in the cerebrospinal fluid. Treatment with the antiretroviral agent azidothymidine decreased the viremia and associated dementia and lowered the cerebrospinal fluid levels of quinolinate. Because cerebrospinal fluid levels of quinolinate are often higher than those found in blood, it has been argued that the origin of this neurotoxin may be via intracerebral synthesis, although this has not yet been documented. The highest reported levels of cerebrospinal fluid quinolinate are often obtained in AIDS patients with opportunistic infections. Elevated levels of cerebrospinal fluid quinolinate are found in human patients and in nonhuman primates with inflammatory neurological diseases, but it is unclear whether the mechanisms underlying the increased levels of quinolinate induced by bacterial and viral infection are similar. It also remains to be determined if the bacteria are actually involved in the synthesis of some of the increased quinolinate. No specific organism has been identified as responsible for elevating quinolinic acid since this neurotoxin is elevated in a variety of bacterial and viral infections. The discovery of an effective inhibitor of quinolinic acid synthesis will help to elucidate the importance of this neurotoxin in inflammatory and infectious disease.

In addition to the multiple ligand-binding and regulatory sites on the NMDA receptor for these relatively small molecules, the NMDA receptor exhibits an extremely high degree of interaction with many other membrane and cytoplasmic proteins. In fact, the complexity of the proteins with which this receptor has demonstrable interactions has established a new form of high-throughput molecular analysis combining the classical forms of two-dimensional gel chromatography with very sensitive methods of peptide sequence identification employing mass fragmentography. Using such "proteomic" methods, the NMDA receptor's known interacting partners increased from a single association with the intrinsic membrane protein PSD95, which was thought to be critical to anchor the receptor to the postsynaptic specialization site, to more than six dozen other proteins. Within this NMDA receptor protein complex, five main classes of proteins have so far been identified: neurotransmitter receptors (including other Glu receptors), cell adhesion proteins, adaptors, signaling enzymes, and cytoskeletal proteins.

Non-NMDA Receptors

Both AMPA and kainic acid (KA) receptors mediate fast excitatory synaptic transmission and are associated primarily with voltage-independent channels that gate a depolarizing current primarily carried by the influx of Na^+ ions. While these receptors are easily distinguished from NMDA receptors, they are more difficult to distinguish from each other. Molecular biological studies have confirmed the existence of AMPA and KA classes of non-NMDA receptors but have revealed a considerable degree of heterogeneity within these two families. Some pharmacological discrimination can be achieved, with AMPA and quisqualate being the preferred agonists for AMPA receptors and domoate and kainate the preferred agonists for KA receptors. The most selective and potent non-NMDA antagonists available are a series of dihydroxyquinoxaline derivatives including 2,3-dihydro-6-nitro-7-sulfamoyl-benzo(F)quinoxaline (NBQX), 6-cyano-7-nitroquinoxaline-2,3-dione (CNQX), and 6,7-dinitroquinoxaline-2,3-dione (DNQX) (see Table 6–3). However, these agents competitively block both types of non-NMDA receptor, although NBQX appears to exhibit the best selectivity for AMPA receptors. Very few KA receptor–selective compounds have been identified. A new class of 2,3-benzodiazepine derivatives (most notably GYKI-52466) has been shown to block AMPA-induced responses noncompetitively and to attenuate ischemic neuronal damage effectively in animal models, highlighting the fact that non-NMDA receptors also play an important role in CNS pathology.

In studies of immature hippocampal neurons in culture, AMPA receptors were surprisingly found almost exclusively within the dendritic cytoplasm rather than in the dendritic membranes and spines. Using special constructs of the mRNA for the AMPA receptor coupled with green fluorescent protein, regular dynamic insertion and removal could be demonstrated for the receptors, which was accelerated by activity. This receptor cycling was quickly advanced as a mechanism for activating silent synapses by recruiting these receptors to the synaptic surfaces. However, whether this explanation would hold for mature neurons is dubious since such neurons exhibit abundant spine and synaptic AMPA receptors. Recent evidence suggests that in mature neurons, it may well be the NMDA receptors that are recruited to the cell surface during depolarizing events.

Metabotropic Glutamate Receptors

The mGluRs constitute a family of EAA receptors that are linked to G proteins and second-messenger systems and are distinct from the ionotropic EAA receptors that form ion channels and are comprised of the NMDA, AMPA, and kainate subtypes discussed above. This more recently characterized group of receptors is coupled to a variety of signal-transduction pathways via G proteins, producing alterations in intercellular second messengers and generating slow synaptic responses. This is in clear contrast to the ionotropic Glu receptors, which are directly coupled to cation-specific ion channels and mediate fast excitatory synaptic responses. *Metabotropic* is a term that was coined to indicate that these receptors, unlike the inotropic receptors which form ion channels, affect cellular metabolic processes. Unfortunately, this nomenclature is very misleading since mGluRs, similar to other G protein–coupled receptors, exert profound effects on neuronal function through the regulation of ion channels, protein phosphorylation, and second-messenger cascades. The widespread distribution of metabotropic receptors in the CNS coupled with the prevalence of Glu as a neurotransmitter indicates that this system is a major modulator of second messengers in the mammalian CNS. Molecular cloning studies have revealed the existence of at least eight different subtypes of mGluR, $mGluR_1$ through $mGluR_8$, which have a common structure of a large extracellular domain preceded by the seven-member spanning domains.

Members of the mGluR family can be divided into three subgroups according to their sequence similarities, signal-transduction properties, and pharmacological profiles to agonists (i.e., relative potencies when expressed in cell lines of Glu, quisqualate, ACPD, and AP-4). The first subgroup, comprised of $mGluR_1$ and $mGluR_5$, is coupled to the stimulation of phosphatidylinositol hydrolysis/Ca^{2+} signal transduction. The second group, $mGluR_2$ and $mGluR_3$, is negatively coupled through adenylyl cyclase to cyclic adenosine monophosphate formation. The third group, $mGluR_4$, $mGluR_6$, $mGluR_7$, and $mGluR_8$, is also negatively linked to adenylyl cyclase activity but shows a different agonist preference from that of $mGluR_2$ and $mGluR_3$. As a group, the mGluRs are widely expressed throughout the brain, but the individual subtypes show some differential distribution. The pharmacology of the individual subtypes expressed in Chinese hamster ovary or in baby hamster kidney cells shows some interesting differences (Table 6–3). LAP4 is a potent agonist of $mGluR_4$, $mGluR_6$, $mGluR_7$, and $mGluR_8$ but has little effect on the other receptor subtypes. L-(2S,3S,4S)-α-(carboxycyclopropyl) Gly (L-CCG-I), however, activates $mGluR_2$ at concentrations that have little or no effect on $mGluR_1$ and $mGluR_4$. No agonist yet identified appears to be specific for any single metabotropic receptor subtype.

Considerable experimental evidence indicates that the mGluRs are involved in the regulation of synaptic transmission in the CNS. However, the lack until recently of specific antagonists has limited the precise characterization of the role of individual mGluRs in glutamatergic transmission and has severely hampered progress in identifying their physiological and pathological roles. The discovery that phenylglycine derivatives are selective antagonists of mGluRs has permitted more rigorous testing of the physiological role of this receptor subclass in brain function and dysfunction. Data are emerging that suggest a role in both synaptic transmission and synaptic plasticity. The most exciting new development concerning MGluRs is the finding by Conquet and collaborators using knockout mice, that $mGluR_5$ is probably an essential factor in cocaine self-administration and locomotor effects. Their studies showed that the reinforcing properties of cocaine are absent in mice lacking $mGluR_5$ and that a selective $mGluR_5$ antagonist, 2-methyl-6-(phenylethynyl)-pyridine, dose-dependently decreased cocaine self-administration (Fig. 6–12). This finding should provide a new incentive for the rapid development of more potent mGluR antagonists with greater subtype specificity and enhanced bioavailability. With the availability of these new subtype-selective antagonists, the next few years should witness major advances

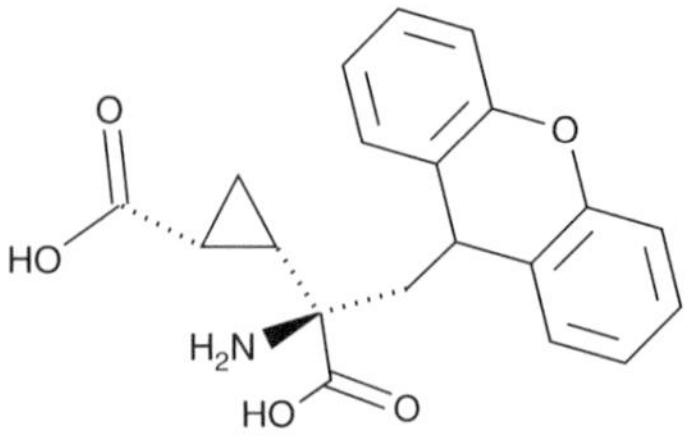

LY344545

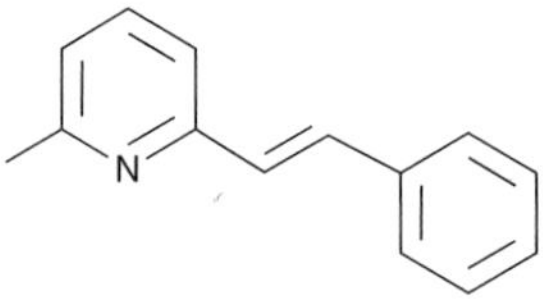

SIB1893

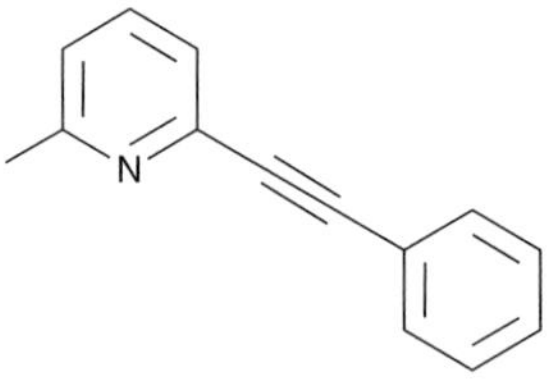

MPEP

FIGURE 6–12. Structures of group 1-selective metabotropic glutamate (mGlu) receptor antagonists. LY344545 is a selective mGlu5 receptor competitive antagonist. SIB1893 and MPEP [2-methyl-6-(phenylethynyl)-pyridine] are two new noncompetitive selective mGlu5 receptor antagonists.

in our knowledge of the roles played by mGluRs in physiological and pathological processes and perhaps even a new treatment of addictive disorders.

The function and distribution of the four classes of EAA receptors are summarized in Table 6–4. The NMDA receptor is an essential component in the generation of LTP. LTP results in an increase in synaptic efficacy that has been proposed as an underlying mechanism involved in memory and learning (see Chapter 13).

In addition to the roles excitotoxic mechanisms may play in various chronic neurodegenerative disorders like Huntington's disease and viral diseases like AIDS, two chronic neurological syndromes have been linked to dietary consumption of amino acid toxins of plant origin. Neurolathyrism, a spastic disorder occurring in eastern Africa and southern Asia, is associated with dietary consumption of the chick pea *Lathyrus sativus*. β-*N*-Oxalylamino-L-alanine (L-BOAA) has been identified as the responsible toxin in this plant. This amino acid acts as an agonist at AMPA receptors. One of the primary effects of L-BOAA toxicity is the inhibition of mitochondrial complex 1 selectively in the motor cortex and lumbar spinal cord. Guam disease, also referred to as amyotrophic lateral sclerosis/parkinsonism/dementia, is thought

TABLE 6–4. Distribution and Function of Excitatory Amino Acid Receptors in the Mammalian Central Nervous System (CNS)

Receptor Type	*Distribution/Function*
NMDA	Widely distributed in mammalian CNS (enriched in hippocampus, cerebral cortex). Demonstrated most easily by pharmacological antagonism under MG^{2+}-free or depolarizing conditions or in binding experiments. Usually recognized as a slow component in repetitive activity generated primarily by non-NMDA receptors. Important in synaptic plasticity.
AMPA	Widespread in CNS; parallel distribution to NMDA receptors. Involved in the generation of fast component of EPSPs in many central excitatory pathways.
Kainate	Concentrated in a few specific areas of CNS, complementary to NMDA/AMPA distribution (e.g., stratum lucidum region of hippocampus). Difficult to distinguish from AMPA receptors pharmacologically due to nonspecificity of kainate in electrophysiological experiments. However, present specificity (in absence of AMPA and NMDA receptors) on dorsal root C fibers.
ACPD (metabotropic)	Linked to IP_3 formation. Activated by glutamate, quisqualate, ibotenate, and *trans*-ACPD but not by AMPA, NMDA, or kainate. Not antagonized by NMDA or non-NMDA antagonists but sensitive to pertussis toxin. May be involved in developmental plasticity.

NMDA, *N*-methyl-D-aspartate; AMPA, α-amino-3-hydroxy-5-methylisoxazole-4-propionic acid; ACPD, 1-aminocyclopentane-1,3-dicarboxylic acid; EPSP, excitatory postsynaptic potential; IP_3, inositol-1,4,5-trisphosphate.

SOURCE: Modified from Watkins et al. (1990).

to be related to the consumption of flour prepared from the seeds of the cycad *Cycas circinalis*, which contains the amino acid β-*N*-methyl-amino-L-alanine (BMAA). Although BMAA is a neutral amino acid that is not directly excitatory or toxic in vitro, in the presence of bicarbonate it becomes excitotoxic and acts as an agonist at AMPA and NMDA receptors.

Neurotoxicity has also been observed following ingestion of domoic acid. Domoic acid is an analog of KA that is about three times as potent. This substance is synthesized by seaweed and can be consumed in toxic amounts by eating mussels that have fed on the seaweed. An outbreak of domoic poisoning occurred in 1987 in Canada. Consumption of this neurotoxin by humans can damage the hippocampus and produce dementia.

With the availability of more specific pharmacological agents, it should be possible to evaluate in more detail the involvement of EAA pathways in normal brain function and in neuropathological conditions. The participation of NMDA receptors in LTP provides a strong link between these systems and the mechanisms of learning and memory. NMDA and other EAA receptors also appear to play a role in cell damage caused by hypoglycemia, hypoxia, seizures, and other disturbances associated with excess EEAs.

Selected References

GABA

Bowery, N. G. (1993). $GABA_B$ receptor pharmacology. *Annu. Rev. Pharmacol. Toxicol.* 33, 109–147.

Jursky, F., S. Tamura, A. Tamura, S. Mandiyan, H. Nelson, and N. Nelson (1994). Structure, function and brain localization of neurotransmitter transporters. *J. Exp. Biol.* 196, 283–295.

Lüddens, H. and W. Wisden (1991). Function and pharmacology of multiple $GABA_A$ receptor subunits. *Trends Pharmacol. Sci.* 12, 49–51.

Macdonald, R. L. and Olsen, R. W. (1994). $GABA_A$ receptor channels. *Annu. Rev. Neurosci.* 17, 569–602.

Mellon, S. H. and L. S. Griffin (2002). Neurosteroids: biochemistry and clinical significance. *Trends Endocrinol. Metab.* 13, 35–43.

Mohler, H., F. Crestani, and U. Rudolph (2001). $GABA_A$ receptor subtypes: a new pharmacology. *Curr. Opin. Pharmacol.* 1, 22–25.

Paul, S. M. (1995). GABA and glycine. In *Psychopharmacology: The Fourth Generation of Progress* (F. E. Bloom and D. J. Kupfer, eds.). Raven Press, New York, pp. 87–94.

Purdy, R. H., A. L. Morrow, P. H. Moore, Jr., and S. M. Paul (1991). Stress-induced elevations of γ-aminobutyric acid type A receptor-active steroids in the rat brain. *Proc. Natl. Acad. Sci. U.S.A.* 88, 4553–4557.

Rudolph, U., F. Crestani, and H. Mohler (2001). $GABA_A$ receptor subtypes: dissecting their pharmacological functions. *Trends in Pharmacol. Sci.* 22, 1188–194.

Scott, R. H., K. G. Sutton, and A. C. Dolphin (1993). Interactions of polyamines with neuronal ion channels. *Trends Neurosci.* 16, 153–160.

Sieghart, W. (2000). Unraveling the function of $GABA_A$ receptor subtypes. *Trends Pharmacol. Sci.* 21, 411–413.

Glycine

Becker, C.-M. (1990). Disorders of the inhibitory glycine receptor: the spastic mouse. *FASEB J.* 4, 2767–2774.

Betz, H. (1992). Structure and function of inhibitory glycine receptors. *Q. Rev. Biophys.* 25, 381–394.
Ottersen, O. P. and J. Storm-Mathisen (eds.) (1990). *Glycine Neurotransmission.* John Wiley & Sons, New York.
Roberts, E. and R. Hammerschlag (1972). *Basic Neurochemistry*. Little, Brown, Boston, pp. 131–163.
Vandenberg, R. J., C. A. Handford, and P. R. Schofield (1992). Distinct agonist- and antagonist-binding sites on the glycine receptor. *Neuron* 9, 491–496.
Vannier, C. and A. Triller (1997). Biology of the postsynaptic glycine receptor. *Int. Rev. Cytol.* 176, 201–244.
Zafra, F., C. Aragon, and C. Gimenez (1997). Molecular biology of glycinergic neurotransmission. *Mol. Neurobiol.* 14, 117–142.

Excitatory amino acids

Bridges, R. J. (2001). The ins and outs of glutamate transporter pharmacology. *Tocris Rev.* 17, 1–5.
Conn, P.J., and J.-P. Pin (1997). Pharmacology and functions of metabotropic glutamate receptors. *Annu. Rev. Pharmacol. Toxicol.* 37, 205–237.
Chiamulera, C., M. P. Epping-Jordan, A. Zocchi, C. Marcon, C. Cottiny, S. Tacconi, M. Corsi, F. Orzi, and F. Conquet (2001). Reinforcing and locomotor stimulant effects of cocaine are absent in $mGluR_5$ null mutant mice. *Nat. Neurosci.* 4, 873–874.
Cotman, C. W., J. S. Kahle, S. E. Miller, J. Ulas, and R. J. Bridges (1995). Excitatory amino acid neurotransmission. In *Psychopharmacology: The Fourth Generation of Progress* (F. E. Bloom and D. J. Kupfer, eds.). Raven Press, New York, pp. 75–85.
Grant, S. N. and W. P. Blackstock (2001). Proteomics in neuroscience: from protein to network. *J. Neurosci.* 21, 8315–8318.
Grosshaus, D. R., D. A. Clayton, S. J. Coultrop, and M. D. Browning (2002). LTP leads to rapid surface expression of NMDA but not AMPA receptors in adult rat CA1. *Nat. Neurosci.* 5, 27–33.
Hayashi, Y., N. Sekiyama, S. Nakanishi, D. E. Jane, D. C. Sunter, E. F. Birse, P. M. Udvarhelyi, and J. C. Watkins (1994). Analysis of agonist and antagonist activities of phenylglycine derivatives for different cloned metabotropic glutamate receptor subtypes. *Neuroscience* 14, 3370–3377.
Husi, H, M. A. Ward, J. S. Choudhary, W. P. Blackstock, and S. G. Grant (2000). Proteomic analysis of NMDA receptor-adhesion protein signaling complexes. *Nat. Neurosci.* 3, 661–669.
Ishii, T., K. Moriyoshi, H. Sugihara, K. Sakurada, H. Kadotani, M. Yokoi, C. Akazawa, R. Shigemoto, N. Mizuno and M. Masu (1993). Molecular characterization of the family of the *N*-methyl-D-aspartate receptor subunits. *J. Biol. Chem.* 268, 2836–2843.
Kemp, J. A. and P. D. Leeson (1993). The glycine site of the NMDA receptor—five years on. *Trends Pharmacol. Sci.* 14, 20–25.
Mothet, J.-P., A. T. Parent, H. Wolosker, R. O. Brady, Jr., D. J. Linden, C. D. Ferris, M. A. Rogawski, and S. H. Snyder (2000). D-Serine is an endogenous ligand for the glycine site of the *N*-methyl-D-aspartate receptor. *Proc. Natl. Acad. Sci. U.S.A.* 97, 4926–4931.

Nicholls, D. G. (1994). *Proteins, Transmitters and Synapses*. Blackwell Science, Cambridge, MA.

Ravindranath, V. (2002). Neurolathyrism: mitochondrial dysfunction in excitotoxicity mediated by L-beta-oxalyl aminoalanine. *Neurochem. Int.* 40(6), 505–509.

Reinhard, J. F., Jr., J. B. Erickson, and E. M. Flanagan (1994). Quinolinic acid in neurological disease: opportunities for novel drug discovery. *Adv. Pharmacol.* 30, 85–127.

Rogawski, M. A. (1993). Therapeutic potential of excitatory amino acid antagonists: channel blockers and 2,3-benzodiazepines. *Trends Pharmacol. Sci.* 14, 325–331.

Sacco, R. L., J. T. DeRosa, E. C. Haley, Jr., B. Levin, P. Ordronneau, S. J. Phillips, T. Rundek, R. G. Snipes, J. L. Thompson and for the GAIN Americas Investigators (2001). Glycine antagonist in neuroprotection for patients with acute stroke GAIN Americas: a randomized controlled trial. *JAMA* 285, 1719–1728.

Scatton, B. (1993). The NMDA receptor complex. *Fundam. Clin. Pharmacol.* 7, 389–400.

Schoepp, D. D., D. E. Jane, and J. A. Monn (1999). Pharmacological agents acting at subtypes of metabotropic glutamate receptors. *Neuropharmacology* 38, 1431–1476.

Seeburg, P. H. (1993). The molecular biology of mammalian glutamate receptor channels. *Trends Pharmacol. Sci.* 14, 297–303.

Shi, S.-H. (2001). AMPA receptor dynamics and synaptic plasticity. *Science* 294, 1851–1852.

Spooren, W. P. S., F. Gasparomo, T. E. Salt, and R. Kuhn (2001). Novel allosteric antagonists shed light on $mGlu_5$ receptors and CNS disorders. *Trends Pharmacol. Sci.* 22, 331–337.

Takamori, S., J. S. Rhee, C. Rosenmund, and R. Jahn (2000). Identification of a vesicular glutamate transporter that defines a glutaminergic phenotype in neurons. *Nature* 407, 189–194.

Valtschanoff, J. G., K. D. Phend, P. S. Bernardi, R. J. Weinberg, and A. Rustioni (1994). Amino acid immunocytochemistry of primary afferent terminals in the rat dorsal horn. *J. Comp. Neurol.* 346, 237–252.

Watkins, J. and G. Collingridge (1994). Phenylglycine derivatives as antagonists of metabotropic glutamate receptors. *Trends Pharmacol. Sci.* 15, 333–342.

7

Acetylcholine

The neurophysiological activity of acetylcholine (ACh) has been known since the turn of the century and its neurotransmitter role since the mid-1920s. With this long history, it is not surprising when students assume that everything is known about the subject. Unfortunately, the delay in developing sophisticated methods for determining the presence of ACh in cholinergic tracts and terminals, which only recently has been overcome, has left this field far behind the biogenic amines. The structural formula of ACh is presented below:

$$(CH_3)_3N^+-CH_2CH_2-O-\underset{\underset{\displaystyle O}{\|}}{C}-CH_3$$

Assay Procedures

ACh may be assayed by its effect on biological test systems or by physiochemical methods. Popular bioassay preparations include the frog rectus abdominis, the dorsal muscle of the leech, the guinea pig ileum, the blood pressure of the rat (or cat), and the heart of *Venus mercenaria*. In general, bioassays tend to be laborious, to be subject to interference by naturally occurring substances, and on occasion to behave in a mysterious fashion (e.g., the frog rectus abdominis is not as sensitive to ACh in the summer months as in the winter). Nevertheless, bioassays currently represent one of the most sensitive (0.01 pmol in the toad lung) and, under properly controlled conditions, the most specific procedures for determining ACh. It is probably fair to say that the neurochemically oriented investigators' natural fear and distrust of a bioassay have hampered progress in elucidating the biochemical and biophysical aspects of ACh. This statement is supported by a consideration of the plethora of information on norepinephrine. This neurotransmitter can also be bioassayed, but it was only after the development of sensitive fluorometric and radiometric procedures for determining components of the adrenergic nervous system that the information explosion occurred.

Until about 1965, physiochemical methods for determining ACh were so insensitive as to be virtually useless for measuring endogenous levels. Since then, however, papers have been published on enzymatic, fluorometric, gas chromatographic, chemiluminescent, and radioimmunoassay techniques that approach the sensitivity and specificity of the bioassays. Currently, the most popular assays are the gas chromatographic–mass spectrometric procedure of Jenden and co-workers, the radiometric procedure of McCaman and Stetzler, the chemiluminescent assay of Israel and Lesbats, and the high-performance liquid chromatographic–electrochemical procedure of Potter and colleagues. Ricny and co-workers have modified and combined several existing methods to develop a highly sensitive (0.2 pmol) determination of the transmitter, and the sensitivity has been further increased in the procedure of Flentge et al.

Synthesis

ACh is synthesized in a reaction catalyzed by choline acetyltransferase (ChAT):

$$\text{Acetyl CoA} + \text{choline} \rightleftharpoons \text{ACh} + \text{CoA}$$

Before entering into a discussion of ChAT, we should take note of Figure 7–1, which depicts the possible sources of acetyl coenzyme A (CoA) and choline. In theory, the acetyl CoA for ACh synthesis may arise from glucose, through glycolysis and the pyruvate dehydrogenase system; from citrate, either by reversal of the condensing enzyme (citrate synthetase) or by the citrate cleavage enzyme (citrate lyase); or from acetate through acetate thiokinase. In brain slices, homogenates, acetone powder extracts, and preparations of nerve-ending particles, glucose, or citrate are the best sources for ACh synthesis, with acetate rarely showing any activity. In lobster axons, the electric organ of the *Torpedo*, corneal epithelium, and frog neuromuscular junction, acetate appears to be the preferred substrate. However, all of these systems are in vitro and do not necessarily reflect the situation in vivo. Regardless of its source, acetyl CoA is primarily synthesized in mitochondria. Since, as detailed below, ChAT appears to be in the synaptosomal cytoplasm, another still unsolved problem is how acetyl CoA is transported out of the mitochondria to participate in ACh synthesis. A probable carrier for acetyl CoA is citrate, which can diffuse into the cytosol and produce acetyl CoA via citrate lyase; a possible carrier is acetyl carnitine, and another possibility is Ca^{2+}-induced leakage of acetyl CoA from mitochondria.

Although choline can be synthesized de novo in brain by successive methylations of ethanolamine, the extent is too minor to be of significance. Rather, choline is transported to the brain both free and in phospholipid form (possibly as phosphatidylcholine) by the blood. Following the hydrolysis of ACh, about 35%–50% of the liberated choline is transported back into the presynaptic terminal by a sodium-dependent, high-affinity active transport system, to be reutilized in ACh synthesis. As outlined in Figure 7–1, the remaining choline may be catabolized or become incorporated into phospholipids, which can again serve as a source of choline. A curious observation is that when brain cortical slices are incubated for 2 hours in a Krebs-Ringer medium, choline accumulates to about 10 times its original concentration. Similarly, a rapid postmortem increase in choline has been observed. The precise source of this choline is unknown; a probable candidate is phosphatidylcholine.

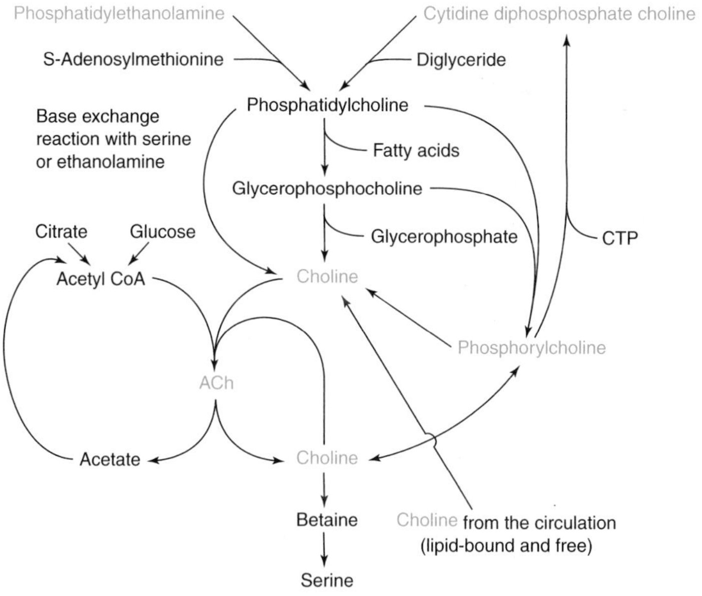

FIGURE 7–1. Acetylcholine (ACh) metabolism. CoA, coenzyme A; CTP, cytidine triphosphate.

Choline Transport

Recent studies of choline transport have produced a number of significant findings.

1. Choline crosses cell membranes by two processes, referred to as *high-affinity* and *low-affinity transport.* High-affinity transport, with a K_m for choline of 1–5 μM, is saturable, carrier-mediated, dependent on sodium, and stimulated by chloride. It is also dependent on the membrane potential of the cell or organelle so that any agent (e.g., K^+) that depolarizes the cell will concurrently inhibit high-affinity transport. Low-affinity choline transport, with a K_m of 40–80 mM, appears to operate by a passive diffusion process, to be linearly dependent on the concentration of choline, and to be virtually nonsaturable.
2. In contrast to the other neurotransmitters, ACh is taken up in terminals only via low-affinity transport; it is only choline that exhibits high-affinity kinetics.
3. Current evidence suggests that the high-affinity transport of choline is specific for cholinergic terminals and is not present in aminergic nerve terminals. Furthermore, it is kinetically (but not physically) coupled to ACh synthesis. About 50%–85% of the choline that is transported by the high-affinity process is utilized for ACh synthesis. Low-affinity transport, however, is found in cell bodies and in tissues such as the corneal epithelium, and it is thought to function in the synthesis of choline-containing phospholipids. Tissues that do not synthesize ACh (e.g., fibroblasts, erythrocytes, photoreceptor cells) exhibit high-affinity choline transport that is coupled to phospholipid synthesis.
4. Hemicholinium-3 is an extremely potent inhibitor of high-affinity transport (K_m of 0.05–1 μM) but a relatively weak inhibitor of low-affinity transport (K_m of 10–120 μM).
5. There are three obvious mechanisms for regulating the level of ACh in cells: feedback inhibition by ACh on ChAT, mass action, and the availability of acetyl CoA and/or choline. Of these three possibilities, the major regulatory factor seems to be high-affinity choline transport. This view derives from early observations that choline is rate-limiting in the synthesis of ACh coupled with findings in a number of laboratories. Using the septal-hippocampal pathway, a known cholinergic tract, Kuhar and associates showed that changes in impulse flow induced via electrical stimulation or pentylenetetrazol administration (both of which increase impulse flow) or via lesioning

or the administration of pentobarbital (both of which decrease neuronal traffic) will alter high-affinity transport of choline into hippocampal synaptosomes. In their studies, procedures that activated impulse flow increased the maximal velocity (V_{max}) of choline transport, while agents that stopped neuronal activity decreased V_{max}. In neither situation was the K_m changed, a result to be expected since the concentration of choline outside the neuron (5–10 m*M*) normally exceeds the K_m for transport (1–5 m*M*). Recent evidence, however, suggests that this relationship between impulse traffic and choline transport does not occur in all brain areas (e.g., in the striatum, where cholinergic interneurons abound). In addition, the endogenous concentration of ACh is implicated in regulating the level of the transmitter in the brain. Thus, in several studies, an increase in choline uptake following depolarization of a preparation has been attributed to the release of endogenous ACh upon depolarization. Other studies, however, suggest that this increased choline uptake is not related to ACh release but rather to an increase in Na-K adenosine triphosphatase (ATPase) activity.

6. Recently, the high-affinity choline transporter has been cloned by the Okuda group, who developed an antibody to map cholinergic neurons.

Choline Acetyltransferase

With respect to the cellular localization of ChAT, the highest activity is found in the interpeduncular nucleus, caudate nucleus, retina, corneal epithelium, and central spinal roots (3000–4000 mg ACh synthesized $\cdot$ g^{-1} $\cdot$ $hour^{-1}$). In contrast, dorsal spinal roots contain only trace amounts of the enzyme, as does the cerebellum.

Intracellularly, after differential centrifugation in a sucrose medium, ChAT is found in mammalian brain, predominantly in the crude mitochondrial fraction. This fraction contains mitochondria, nerve-ending particles (synaptosomes) with enclosed synaptic vesicles, and membrane fragments. When this fraction is subjected to sucrose density gradient centrifugation, the bulk of the ChAT is associated with nerve-ending particles. When these synaptosomes are ruptured by hypo-osmotic shock, synaptosomal cytoplasm can be separated from synaptic vesicles. In a solution of low ionic strength, ChAT is adsorbed to membranes and to vesicles; but in the presence of salts at physiological concentration, the enzyme is solubilized and remains in the cytoplasm. In vivo, the enzyme is most likely present in the cytoplasm of the nerve-ending particle. However, a particulate form of the enzyme has been found at nerve terminals and, more recently, at synaptic vesicle membranes. This ChAT, which when solubilized has the same kinetic characteristics as the soluble form, occurs in a much lower concentration but with a higher specific activity. Since kinetic coupling has been observed between uptake of choline and acetylation, it is conceivable that this membrane-bound ChAT is the physiologically relevant form. Support for this contention derives from three experimental findings: *(1)* homocholine cannot be acetylated with purified soluble ChAT but can form acetylhomocholine when a lysed synaptosomal preparation is used; *(2)* choline mustard aziridinium ion, an inhibitor of choline transport as well as of ChAT, is a much more potent inhibitor of membrane-bound ChAT than the soluble form of the enzyme; *(3)* a monoclonal antibody raised against *Torpedo* terminal membranes inhibits ChAT and ACh release. Regardless of these observations, most cholinergic experts favor the soluble form of ChAT as the major source of ACh.

A cell-free system of ChAT was first described by Nachmansohn and Machado in 1943. Since that time, the enzyme from squid head ganglia, human placenta, *Drosophila melanogaster*, *Torpedo californica*, and brain has been purified and some of its characteristics have been defined.

When highly purified from rat brain, ChAT has a molecular weight of 67–75 kDa: it has an apparent Michaelis constant (K_m) for choline of 7.5×10^{-4} M and for acetyl CoA of 1.0×10^{-5} M. Recent estimates suggest an equilibrium constant of 13. The enzyme is activated by chloride and inhibited by sulfhydryl reagents. A variety of studies on the substrate specificity of the enzyme indicate that various acyl derivatives of both CoA and ethanolamine can be utilized. The major gap in our knowledge of ChAT is that as yet we do not know of any useful (i.e., potent and specific) direct inhibitor. Styrylpyridine derivatives inhibit it but suffer from the fact that they are light-sensitive, somewhat insoluble, and possess varying degrees of anticholinesterase activity. Hemicholinium inhibits the synthesis of ACh indirectly by preventing the transport of choline across cell membranes. The genomic aspects of ChAT have been reviewed by Wu and Hersh (1994). Dobransky et al. (2001) have described the results of phosphorylation of the enzyme.

Acetylcholinesterase

Everybody agrees that ACh is hydrolyzed by cholinesterases, but nobody is sure just how many cholinesterases exist in the body. All cholinesterases will hydrolyze not only ACh but other esters as well. Conversely, hydrolytic enzymes such as arylesterases, trypsin, and chymotrypsin will not hydrolyze choline esters. The problem in determining the number of cholinesterases that exist is that different species and organs sometimes exhibit maximal activity with different substrates. For our purposes, we will divide the enzymes into two rigidly defined classes: acetylcholinesterase (also called "true" or specific cholinesterase) and butyrylcholinesterase (also called "pseudo" or nonspecific cholinesterase; the term *propionylcholinesterase* is sometimes used since in some tissues propionylcholine is hydrolyzed more rapidly than butyrylcholine). Although their molecular forms are similar, the two enzymes are distinct entities, encoded by specific genes. Current evidence suggests that in lower forms butyrylcholinesterase predominates, gradually giving way to acetylcholinesterase with evolution. When distinguishing between the two types of cholinesterase, at least two criteria should be used because of the aforementioned species or organ variation.

The first criterion is the optimum substrate. Acetylcholinesterase hydrolyzes ACh faster than butyrylcholine, propionylcholine, or tributyrin; the reverse is true with butyrylcholinesterase. In addition, acetyl-*N*-methyl choline (methacholine) is split only by acetylcholinesterase. That this criterion is not inviolate and must be used along with other indices is illustrated by the fact that chicken brain acetylcholinesterase will hydrolyze acetyl-β-methyl choline but will also hydrolyze propionylcholine faster than ACh. Also, the beehead enzyme will not hydrolyze either ACh or butyrylcholine but will split acetyl-β-methyl choline.

The second criterion is the substrate concentration versus activity relationship. Acetylcholinesterase is inhibited by high concentrations of ACh so that a bell-shaped substrate concentration curve results. This is observed also when butyrylcholine or propionylcholine is used. In contrast, butyrylcholinesterase is not inhibited by high substrate concentrations so that the usual Michaelis-Menten type of substrate concentration curve is obtained. The reason for this difference is that in acetylcholinesterase there is at least a two-point attachment of substrate to enzyme, whereas with butyrylcholinesterase the substrate is attached at only one site.

The type of cholinesterase found in a tissue is often a reflection of the tissue. This fact is used as a discriminating index between cholinesterases. In general, neural tissue contains acetylcholinesterase, while glial cells and nonneural tissue usually contain butyrylcholinesterase. However, this is a generalization, and some neural tissue (e.g., autonomic ganglia) contains both esterases, as do some extraneural organs (e.g., liver, lung). In the blood, erythrocytes contain only acetylcholinesterase, while plasma contains butyrylcholinesterase. However, plasma has primary substrates varying from species to species. Because of its ubiquity, cholinesterase activity cannot be used as the sole indicator of a cholinergic system in the absence of additional supporting evidence. To generalize on this point, until neuron-specific, transmitter-degrading enzymes are discovered, it is a neurochemical commandment that, to delineate a neuronal tract, one must always assay an enzyme involved in the synthesis of a neurotransmitter and not one concerned with catabolism.

A final criterion that may be applied to differentiate between the esterases is their susceptibility to inhibitors. Thus, the organophosphorous anticholinesterases, such as diisopropyl phosphorofluoridate and iso-octamethyl pyrophosphoramide, are more potent inhibitors of butyrylcholinesterase whereas WIN8077 (Ambinonium) is about 2000 times better an inhibitor of acetylcholinesterase. The compound BW284C51 [1,5-bis-(4-allyldimethylammoniumphenyl)penta-3-one dibromide] is presumed to be a specific reversible inhibitor of acetylcholinesterase.

In discussing the various techniques used to classify the cholinesterases, we touched on some aspects of the molecular properties of the enzymes. Because very little work has been done on butyrylcholinesterase and no physiological role for this enzyme (or enzymes) has been demonstrated, we will focus our attention on acetylcholinesterase. In sucrose homogenates of mammalian brain subjected to differential centrifugation, acetylcholinesterase is found in both the mitochondrial and the microsomal fractions. The latter, consisting of endoplasmic reticulum and plasma cell membranes, exhibit a higher specific activity. This localization of the enzyme is supported by electron microscopic and histochemical studies that fix the activity at membranes of all kinds in both the CNS and the peripheral nervous system.

Both cholinesterases occur in several molecular forms, which are classified as either globular or asymmetric. The globular forms, G1, G2, and G4 (Fig. 7–2), exist as monomers, dimers, and tetramers. Elongated forms, which contain as many as 12 subunits and are attached to a collagen tail, are classified as asymmetric. Regardless of the form, both cholinesterases occur in a water-soluble and a membrane-bound state.

With its turnover time of 150 milliseconds, equivalent to hydrolyzing 5000 molecules of ACh per molecule of enzyme per second, acetylcholinesterase ranks as one of the most efficient enzymes extant.

With respect to the topography of the enzyme, the twin-hatted diagram of the anionic and esteratic sites has been reproduced countless times and need not be presented again here. However, some discussion is in order since this was the first enzyme to be dissected at the molecular level. For this initiation into molecular biology, we owe a debt of gratitude to Nachmansohn and colleagues, particularly Wilson.

The active center of acetylcholinesterase has two main subsites. The first is an anionic site, which attracts the positive charge in ACh; the second, about

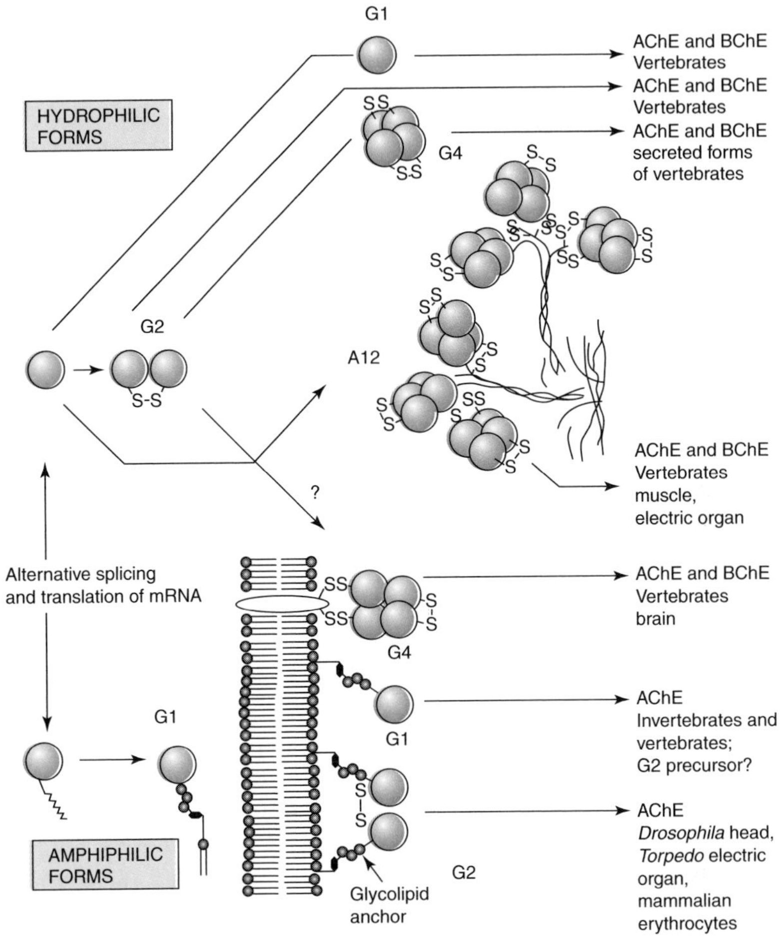

FIGURE 7–2. Schematic model of the molecular polymorphism of acetylcholinesterase (AChE) and butyrylcholinesterase (BChE). Open circles designate catalytic subunits. Disulfide bonds are indicated by S-S. Hydrophilic forms are G_1, G_2, and G_4. The asymmetric A_{12} forms have three hydrophilic G_4 heads linked to a collagen tail via disulfide bonds. The G_4 amphiphilic forms of brain are anchored into a phospholipid membrane through a 20 kDa anchor. The G_2 amphiphilic forms of erythrocytes have a glycolipid anchor. In *Torpedo*, AChE hydrophilic forms and amphiphilic G_2 forms are produced by alternative splicing so that the proteins are identical at 535 amino acids but non-identical at their C termini (From Chatonnet and Lockridge, 1989.)

5 Å distant, is an esteratic site, which binds the carbonyl carbon atom of ACh. Current information suggests that the anionic site contains at least one carboxyl group, possibly from glutamate, and the esteratic site involves a histidine residue adjacent to serine. The overall reaction is written as follows:

$$\text{E-OH} + \text{ACh} \underset{k_{-1}}{\overset{k_1}{\rightleftharpoons}} (\text{E-OH-Ach}) \xrightarrow{k_2}$$

$$\text{E-OAC} + \text{choline} \underset{H_2O}{\xrightarrow{k_3}} \text{E-OH} + \text{ACO}^- + \text{H}^+$$

Information on the architecture of the active center has been derived not only from kinetic studies using model compounds but also from a group of inhibitors known as the anticholinesterases. (The pharmacology of these agents will be discussed below, see ACh in Disease States.) The anticholinesterases are classified as reversible and irreversible inhibitors of the enzyme. Like ACh, both types of inhibitor acylate the enzyme at the esteratic site. However, in contrast to ACh or to a reversible inhibitor such as physostigmine, the irreversible inhibitors, which are organophosphorous compounds, irreversibly phosphorylate the esteratic site. This phosphorylation occurs on the hydroxyl group of serine when diisopropylfluorophosphonate (DFP) is incubated with purified acetylcholinesterase. Although the organophosphorous agents (referred to as nerve gases, although they are actually oils) are classed as irreversible anticholinesterases, there is a slow detachment of the compounds from the enzyme. Wilson observed that hydroxylamine speeded up this dissociation and regenerated active enzyme. He then designed a nucleophilic agent with a spatial structure that would fit the active center of acetylcholinesterase and produced a compound that is very active in displacing the inhibitor. This is 2-pyridine aldoxime methiodide (PAM), which has been used with moderate success in treating poisoning from the organophosphorous compounds in insecticides. PAM, with its quaternary ammonium group, does not penetrate the blood–brain barrier well enough to overcome central actions of the anticholinesterase. For this reason, atropine is usually used as an antidote along with PAM. Atropine blocks the effect of ACh at neuroeffector sites and has nothing to do with acetylcholinesterase. Currently, the oxime of choice in dephosphorylating organophosphorous compound–inactivated cholinesterase is HI-6 (1 [2-{hydroxyimino}methylpyridinium]-2-[4-carboxyamidopyridinium] dimethyl ether dichloride).

A curious observation (useful for cocktail party conversation) is that heroin is deacetylated to 6-monoacetylmorphine by both serum butyrylcholinesterase and erythrocyte acetylcholinesterase, but only acetylcholinesterase will hydrolyze the monoacetylmorphine to morphine. Interestingly, brain acetylcholinesterase will not deacetylate heroin. Another interesting and potentially useful observation is that serum butyrylcholinesterase will hydrolyze and thus inactivate cocaine. It has been suggested that mutants of the enzyme can be prepared that will inactivate the drug at a faster rate, to be useful in cocaine overdosage. Reviews of the cholinesterases by Taylor et al. and by Soreq and Seidman are recommended.

Uptake, Synthesis, and Release of ACh

Superior Cervical Ganglion, Brain, and Skeletal Muscle

To date, the only major, thorough studies of ACh turnover in nervous tissue were done originally by MacIntosh and colleagues and subsequently by Collier using the superior cervical ganglion of the cat. Using one ganglion to assay the resting level of ACh and perfusing the contralateral organ, these investigators determined the amount of transmitter synthesized and released under a variety of experimental conditions, including electrical stimulation, addition of an anticholinesterase to the perfusion fluid, and perfusion media of varying ionic composition. Their results may be summarized as follows:

1. During stimulation, ACh turns over at a rate of 8%–10% of its resting content every minute (i.e., about 24–30 ng/minute). At rest, the turnover rate is about 0.5 ng/minute. Since there is no change in the ACh content of the ganglion during stimulation at physiological frequencies, it is evident that electrical stimulation not only releases the transmitter but also stimulates its synthesis.
2. Choline is the rate-limiting factor in the synthesis of ACh.
3. In the perfused ganglion, Na^+ is necessary for optimum synthesis and storage and Ca^{2+} is necessary for release of the neurotransmitter.
4. Newly synthesized ACh appears to be more readily released upon nerve stimulation than depot or stored ACh.
5. About half of the choline produced by cholinesterase activity is reutilized to make new ACh.
6. At least three separate stores of ACh in the ganglion are inferred from these studies: *surplus* ACh, considered to be intracellular, which accumulates only in an eserine-treated ganglion and is not released by nerve stimulation but is released by K depolarization; *depot* ACh,

which is released by nerve impulses and accounts for about 85% of the original store; and *stationary* ACh, which constitutes the remaining 15% that is nonreleasable.

7. Choline analogs, such as triethylcholine, homocholine, and pyrrolcholine, are released by nerve stimulation only after they are acetylated in the ganglia.
8. Increasing the choline supply in the plasma during perfusion of the ganglion only transiently increases the amount of ACh that is releasable with electrical stimulation, despite accumulation of the transmitter in the ganglion.
9. The compound AH5183 (vesamicol), shown by the Parsons lab to inhibit ACh transport into synaptic vesicles, ultimately blocks release of ACh from the stimulated ganglia. This finding supports the contention that there is vesicular release of the transmitter in the periphery.

As stated above, this work on the superior cervical ganglion represents the most complete information on the turnover of ACh in the nervous system. With respect to the regulation of ACh turnover in cholinergic terminals of skeletal muscle, Vaca and Pilar, using chick iris, have elaborated on the work of Potter with the rat phrenic nerve-diaphragm preparation. With the iris and the diaphragm preparation, the same relationship of high-affinity choline uptake, ACh synthesis, and regulation by endogenous ACh has been demonstrated as previously described with CNS and autonomic nervous system preparations.

As noted earlier, ACh is not taken up into cholinergic terminals by a high-affinity transport system. However, as first described by Parsons' laboratory, ACh is transported into synaptic vesicles via a proton-pumping ATPase activity. A glycosylated ATPase pumps protons out of vesicles and drives ACh via a separate transporter into vesicles in exchange for the protons. This uptake is blocked by vesamicol. Interestingly, this vesicular transporter and ChAT arise from the same gene locus on chromosome 10.

Brain Slices, Nerve-Ending Particles (Synaptosomes), and Synaptic Vesicles

In 1939 Mann, Tennenbaum, and Quastel demonstrated the synthesis and release of ACh in cerebral cortical slices. Since then, these observations have been repeatedly confirmed but only moderately extended. The major finding of interest in all of these studies is that in the usual incubation medium the level of ACh in the slices reaches a limit and cannot be raised. In a high K^+ medium, the total ACh is increased substantially because much of it leaks into the medium from the slices. The experiments again suggest that the intracellular concentration of the neurotransmitter plays a role in regulating its rate of synthesis, in addition to the high-affinity uptake system for choline. This concept of a feedback mechanism is supported by the findings that the administration of drugs such as morphine, oxotremorine, and the anticholinesterases at the most succeed only in doubling the original level of ACh in the brain. Regardless of the dose of the drug, no higher level can be obtained.

Much of the current neurochemical work on the release of ACh from the brain involves the use of nerve-ending particles (synaptosomes). This preparation, independently developed by DeRobertis and by Whittaker, is derived from sucrose density gradient centrifugation of a crude mitochondrial fraction of brain. Although synaptosomes represent presynaptic terminals with enclosed vesicles and mitochondria, some postsynaptic fragments are often attached to them. A disadvantage of the preparation is its heterogeneity; the usual synaptosome fraction is a mixture of cholinergic, noradrenergic, serotonergic, and other terminals. In addition, as judged by electron microscopy and enzyme markers, the purity is around 60%; the contaminants may include glial cells, ribosomes, and membrane fragments that may be axonal, mitochondrial, or perikaryal. However, the advantage of these preparations is that they can be isolated easily and that synaptic vesicles can be collected by hypo-osmotically shocking the synaptosomes, followed by centrifugation. With respect to the disposition of ACh in synaptosomes, roughly half of the transmitter is found in vesicles and the other half in synaptosomal cytoplasm. The cytosolic localization could be artifactual, resulting from the preparation methods.

Following the discovery of these presynaptically localized vesicles that contained ACh, the conclusion was almost unavoidable that these organelles are the source of the quantal release of transmitter as described in the neurophysiological experiments of Katz and collaborators. Thus, the obvious interpretation has been that as the nerve is depolarized Ca^{2+} enters the terminal, vesicles in apposition to the terminal fuse with the presynaptic membrane, and ACh is released into the synaptic cleft to interact with receptors on the postsynaptic cell to change ion permeability. Synapsin I, a phosphoprotein that is localized in vesicles, may mediate the translocation of vesicles to the plasma membrane. Other vesicular proteins that have been implicated in the exocytotic process are the synaptotagmins, synaptophysins, and synaptobrevins (see Chapter 3). Synaptobrevin is of particular interest in that both tetanus toxin and botulinum toxin type B, which are zinc endopeptidases, inhibit ACh release by cleaving it. Synaptotagmin has been implicated as a Ca^{2+}-sensor in the release process. The subsequent sequence of events is not clear, but in some fashion the presynaptic membrane is pinocytotically recaptured and vesicles are resynthesized and simultaneously or subsequently repleted with ACh. This endocytotic event is apparently triggered by calcineurin, a Ca^{2+}-dependent protein phosphatase. Another, though minor, possibility to explain ACh release comes from the laboratories of Israel and Dunant. The Israel group isolated a lipoprotein from presynaptic terminals of the *Torpedo* electric organ, which they refer to as a *mediatophore*. A 15 kDa subunit of this protein exhibits vacuolar ATPase activity. When the mediatophore is incorporated into ACh-loaded proteoliposomes, addition of calcium and the calcium ionophore A23187 triggers the release of ACh. The Dunant group transfected mediatophore cDNA into neuroblastoma cells and restored quantal ACh release. Dunant and Israel suggest that the vesicles release ACh to the cytosol and that the mediatophore, acting as a gate around a calcium channel, will release the transmitter. The action of calcium is terminated by uptake into synaptic vesicles, where it is ultimately excreted.

Cholinergic Pathways

The identification of cholinergic synapses in the peripheral nervous system has been relatively easy, and we have known for a long time now that ACh is the transmitter at autonomic ganglia, at parasympathetic postganglionic synapses, and at the neuromuscular junction. In the CNS, however, until relatively recently, technical difficulties have limited our knowledge of cholinergic tracts to the motoneuron collaterals to Renshaw cells in the spinal cord. With respect to the aforementioned technical difficulties, the traditional approach has been to lesion a suspected tract and then assay for ACh, ChAT, or high-affinity choline uptake at the presumed terminal area. Problems with lesioning include making discrete, well-defined lesions and interrupting fibers of passage. This latter problem is illustrated by the discovery that a habenula-interpeduncular nucleus projection that, based on lesioning of the habenula, was always described as a cholinergic pathway is not: it turned out that what was lesioned were cholinergic fibers that passed through the habenula. Thus, although the interpeduncular nucleus has the highest choline uptake and ChAT activity of any area in the brain, the origin of this innervation remains largely unknown. A quantum leap in technology for tracing tracts in the CNS has occurred in the past several years. Through the use of histochemical techniques (originally developed by Koelle and co-workers) that stain for regenerated acetylcholinesterase after DFP treatment (Butcher, Fibiger), autoradiography with muscarinic receptor antagonists (Rotter, Kuhar), and immunohistochemical procedures with antibodies to ChAT (McGeer, Salvaterra, Cuello, Wainer), a clear picture of cholinergic tracts in the CNS is now emerging. The well-documented tracts are depicted in Figure 7–3. There is additional information that in the striatum and the nucleus accumbens septi, only cholinergic interneurons are found. Also, intrinsic cholinergic neurons have been reported to exist in the cerebral cortex, colocalized with vasoactive intestinal polypeptide and often in close proximity to blood vessels.

With respect to other neurotransmitter functions, ACh may participate in circuits involved with pain reception. Thus, the findings that nettles (*Urtica dioica*) contain ACh and histamine, that high concentrations of ACh injected into the brachial artery of humans result in intense pain, and that ACh applied to a blister produces a brief but severe pain indicate a relationship between ACh and pain. That ACh may act as a sensory transmitter in thermal receptors, taste fiber endings, and chemoreceptors has also been suggested, based on the excitatory activity of the compound on these sensory nerve endings.

Cellular Effects

A variety of actions of ACh that may be viewed as cellular effects rather than neurotransmitter activity have been described. These include ciliary movement in the gill plates of *Mytilus edulis*, ciliary motility of mammalian respiratory and esophageal tracts, water resorption and photosynthesis in plants, a hyperpolarizing effect on atrial muscle, limb regeneration in salamanders, protein production in the silk gland of spiders, induction of sporulation in the fungus *Trichoderma*, protoplasmic streaming in slime molds, and photic control of circadian rhythms and seasonal reproductive cycles.

There are also a variety of tissues and organisms, such as human placenta, immune cells, *Lactobacillus plantarum*, *Trypanosoma rhodesiense*, and the fungus *Claviceps purpurea* in which ACh is found but nothing is known of its action. One of the most interesting situations is the corneal epithelium, which contains the highest concentrations of ACh of any tissue in the body. ACh may be involved in sodium transport in this tissue.

All of the activities of ACh and its localization in nonnervous tissue that we have noted above suggest that this agent may be a hormone as well as a neurotransmitter. All known neurotransmitters may possess this dual func-

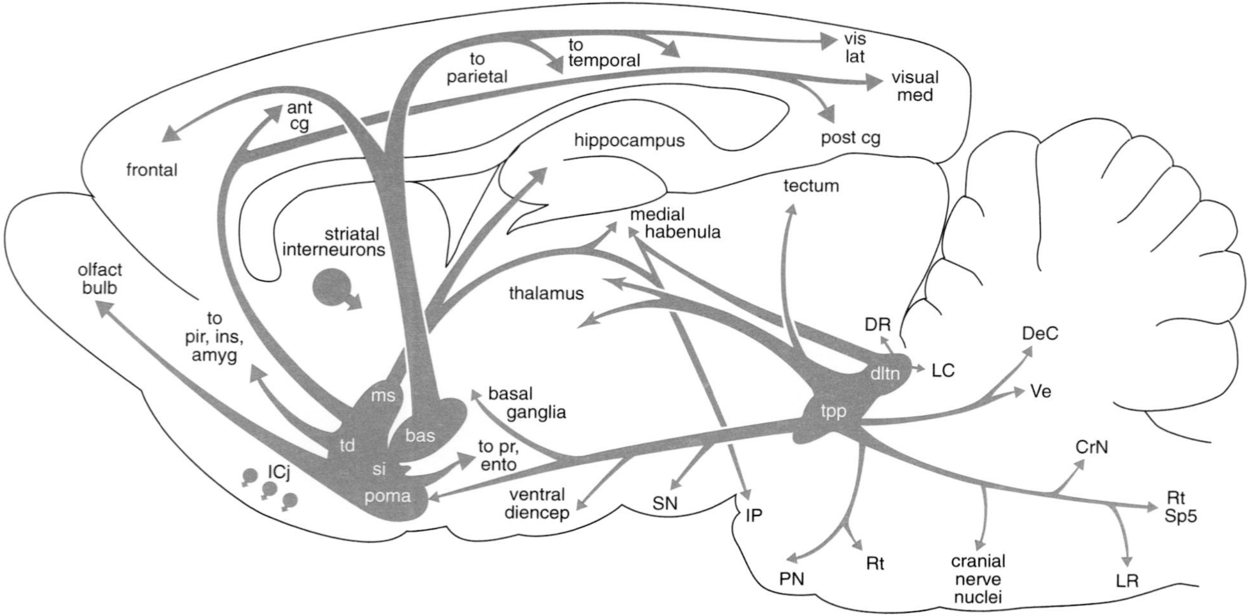
vis
lat
to
temporal
to
parietal
visual
med
ant
cg
post cg
frontal
hippocampus
tectum
medial
habenula
striatal
interneurons
olfact
bulb
thalamus
to
pir, ins,
amyg
DR
DeC
dltn
LC
Ve
ms
basal
ganglia
tpp
bas
td
to pr,
ento
CrN
si
ICj
poma
ventral
diencep
SN
IP
Rt
Sp5
PN
Rt
cranial
nerve
nuclei
LR

Figure 7–3. Schematic representation of the major cholinergic systems in the mammalian brain. Central cholinergic neurons exhibit two basic organizational schemata: local circuit cells (i.e., those that morphologically are arrayed wholly within the neural structure in which they are found), exemplified by the interneurons of the caudate-putamen nucleus, nucleus accumbens, olfactory tubercle, and Islands of Calleja complex (ICj), and projection neurons (i.e., those that connect two or more different regions). Of the cholinergic projection neurons that interconnect central structures, two major subconstellations have been identified: the basal forebrain cholinergic complex composed of choline acetyltransferase (ChAT)-positive neurons in the medial septal nucleus (ms), diagonal band nuclei (td), substantia innominata (si), magnocellular preoptic field (poma), and nucleus basalis (bas) and projecting to the entire nonstriatal telencephalon and the pontomesencephalotegmental cholinergic complex composed of ChAT-immunoreactive cells in the pendunucloppontine (tpp) and laterodorsal (dltn) tegmental nuclei and ascending to the thalamus and other diencephalic loci and descending to the pontine and medullary reticular formations (Rt), deep cerebellar (DeC) and vestibular (Ve) nuclei, and cranial nerve nuclei. Not shown are the somatic and parasympathetic cholinergic neurons of cranial nerves III–VII and IX–XII and the cholinergic α and γ motor and autonomic neurons of the spinal cord. amyg, amygdada; ant cg, anterior cingulate cortex; CrN, dorsal cranial nerve nuclei; diencep, diencephalon; DR, dorsal raphe nucleus; ento, entorhinal cortex; frontal, frontal cortex; IP, interpeduncular nucleus; ins, insular cortex; LC, locus ceruleus; LR, lateral reticular nucleus; olfact, olfactory; pir, piriform cortex; PN, pontine nuclei; pr, perirhinal cortex; parietal, parietal cortex; post cg, posterior cingulate cortex; SN, substantia nigra; Sp5, spinal nucleus of cranial nerve V. (From Butcher and Woolf, 1986, and Woolf and Butcher, 1989.)

tion. These two activities have already been shown to occur with the biogenic amines. Even when a certified neurotransmitter is found in nervous tissue, its action may satisfy the criteria for defining a hormone or a modulator rather than the currently strict criteria for a "classic" neurotransmitter.

Cholinergic Receptors

As noted in Chapter 4, cholinergic receptors fall into two classes, muscarinic (Table 7–1) and nicotinic (Table 7–2). At last count, five muscarinic receptors (M_1–M_5) had been cloned. All of them exhibit a slow response time (100–250 milliseconds), are coupled to G proteins, and either act directly on ion channels or are linked to a variety of second-messenger systems. M_1, M_3, and M_5 via Gq are coupled to phosphatidylinositol hydrolysis; M_2 and M_4 via G_i are coupled to cyclic adenosine monophosphate (cAMP). When activated, the final effect can be to open or close K channels, Ca channels, or Cl

TABLE 7–1. Muscarinic Receptors

Currently Accepted Name	*M_1*	*M_2*	*M_3*	*M_4*	*M_5*
Molecular biology classification	m1	m2	m3	m4	m5
Structural information	460 aa (human)	466 aa (human)	590 aa (human)	479 aa (human)	532 aa (human)
Subtype-selective agonists	McN-A343 (ganglion) Pilocarpine (relative to M_3 and M_5) L-689,660 Xanomeline CDD-0097	Bethanechol (relative to M_4)	L-689,660	McN-A343 (relative to M_2)	None known
Subtype-selective antagonists	Pirenzepine Telenzepine	Methoctramine AF-DX-116 AF-DX-384 Gallamine (non-competitive) Himbacine Tripitramine	Hexahydro-sila-difenidol *p*-Fluorohexahydro-sila-difenidol 4-DAMP	Tropicamide Himbacine AF-DX-384	None known

Receptor-selective agonists	Bethanechol Metoclopramide Muscarine Pilocarpine Oxotremorine M	Bethanechol Metoclopramide Muscarine Pilocarpine Oxotremorine M	Bethanechol Metoclopramide Muscarine Pilocarpine Oxotremorine M	Bethanechol Metoclopramide Muscarine Pilocarpine Oxotremorine M	Bethanechol Metoclopramide Muscarine Pilocarpine Oxotremorine M
Receptor-selective antagonists	Scopolamine QNB, (±) QNB, R(−) Atropine	Scopolamine QNB, (±) QNB, R(−) Atropine	Scopolamine QNB, (±) QNB, R(−) Atropine	Scopolamine QNB, (±) QNB, R(−) Atropine	Scopolamine QNB, (±) QNB, R(−) Atropine
Signal-transduction mechanisms	$G_{q/11}$ (increase IP_3/DAG) NO	G_i (cAMP modulation) ↑ K^+ (G)	$G_{q/11}$ (increase IP_3/DAG) NO	G_i (cAMP modulation) ↑ K^+ (G)	$G_{q/11}$ (increase IP_3/DAG) NO
Radioligands of choice	[^{3}H]-Pirenzepine [^{3}H]-Telenzepine [^{3}H]-QNB	[^{3}H]-AF-DX-384 [^{3}H]-QNB	[^{3}H]-4-DAMP [^{3}H]-QNB	[^{3}H]-AF-DX-384 [^{3}H]-QNB	[^{3}H]-QNB [^{3}H]-NMS

AF-DX-116, 11-([2-[(Diethylamino)methyl]-1-piperdinyl]acetyl)-5,11-dihydro-6-pyridol[2,3-b][1,4]benzodiazepin-6-one; AF-DX-384, 5,11-dihydro-11-[2-[2-[*N*,*N*-dipropylaminomethyl)piperidin-1-yl]ethylamino]-carbonyl]6H-pyridol[2,3-b][1,4]benzodiazepin-6-one; CDD-0097, 5-propargyloxycarbonyl-1,4,5,6-tetrahydropyrimidine; 4-DAMP, 4-diphenylacetoxy-*N*-methylpiperidine methiodide; L-689,660, 1-azabicyclo [2,2,2]octane,3-(6-chloropyrazinyl)maleate; McN-A343, 4-(*N*-[3-chlorophenyl]carbamoyloxy)-2-butynyltrimethylammonium chloride; NMS, *N*-methylscopolamine; QNB, quinuclidinyl-α-hydroxydiphenylacetate (quinuclidinylbenzylate); NO, nitric oxide; IP_3/DAG, inositol triphosphate/diacylglycerol; cAMP, cyclic adenosine monophosphate.

SOURCE: Modified from Watling (2001).

TABLE 7–2. Nicotinic Receptors

Currently used name	Neuronal (CNS) (α-bungarotoxin-insensitive)	Neuronal (CNS) (α-bungarotoxin-sensitive)	Neuronal (autonomic ganglia)	Muscular
Subunits (arranged as pentamers)	$\alpha4\beta2$ (major) $\alpha3\beta4$ $\alpha2?\alpha3?$	$\alpha7$ homomers $\alpha8$? $\alpha9$? $\alpha9/\alpha10$	$\alpha7$ homomers $\alpha3\alpha5\beta4$ $\alpha3\alpha5\beta2\beta4$	$\alpha1\beta1\delta\gamma(\varepsilon)$
Receptor-selective agonists	Cytisine RJR-2403 Epibatidine Anatoxin A ABT-418 A-85380 DBO-83	Anatoxin A DMAC GTS-21 AR-R-17779	Epibatidine SIB-1553A DMPP	Epibatidine Anatoxin A TMA
Receptor-selective antagonists	Mecamylamine Dihydro-β-erythroidine Erysodine α-Conotoxin AuIB ($\alpha3\beta4$) α-Conotoxin MII ($\alpha3\beta2$)	Methyllycaconitine α-Bungarotoxin α-Conotoxin IMI	Hexamethonium Chlorisondamine ? Mecamylamine ? κ-Bungarotoxin	α-Bungarotoxin

Signal-transduction mechanisms	Modulation of cation channel conductance/permeability properties	Modulation of cation channel conductance/permeability properties	Modulation of cation channel conductance/permeability properties	Modulation of cation channel conductance/permeability properties
Radioligands of choice	[^{3}H]-Nicotine [^{3}H]-Epibatidine [^{3}H]-Cytisine	[^{125}I]-α-Bungarotoxin [^{3}H]-Methyllycaconitine	[^{3}H]-Epibatidine [^{125}I]-α-Bungarotoxin [^{3}H]-Methyllycaconitine	[^{125}I]-α-Bungarotoxin

A-85380, 3-(2[S]-Azetidinylmethoxy)pyridine; ABT-418, (S)-3-methyl-5-(1-methyl-2-pyrrolidinyl)isoxazole; AR-R-17779, (−)-spirol[1-azabicyclo[2.2.2]octane-3,5′-oxazolidin-2′-one (4a); DBO-83, 3-(6-Cl-3-yl)-diazabicyclo(3.2.1)octane; DMAC, 3-(4)-dimethylaminocinnamylidine anabaseine; DMPP, *N,N*-dimethyl-*N′*-phenyl-piperazinium iodide; GTS-21, [3-(2,4-dimethoxybenzylidene)-anabaseine; RJR-2403, *N*-methyl-4-(3-pyridinyl]-3-buten-1-amine; SIB-1553A, 4-[[2-(1-methyl-2-pyrrolidinyl)ethyl]thio]phenyl hydrochloride; TMA, tetramethylammonium.

SOURCE: Modified from Watling (2001).

channels, depending on the cell type. With this array of channel activity, therefore, stimulation of muscarinic receptors will lead to either depolarization or hyperpolarization. As noted in Table 7–1, second-messenger systems have been described following activation of the muscarinic receptors. Knowing the messengers is fine, but it does not tell us anything about the message; that is, what the ultimate physiological effect is.

The pharmacological antagonists that have been used to define three of the muscarinic subtypes are pirenzepine, which has a high affinity for M_1; AFDX-116 and methoctramine with a high affinity for M_2; and 4-diphenylacetoxy-*N*-methylpiperidine methiodide, which exhibits the highest affinity for M_3. Most antagonists do not show more than a fivefold selectivity for one subtype over all other subtypes. The two classic muscarinic antagonists atropine and quinuclidinylbenzylate do not distinguish the subtypes but block all equally well. What is needed are selective muscarinic agonists that will define the various subtypes. More importantly, it would be a major advance if we knew whether each subtype subserved a specific function, such as bradycardia or smooth muscle contractibility. If this were the case (and it probably is), it could lead to the development of specific drugs devoid of side effects.

Until relatively recently, the identification of nicotinic cholinergic receptors in the CNS was an enigma. Using labeled α-bungarotoxin, nicotine, mecamylamine, or dihydro-β-erythroidine, each investigation yielded mystifying results in which the antagonist could not be easily displaced by ACh or by unlabeled ganglionic or neuromuscular antagonists but on occasion was displaced by muscarinic agonists and antagonists. A major problem has been the low density of nicotinic compared to muscarinic receptors in the brain.

Conversely, much is known about the properties of the nicotinic cholinergic receptor of *Torpedo* and *Electrophorus electricus* organs. This reflects the abundance of the receptor in this tissue and the availability of two snake toxins, α-bungarotoxin and *Naja naja siamensis*, that specifically bind to the receptor and have facilitated its isolation and purification. In the past 10 years, however, research with monoclonal antibodies and cDNA has yielded considerable information about the mammalian nicotinic receptors (Table 7–2). At least nine different functional receptors have been identified and can be tentatively differentiated by CNS, ganglionic, and muscle types as well as pre- and postsynaptic localizations in the CNS. Cholinergic nicotinic receptors from muscle or electric organ contain four different subunits (α, β, γ and δ), with a stoichiometry of two α subunits and each of the other three. In contrast, neuronal nicotinic receptors contain only two kinds of subunit (α and β), with the α occurring in at least eight different forms and the β in three. When one considers the variety of combinations of these α and β subunits that are theoretically possible, there appears to be a surfeit of nicotinic receptors in the brain. Current and future electrophysiological and behavioral studies are attempting to assign a specific function to each of them. A review of this problem by Klink et al. and a review of neuronal nicotinic receptors by Clementi et al. are recommended.

Meanwhile, considerable attention is being devoted to nicotine since tobacco smoking has been shown to reduce the incidence of Parkinson's disease and Alzheimer's disease. Thus, nicotine patches have been shown to increase cognition in Alzheimer patients and α_7 nicotine receptor agonists specifically appear to be involved in regulating neuronal growth during development (Reviewed by Picciotto et al., 2000). The $\alpha_4\beta_2$ receptor has been implicated in learning deficiencies. Finally, nicotine antagonists such as dihydro-β-erythroidine impair working memory. Obviously, pharmaceutical chemists have more than a casual interest in pursuing this topic. A currently promising compound is galantamine, an anticholinesterase that sensitizes nicotinic receptors.

As noted above, the nicotinic ACh receptor from electric organs is a pentameric integral membrane protein composed of four glycosylated polypeptide chains designated α, β, γ, and δ, with a stoichiometry of two α subunits to one each of the other three (Fig. 7–4). All subunits traverse the membrane and, when viewed face on, resemble a five-petal rosette with a central pit. ACh binds to the α subunits and produces a conformational change in the channel that selectively allows cations rather than anions with a diameter of about 0.65 nm to pass through. The molecular weight of the receptor complex is about 255,000. Desensitization of the receptor increases when it is phosphorylated by cAMP protein kinase (protein kinase A), protein kinase C, or tyrosine kinase.

An unexpected windfall resulting from the isolation of the neuromuscular nicotinic cholinergic receptor has been the demonstration that when this lipoprotein, isolated from *Electrophorus*, is injected into rabbits, all signs of myasthenia gravis appear. This finding, coupled with autoradiographic studies of muscle biopsies of myasthenics using labeled α-bungarotoxin as a tag where a marked deficiency of receptors is observed, has led to a better understanding of the disease at the molecular level. It is now clear that myasthenia gravis is an autoimmune disease in which a circulating antibody appears to be involved in an increased rate of degradation and damage, as well as antagonism of the ACh receptor. Since antibodies produced from *Electrophorus* or *Torpedo* ACh receptor result in myasthenic signs in rabbits, guinea pigs, and monkeys, this work also carries the implication that the ACh receptor is a phylogenetically conserved protein that can exhibit immunological cross-reactivity.

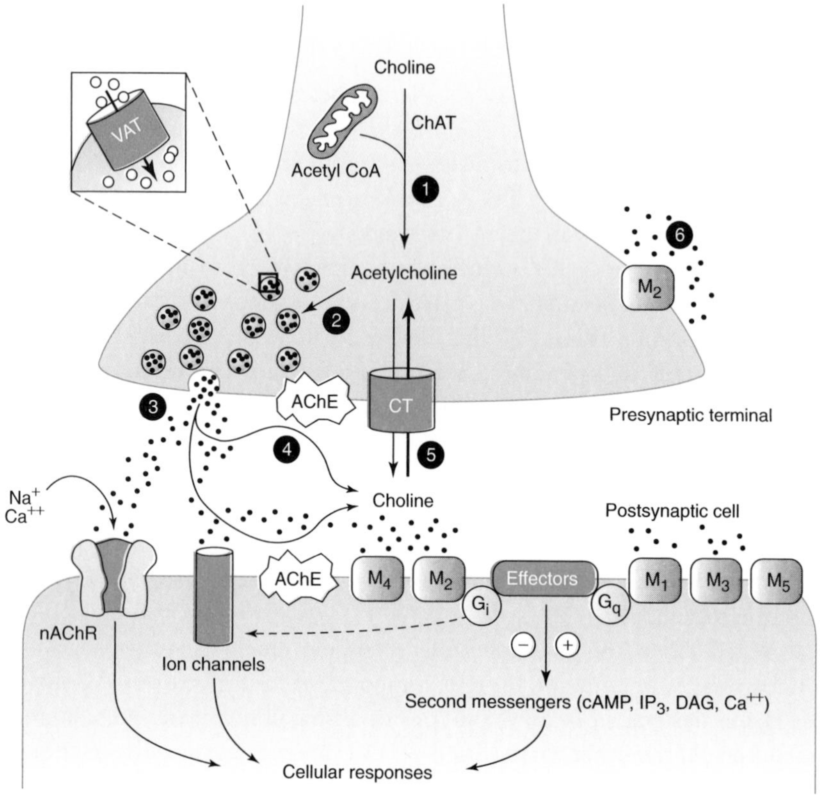

FIGURE 7–4. Model of a acetylcholine synapse illustrating the presynaptic and postsynaptic molecular entities involved in the synthesis, storage, release, reuptake, and signaling of acetylcholine. Choline is transported into the presynaptic terminal by an active uptake mechanism and converted to acetylcholine by a single enzymatic step. The acetyl coenzyme A required for acetylcholine synthesis is provided by presynaptic mitochondria. Muscarinic M_2 autoreceptors in the presynaptic terminal modulate the release of acetylcholine. In contrast to monoamine synapses, the plasma membrane transporter of an acetylcholine synapse does not return the neurotransmitter to the presynaptic terminal but rather recycles precursor choline. Acetylcholine is metabolized by acetylcholine esterase (AchE) present on both pre- and postsynaptic membranes, which serves to terminate its action. Both G protein–coupled muscarinic receptors and ligand-gated ion channels (nicotinic receptors) may be present as depicted on the postsynaptic cell. Sites of drug action:

Site 1: ACh synthesis can be blocked by styryl pyridine derivatives such as NVP.

Site 2: ACh transport into vesicles is blocked by vesamicol (AH5183).

Site 3: Release is promoted by β-bungarotoxin, black widow spider venom, and La^{3+}. Release is blocked by botulinum toxin, cytochalasin B, collagenase pretreatment, and Mg^{2+}.

ACh in Disease States

Aside from myasthenia gravis and other autoimmune diseases, such as the Lambert-Eaton myasthenic syndrome (a presynaptic problem involving diminished release of ACh), the role of ACh in nervous system dysfunction is unclear. Certainly, a strong case can be made for familial dysautonomia, an autosomal recessive condition affecting Ashkenazi Jews that is diagnosed by a supersensitivity of the iris to methacholine. Huntington's disease, involving a degeneration of Golgi type 2 cholinergic interneurons in the striatum, is partially ameliorated by physostigmine. Administration of physostigmine to patients with tardive dyskinesia has produced mixed results.

In Alzheimer's disease, characterized behaviorally by a severe impairment in cognitive function and neuropathologically by the appearance of neuritic plaques and neurofibrillary tangles, a cholinergic dysfunction has been implicated. As discussed more fully in Chapter 13, this is based on the fact that a variety of cholinergic abnormalities have been found. However, efforts to treat patients with choline, lecithin, or anticholinesterases such as tetrahydroaminoacridine (tacrine) have met with little success. In animal models with imposed cognitive deficits, memory has been restored with synthetic cholinergic agonists such as YM796, DuP996, L-689,660, and WEB-1881-FU. Since there are no animal models of Alzheimer's disease, it is still un-

---

Site 4: Acetylcholinesterase is inhibited reversibly by physostigmine (eserine) or irreversibly by DFP, or soman.

Site 5: Choline uptake competitive blockers include hemicholinium-3, troxypyrrolium tosylate, or AF64A (noncompetitive).

Site 6: Presynaptic muscarinic receptors may be blocked by AFDX-116 (an M_2 antagonist), atropine, or quinuclidinyl benzilate. Muscarinic agonists (e.g., oxotremorine) will inhibit the evoked release of ACh by acting on these receptors.

Site 7: Postsynaptic receptors are activated by cholinomimetic drugs and anticholinesterases. Nicotinic receptors, at least in the peripheral nervous system, are blocked by rabies virus, curare, hexamethonium, or dihydro-β-erythroidine; n-methylcarbamylcholine and dimethylphenyl piperazinium are nicotinic agonists. Muscarinic receptors are blocked by atropine, pirenzepine, and quinuclidinyl benzilate.

AC, adenylyl cyclase; AChE, acetylcholine esterase; ChAT, choline acetyltransferase; CT, plasma membrane choline transporter; DAG, diacylglycerol; IP3, inositol triphosphate; M, muscarinic receptors; nAChR, nicotinic acetylcholine receptors; PLC, phospholipase C; VAT, vesicular acetylcholine transporter (not yet isolated). (Modified from Nestler et al., *Molecular Neuropharmacology*)

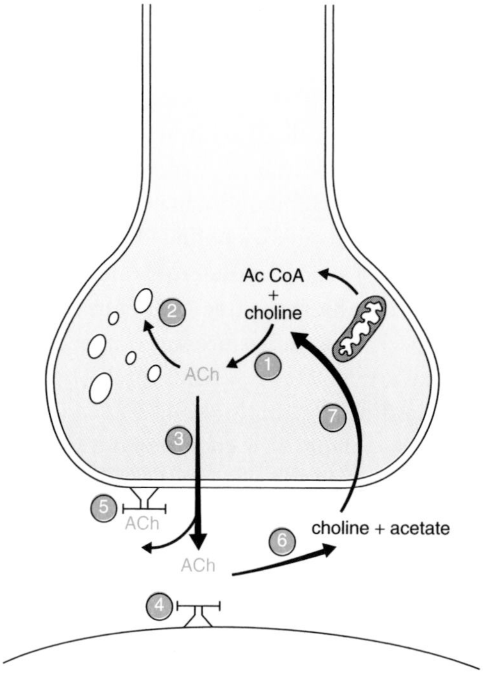

FIGURE 7–5. Sites of drug action at cholinergic synapses.

Site 1: Acetylcholine (ACh) synthesis can be blocked by styryl pyridine derivatives such as naphylvinylpyridine (NVP).

Site 2: ACh transport into vesicles is blocked by vesamicol (AH5183).

Site 3: Release is promoted by β-bungarotoxin, black widow spider venom, and La^{3+}. Release is blocked by botulinum toxin, cytochalasin B, collagenase pretreatment, and Mg^{2+}.

Site 4: Postsynaptic receptors are activated by cholinometic drugs and anticholinesterases. Nicotinic receptors, at least in the peripheral nervous system, are blocked by rabies virus, curare, hexamethonium, or dihydro-β-erythroidine; n-methylcarbamylcholine and dimethylphenyl piperazinium are nicotinic agonists. Muscarinic receptors are blocked by atropine, pirenzepine, and methoctramine.

Site 5: Presynaptic muscarinic receptors may be blocked by AFDX-116 (an M_2 antagonist), atropine, or quinuclidinyl benzilate. Muscarinic agonists (e.g., oxotremorine) will inhibit the evoked release of ACh by acting on these receptors.

Site 6: Acetylcholinesterase is inhibited reversibly by physostigmine (eserine) or irreversibly by diisopropylfluorophosphate or sarin.

Site 7: Choline uptake competitive blockers include hemicholinium-3, troxypyrrolium tosylate, or AF64A (noncompetitive).

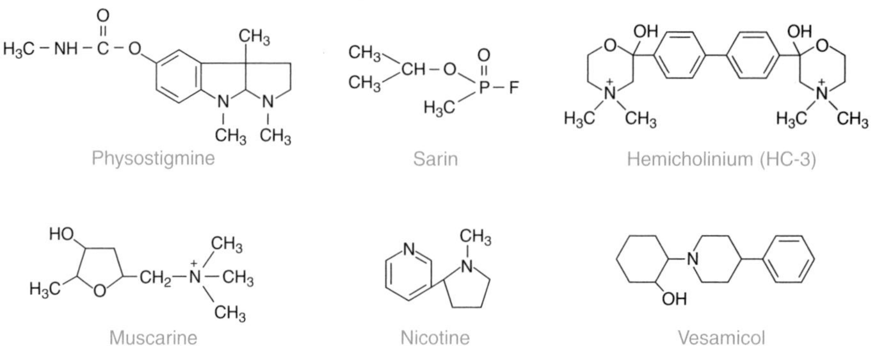

FIGURE 7–6. Structure of some drugs that affect the cholinergic nervous system.

known if any of these agents will produce a clinical benefit in humans. At any rate, the major focus currently is on the β-amyloid peptide, which is found in the plaques of Alzheimer's disease brain. Considerable evidence suggests that aberrant proteolytic processing of this β-amyloid protein is associated with degenerating nerve terminals. Thus, current research is directed to developing specific protease inhibitors as well as exploring techniques for delivering neurotrophic factors such as nerve growth factor (incidentally stimulated by nicotine) to the brain. Sites of drugs that affect the synthesis, release, and neurotransmitter activity of ACh are shown in Figure 7–5. Structures of drugs that affect the cholinergic nervous system are depicted in Figure 7–6.

SELECTED REFERENCES

Allen, D. D. and Q. R. Smith (2001). Characterization of the blood–brain barrier choline transporter using the in situ rat brain perfusion technique. *J. Neurochem.* 76, 1032.

Benishin, C. G. and P. T. Carroll (1983). Multiple forms of choline-*O*-acetyltransferase in mouse and rat brain: solubilization and characterization. *J. Neurochem.* 41, 1016.

Bloc, A., E. Bugnard, Y. Dunant, J. Falk-Vairant, M. Israel, F. Loctin, and E. Roulet (1999). Acetylcholine synthesis and quantal reconstituted by transfection of mediatophore and choline acetyltransferase cDNAs. *Eur. J. Neurosci.* 11, 1523.

Butcher, L. L. and N. J. Woolf (1986). Central cholinergic systems: synopsis of anatomy and overview of physiology and pathology. In *The Biological Substrates of Alzheimer's Disease* (A. B. Scheibel and A. F. Wechsler, eds.). Academic Press, New York, pp. 73–86.

Chatonnet, A. and O. Lockridge (1989). Comparison of butyrylcholinesterase and acetylcholinesterase. *Biochem. J.* 260, 625.

Clementi, F., D. Fornasari, and C. Gotti (2000). Neuronal nicotine acetylcholine receptors: from structure to therapeutics. *Trends Pharmacol. Sci.* 21, 33.

Collier, B., A. Tandon, M. A. M. Prado, and M. Bachoo (1993). Storage and release of acetylcholine in a sympathetic ganglion. *Prog. Brain Res.* 98, 183.

Cooper, J. R. (1994). Unsolved problems in the cholinergic nervous system. *J. Neurochem.* 63, 395.

Cordell, B. (1994). β-Amyloid formation as a potential therapeutic target for Alzheimer's disease. *Annu. Rev. Pharmacol. Toxicol.* 34, 69.

Dobransky, T., W. L. Davis, and F. J. Rylett (2001). Functional characterization of phosphorylation of 69-kDa human 440 by protein kinase C. *J. Biol. Chem.* 276, 222.

Eder-Colli, L., P. A. Briand, M. Pellegrini, and Y. Dunant (1989). A monoclonal antibody raised against plasma membrane of cholinergic nerve terminals of the *Torpedo* inhibits choline acetyltransferase activity and acetylcholine release. *J. Neurochem.* 53, 1419.

Eiden, L.E. (1998). The cholinergic gene locus. *J. Neurochem.* 70, 2227.

Ericson, J. P., H. Varoqui, M. K. H. Schafer, W. Modi, M. F. Diebler, E. Weihe, J. Rand, L. Eiden, T. I. Bonner, and T. B. Usdin (1994). Functional identification of a vesicular acetylcholine transporter and its expression from a cholinergic gene locus. *J. Biol. Chem.* 269, 21929.

Flentge, F., K. Venema, T. Koch, and J. Korf (1992). An enzyme-reactor for electrochemical monitoring of choline and acetylcholine: applications in high-performance liquid chromatography, brain tissue, microdialysis, and cerebrospinal fluid. *Anal. Biochem.* 204, 305.

Israel, M. and Y. Dunant (1998). Acetylcholine release and the cholinergic genomic locus. *Mol. Neurobiol.* 16, 1.

Israel, M. and B. Lesbats (1981). Chemiluminescent determination of acetylcholine and continuous detection of its release from *Torpedo* electric organ synapses and synaptosomes. *Neurochem. Int.* 3, 81.

Jenden, D. J., M. Roch, and R. A. Booth (1973). Simultaneous measurement of endogenous and deuterium-labeled tracer variants of choline and acetylcholine in subpicomole quantities by gas chromatography/mass spectrometry. *Anal. Biochem.* 55, 438.

Kawashima, K., T. Hayakawa, Y. Icashima, T. Suzuki, K. Jufimoto, and H. Oohata (1991). Determination of acetylcholine release in the striatum of anesthetized rats using in vivo microdialysis and a radioimmunoassay. *J. Neurochem.* 57, 882.

Klink, R., A. de Kerchove d'Exaerde, M. Zoli, and J.P. Changeux (2001). Molecular and physiological diversity of nicotinic acetylcholine receptors in the midbrain dopaminergic nuclei. *J. Neurosci.* 21, 1452.

Massouli, J., L. Pezzementi, S. Bon, E. Krejci, and F. M. Vallette (1993). Molecular and cellular biology of cholinesterases. *Prog. Neurobiol.* 41, 39.

McCaman, R. E. and J. Stetzler (1977). Radiochemical assay for ACh: modifications for sub-picomole measurements. *J. Neurochem.* 28, 669.

Misawa, H., K. Nakata, J. Matsura, M. Nagao, T. Okuda, and T. Hagg (2001). Distrubution of the high-affinity choline transporter in the central nervous system of the rat. *Neuroscience* 105, 87.

Okuda, T., T. Haga, T. Kanai, H. Endou, T. Ishihara, and I. Katsura (2000). Identification and characterization of the high-affinity choline transporter. *Nat. Neurosci.* 3, 120.

Parsons, S. M., C. Prior, and I. G. Marshall (1993). Acetylcholine transport, storage, and release. *Int. Rev. Neurobiol.* 35, 279.

Picciotto, M.R., B.J. Calderone, S.L. King, and V. Zachariou (2000). Nicotinic receptors in the brain: links between molecular biology and behavior. *Neuropsychopharmacology* 22, 451.

Potter, L. T. (1970). Synthesis, storage and release of [^{14}C]acetylcholine in isolated rat diaphragm muscles. *J. Physiol.* 206, 145.

Potter, P. E., M. Hadjiconstantinou, J. L. Meek, and N. H. Neff (1984). Measurement of acetylcholine turnover rate in brain: an adjunct to a simple HPLC method for choline and acetylcholine. *J. Neurochem.* 43, 288.

Ricny, J., J. Coupek, and S. Tucek (1989). Determination of acetylcholine and choline by flow-injection with immobilized enzymes and fluorometric or luminometric detection. *Anal. Biochem.* 176, 221.

Roghani, A., J. Feldman, S. A. Kohan, A. Shirzadi, C. B. Gundersen, N. Brecha, and R. H. Edwards (1994). Molecular cloning of a putative vesicular transporter for acetylcholine. *Proc. Natl. Acad. Sci. U.S.A.* 91, 10620.

Rotter, A., N. J. Birdsall, A.S.V. Burgen, P. M. Field, E. C. Hulme, G. Sastry, and C. Sadarangivivad (1979). Cholinergic systems in nonnervous tissue. *Pharmacol. Rev.* 30, 65.

Simon, R., S. Atweh, and M. J. Kuhar (1976). Sodium-dependent high affinity choline uptake: a regulator step in the synthesis of acetylcholine. *J. Neurochem.* 26, 909.

Soreq, H. and S. Seidman (2001). Acetylcholinesterase—new roles for an old actor. *Nat. Rev. Neurosci.* 2, 294.

Taylor, P., Z.D. Luo, and S. Camp (2000). *Cholinesterase and Cholinesterase Inhibitors.* (E. Giacobini, ed.) Martin Dynitz, London.

Tucek, S. (1993) Short-term control of the synthesis of acetylcholine. *Prog. Biophys. Mol. Biol.* 60, 59.

Vaca, K. and G. Pilar (1979). Mechanisms controlling choline transport and acetylcholine synthesis in motor nerve terminals during electrical stimulation. *J. Gen. Physiol.* 73, 605.

Wainer, B. H., A. I. Levey, E. J. Mutson, and M.-M. Mesulam (1984). Cholinergic systems in mammalian brain identified with antibodies against choline acetyltransferase. *Neurochem. Int.* 6, 163.

Watling, R. S. (2001). *The Sigma-RBI Handbook of Receptor Classification and Signal Transduction.* Sigma-RBI, Natick, Mass.

Wessler, I., H. Kilbinger, F. Bittinger, and C. J. Kirkpatrick (2001). The biological role of non-neuronal acetylcholine in plants and humans. *Jpn. J. Pharmacol.*, 85, 2.

Woolf, N. J. and L. L. Butcher (1989). Cholinergic systems: synopsis of anatomy and overview of physiology and pathology. In *The Biological Substrates of Alzheimer's Disease* (A. B. Scheibel and A. F. Wechsler, eds.). Academic Press, New York, pp. 73–86.

Wu, D. and L. B. Hersh (1994). Choline acetyltransferase: celebrating its fiftieth year. *J. Neurochem.* 62, 1653.

8

Norepinephrine and Epinephrine

Norepinephrine (NE) and epinephrine (E) are chemically catecholamines. The term *catecholamine* refers generically to all organic compounds that contain a catechol nucleus (a benzene ring with two adjacent hydroxyl substituents) and an amine group (Fig. 8–1). In practice, the term usually means dihydroxyphenylethylamine (dopamine, DA) and its metabolic products NE and E. Major advances in the understanding of the biochemistry, physiology, and pharmacology of these compounds have come mainly through the development of sensitive assay techniques and methods for visualizing catecholamine neurons, their connections and their metabolic enzymes and receptors in vivo and in vitro.

In 1946, it was demonstrated by von Euler in Sweden, and shortly thereafter by Holtz in Germany, that mammalian peripheral sympathetic nerves use NE as a transmitter instead of E (which is the sympathetic transmitter in the frog). With only minimal evidence based on tissue content, von Euler predicted that NE was highly concentrated in the nerve terminal region from which it was released to act as a neurotransmitter. This prediction was conclusively documented some 10 years later with the development of techniques for visualizing catecholamines in freeze-dried tissue sections, which helped to define the anatomy of the peripheral noradrenergic neuron.

Shortly after NE was established as the neurotransmitter substance of adrenergic nerves in the peripheral nervous system, Holtz identified it as a normal constituent of mammalian brain. For some years, however, it was thought that the presence of NE in the mammalian brain only reflected vasomotor

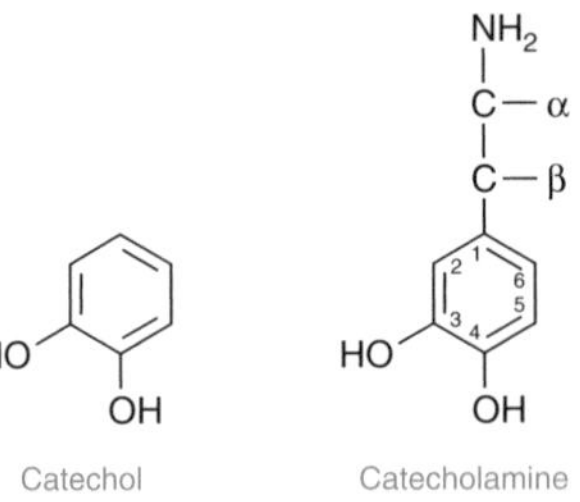

FIGURE 8–1. Catechol and catecholamine structure.

innervation to the cerebral blood vessels. In 1954, Vogt demonstrated that NE was not uniformly distributed in the central nervous system (CNS) and that this nonuniform distribution did not in any way coincide with the density of blood vessels found in a given brain region. This regional localization within the mammalian brain suggested that NE might subserve some specialized function, perhaps as a central neurotransmitter. The relative regional distribution of NE in the brain is quite similar in most mammalian species.

Morphology of the Sympathetic Neuron

The morphology of the peripheral noradrenergic neuron became clear following the development by Falck and Hillarp of the fluorescent histochemical method for the visualization of catecholamines by condensation with dry formaldehyde vapor at 60°C–80°C. Under the fluorescent microscope, all parts of the adrenergic neuron can be visualized (cell bodies, dendrites, axons, and nerve terminals), but the highest concentration of NE and the strongest fluorescence are found in the nerve terminal varicosities. A diffuse, widespread innervation pattern is characteristic of the sympathetic nervous system and its noradrenergic nerves. A single neuron can give rise to nerve terminal branches with lengths on the order of 10–20 cm, possessing several thousand nerve terminal varicosities (Fig. 8–2). The localization of catecholamines and their precursors within morphologically recognizable microscopic structures by fluorescence and electron microscopy has been a great advantage for investigators studying noradrenergic mechanisms where fluorescence intensity, vesicle contents, and amine content correlate. The lack of a suitable histochemical technique for the visualization of acetylcholine has been a serious handicap by comparison for those interested in cholinergic mechanisms (see Chapter 7).

DA is also present in the mammalian CNS, and its distribution differs markedly from that of NE, an early indication that DA functions as more than a precursor of NE in the CNS (see Chapter 9). DA is also present in the carotid body and superior cervical ganglion, where it also likely plays a role independent of NE. The superior cervical ganglion appears to have at least three distinct populations of neurons: cholinergic neurons, noradrenergic neurons, and, small, intensely fluorescent cells that contain DA but whose functional significance is unclear.

The endogenous occurrence of E in the mammalian CNS at relatively low levels (approximately 5%–10% by bioassay of the NE content) was reported in the early 1960s. Many investigators suggested that these original estimates

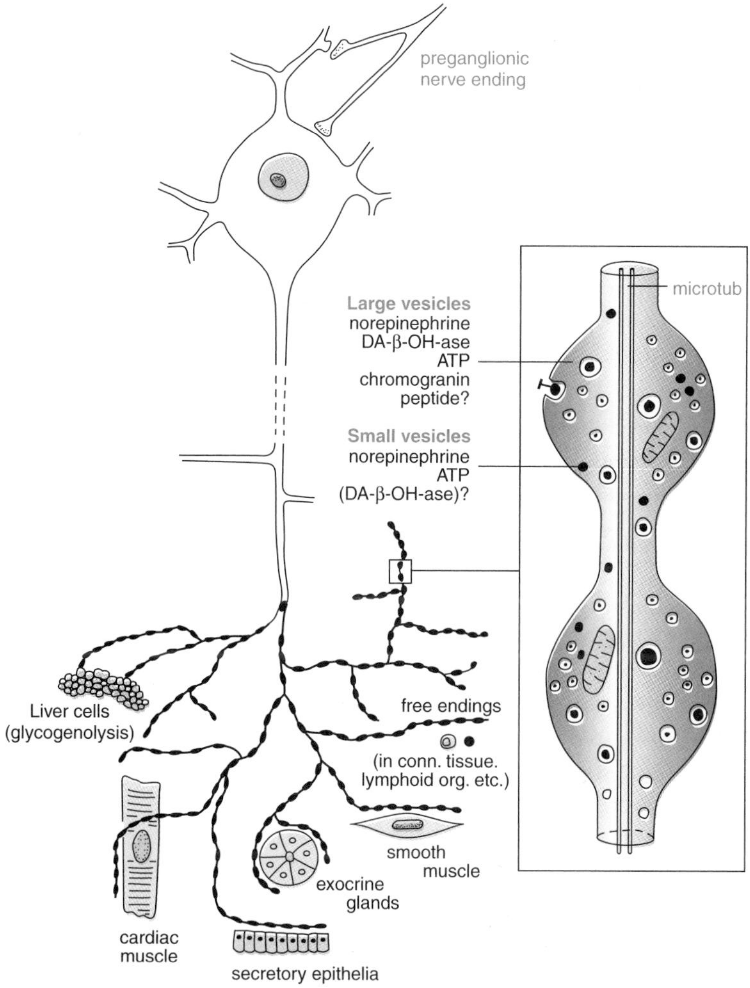

FIGURE 8–2. Schematic drawing of a noradrenergic neuron in the peripheral autonomic nervous system. Examples of innervated structures are illustrated. (Modified from Dahlstrom and Carlsson, 1986.)

are subject to error and in the past have discounted the importance of E in the mammalian brain. However, its presence has now been documented by more sophisticated analytical techniques, such as gas chromatography–mass spectrometry and liquid chromatography coupled with electrochemical detection and confirmed by immunohistochemical techniques.

The detailed topographical survey of brain catecholamines at different levels of organization within the CNS has provided a framework for organizing and conducting logical experiments concerning the possible functions of these amines. The anatomy, biochemistry, and pharmacology of CNS NE and E systems are discussed in detail in the latter half of this chapter.

Life Cycle of the Catecholamines

Biosynthesis

Catecholamines are formed in the brain, chromaffin cells, sympathetic nerves, and sympathetic ganglia from their amino acid precursor tyrosine by a sequence of enzymatic steps first postulated by Blaschko in 1939. This pathway was confirmed in 1964 by Nagatsu and co-workers, who demonstrated that tyrosine hydroxylase (TH) converts L-tyrosine to 3,4-dihydroxyphenylalanine (DOPA). Tyrosine is normally present in the circulation at a concentration of about 5 to 8×10^{-5} M. It is taken up from the bloodstream and concentrated within the brain and presumably in other sympathetically innervated tissue by an active transport mechanism. Once inside the peripheral neuron, tyrosine undergoes a series of chemical transformations, resulting in the formation of NE or, in the brain or chromaffin cell, NE, DA, or E, depending on the availability of the required downstream synthetic enzymes (see Fig. 8–3). Both phenylalanine and tyrosine are normal constituents of the mammalian brain, present in a free form at a concentration of about 5×10^{-5} M. However, NE biosynthesis is usually considered to begin with tyrosine, which represents a branch point for many important biosynthetic processes in animal tissues. The percentage of tyrosine used for catecholamine biosynthesis as opposed to other biochemical pathways is minimal (<2%).

Tyrosine Hydroxylase

The first enzyme in the biosynthetic pathway, TH, was the last enzyme in this series of reactions to be identified. It was demonstrated by Udenfriend and colleagues in 1964, and its properties have been reviewed repeatedly. It is present in the adrenal medulla, brain, and all sympathetically innervated tissues as a unique constituent of catecholamine-containing neurons and chromaffin cells. The enzyme is stereospecific; requires molecular O^2, Fe^{2+}, and a tetrahydropteridine cofactor; and shows a fairly high degree of substrate specificity for L-tyrosine and, to a smaller extent, L-phenylalanine. The single human gene for TH has been cloned and found to encode multiple

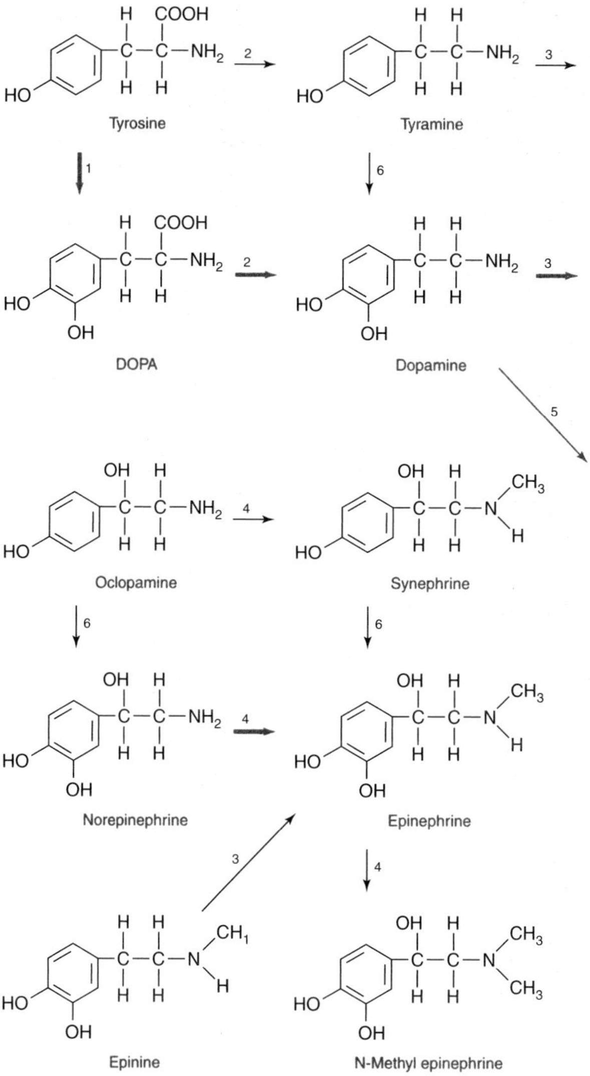

FIGURE 8–3. Primary and alternative pathways in the formation of catecholamine: *(1)* tyrosine hydroxylase; *(2)* aromatic amino acid decarboxylase; *(3)* dopamine-β-hydroxylase; *(4)* phenylethanolamine-*N*-methyltransferase; *(5)* nonspecific *N*-methyltransferase in lung and folate-dependent *N*-methyltransferase in brain; *(6)* catechol-forming enzyme.

mRNAs that are heterogeneous at the 5′ end of the coding region. The functional significance of the varient messages remains to be determined.

Tyrosine hydroxylation is the rate-limiting step in the biosynthesis of NE in the peripheral nervous system and of NE and DA in the brain. In most sympathetically innervated tissues, including the brain, the activity of DOPA decarboxylase and that of dopamine-β-hydroxylase have a magnitude 100–1000 times that of TH. Further proof that this enzyme is the rate-limiting step in catecholamine biosynthesis is that pharmacological intervention at this step reduces NE biosynthesis while blockade of the last two steps in the synthesis of NE does not. Inhibitors of TH markedly reduce endogenous NE and DA in the brain and NE in the heart, spleen, and other sympathetically innervated tissues. Effective inhibitors of this enzymatic step can be categorized into four main groups: *(1)* amino acid analogues, *(2)* catechol derivatives, *(3)* tropolones, and *(4)* selective iron chelators. Some effective amino acid analogues include α-methyl-*p*-tyrosine and its ester, α-methyl-3-iodotyrosine, 3-iodotyrosine, and α-methyl-5-hydroxytryptophan. In general, α-methyl-amino acids are more potent than the unmethylated analogues and a marked increase in activity in the case of the tyrosine analogues can also be produced by substituting a halogen at the 3 position of the benzene ring. Most of the agents in this category act as competitive inhibitors of the substrate tyrosine. α-methyl-*p*-tyrosine and its methyl ester have been the inhibitors most widely used to demonstrate the effects of exercise, stress, and various drugs on the turnover of catecholamines and to lower NE formation in patients with pheochromocytoma and malignant hypertension.

Dihydropteridine Reductase

Although not directly involved in catecholamine biosynthesis, dihydropteridine reductase is intimately linked to the TH step. This enzyme catalyzes the reduction of the quinonoid dihydropterin, which has been oxidized during the hydroxylation of tyrosine to DOPA. Since reduced pteridines are essential for tyrosine hydroxylation, alterations in the activity of dihydropteridine reductase affect the activity of TH. Dihydropteridines with amine substitution at positions 2 and 4 are effective inhibitors of this enzyme, while folic acid antagonists such as aminopterin and methotrexate are relatively ineffective. The distribution of dihydropteridine reductase is quite widespread, the highest activity being found in the liver, brain, and adrenal gland. The distribution of this enzyme activity in the brain extends well beyond catecholamine or serotonin innervation, suggesting that reduced pterins most likely participate in other reactions besides the hydroxylation of tyrosine and tryptophan. In fact, reduced pteridines are critical for nitric oxide (NO) synthetase.

Dihydroxyphenylalanine Decarboxylase

The second enzyme involved in catecholamine biosynthesis is DOPA-decarboxylase, which was actually the first catecholamine-synthesis enzyme to be discovered. Although originally believed to remove carboxyl groups only from L-DOPA, a study of purified enzyme preparations and specific inhibitors demonstrated that this DOPA-decarboxylase acts on all naturally occurring aromatic L-amino acids, including histidine, tyrosine, tryptophan, and phenylalanine as well as both DOPA and 5-hydroxytryptophan. Therefore, this enzyme is more appropriately referred to as "L-aromatic amino acid decarboxylase." There is no appreciable binding of this enzyme to particles within the cell since, when tissues are disrupted and the resultant homogenates centrifuged at high speeds, the decarboxylase activity remains associated largely with the supernatant fraction. The exception to this is in the brain, where some of the decarboxylase activity is associated with synaptosomes. Since synaptosomes are in essence pinched-off nerve endings, however, they would be expected to retain entrapped cytoplasm as well as other intracellular organelles. The DOPA-decarboxylase found in synaptosomal preparations is thought to be present in the entrapped cytoplasm. DOPA-decarboxylase is, relative to other enzymes in the biosynthetic pathway for NE formation, very active and requires pyridoxal phosphate (vitamin B_6) as a cofactor. The apparent K_m value for this enzyme is 4×10^{-4} M. The high activity of this enzyme may explain why it has been difficult to detect endogenous DOPA in sympathetically innervated tissue and brain. It is rather ubiquitous in nature, occurring in the cytoplasm of most tissues, including the liver, stomach, brain, and kidney in high levels, suggesting that its function in metabolism is not limited solely to catecholamine biosynthesis. Although decarboxylase activity can be reduced by vitamin B_6 deficiency in animals, this does not usually result in significant reduction of tissue catecholamines, although it appears to interfere with the rate of repletion of adrenal catecholamines after insulin depletion. In addition, potent decarboxylase inhibitors have very little effect on endogenous levels of NE in tissue. However, these inhibitors have been useful as pharmacological tools (e.g., DOPA accumulation following administration of a decarboxylase inhibitor as an in vivo index of tyrosine hydroxylation).

Dopamine-β-Hydroxylase

Although it has been known for many years that the brain, sympathetically innervated tissue, sympathetic ganglia, and adrenal medulla can transform DA into NE, it was not until 1960 that the enzyme responsible for this conversion was isolated from the adrenal medulla. This enzyme, called dopamine-β-hydroxylase, is, like TH, a mixed-function oxidase. It requires molecular oxygen and utilizes ascorbic acid as a cofactor. Its K_m for its substrate DA is about 5×10^{-3} M. Dicarboxylic acids such as fumaric acid are not absolute requirements, but they stimulate the reaction. Dopamine-β-hydroxylase is a Cu^{2+}-containing protein, with about 2 moles of cupric ion per 1 mole of enzyme, associated with the particulate fraction from the heart, brain, sympathetic nerve, and adrenal medulla, localized primarily in the membrane of the amine storage granules. The mRNA is expressed only in noradrenergic neurons or adrenal chromaffin tissue. Dopamine-β-hydroxylase does not show a high degree of substrate specificity and acts in vitro on a variety of substrates besides DA, oxidizing almost any phenylethylamine to its corresponding phenylethanolamine (i.e., tyramine to octopamine, α-methyldopamine to α-methylnorepinephrine). A number of the resultant structurally analogous metabolites can replace NE at the noradrenergic nerve endings and function as "false" neurotransmitters.

Dopamine-β-hydroxylase can be inhibited by a variety of compounds, including the copper chelators: D-cysteine and L-cysteine, glutathione, mercaptoethanol, and coenzyme A. Inhibition can be reversed by addition of N-ethylmaleimide, which reacts with sulfydryl groups amd interferes with the chelating properties of these substances. Copper-chelating agents such as diethyldithiocarbamate(disulfuram) and bis(1-methyl-4-homopiperazinyl-thiocarbonyl)-disulfide (FLA-63) have proved to be effective inhibitors both in vivo and in vitro. Thus, it has been possible to treat animals with these agents and produce a reduction in brain NE and an elevation of brain DA.

Dopamine-β-hydroxylase purified from the bovine adrenal medulla will produce a specific antibody to the enzyme that will inactivate bovine dopamine-β-hydroxylase but will not cross-react with either DOPA-decarboxylase or TH. However, anti-bovine dopamine-β-hydroxylase will cross-react with human, guinea pig, and dog dopamine-β-hydroxylase, indicating that the enzymes from these various sources are probably structurally related. By coupling immunochemical techniques with fluorescence and electron microscopy, this antibody has already proved useful in the localization of the enzyme in intact peripheral and central tissue. Dopamine-β-hydroxylase has been cloned, and further studies on its molecular structure and expression should yield interesting information.

Phenylethanolamine-N-Methyltransferase

In the adrenal medulla, NE is *N*-methylated by the enzyme phenylethanolamine-*N*-methyltransferase (PNMT) to form E. This enzyme is largely restricted to the adrenal medulla, although low levels of activity exist in heart and mammalian brain . Like the decarboxylase, this enzyme appears in the supernatant of homogenates. Demonstration of activity requires the presence of the methyl donor *S*-adenosylmethionine. Interest in the biosynthetic pathway for catecholamines has also led to the cloning of a single PNMT gene with three exons. The transcript is present in the adrenal medulla, heart, and brain stem; and PNMT is found in these tissues. Regulation of this enzyme in the brain has not been extensively studied, but glucocorticoids are known to regulate the activity of PNMT in the adrenal gland.

Synthesis Regulation

It has been known for a long time that the degree of sympathetic activity does not influence the endogenous levels of tissue NE, and it has been speculated that there must be some homeostatic mechanism whereby the level of transmitter is maintained relatively constant in the sympathetic nerve endings despite the additional losses assumed to occur during enhanced sympathetic activity.

More than 40 years ago, von Euler hypothesized, on the basis of experiments carried out in the adrenal medulla, that during periods of increased functional activity, the sympathetic neuron must also increase the synthesis of its transmitter substance NE to meet the increased demands placed on the neuron. If the sympathetic neuron had the ability to increase transmitter synthesis, this would enable the neuron to maintain a constant steady-state level of transmitter despite substantial changes in transmitter utilization. Some years later, experiments carried out by several laboratories on peripheral sympathetically innervated tissues as well as brain directly demonstrated that this was, in fact, the case. Electrical stimulation of sympathetic nerves, median forebrain bundle, or locus ceruleus both in vivo and in vitro resulted in increased formation of NE in the tissues innervated by these NE neurons. Further studies demonstrated that the observed acceleration of NE biosynthesis produced by enhanced noradreneregic activity was due to an increase in the activity of the rate-limiting enzyme involved in catecholamine biosynthesis, TH. Emphasis then shifted toward defining the mechanisms by which impulse flow changed enzyme activity. It is now well appreciated that the function of TH is determined by two major factors: changes in enzyme activity (the rate at which the enzyme converts the precursor into its product) and changes in the amount of enzyme protein. One major determinant of TH activity is it state of phosphorylation, which occurs at four different serine sites at the N terminus of the TH protein. These four serine residues are differentially phosphorylated by various kinases. Protein kinase C, (cAMP)-dependent protein kinase (protein kinase A), and calcium–calmodulin-dependent kinase phosphorylate TH, producing kinetic activation of the enzyme; the question of which of these mechanisms is operative physiologically remains uncertain. The most recent studies favor the involvement of Ca^{2+}–calmodulin-dependent phosphorylation in the impulse-dependent activation process (Fig. 8–4). α_2-Adrenergic or D_2 dopaminergic autoreceptor–mediated reduction of TH may also be achieved via regulation of the cAMP or Ca^{2+}–calmodulin-dependent protein phosphatases.

A second means of regulating the activity of the enzyme is through end-product inhibition: catecholamines can inhibit the activity of TH through competition for a required pterin cofactor for the enzyme. Finally, the amount of reduced pteridine cofactor (tetrahydobiopterin, BH_4) can also influence TH activity because the levels of the cofactor are not saturating under basal conditions.

In general, very rapid, short-term upregulation of TH occurs primarily by these posttranslational changes, whereas longer-term changes in TH occur through transcriptional upregulation of the TH gene. However the degree to which increases in catecholamine synthesis depend on de novo synthesis of new enzyme protein or changes in enzymatic activity appears to differ between CNS noradrenergic and dopaminergic neurons. For example, increased synthetic demand in noradrenergic neurons of the brain stem nucleus locus

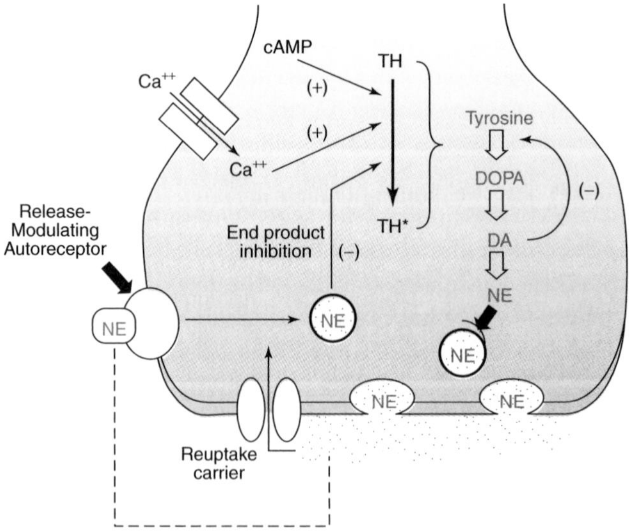

FIGURE 8–4. Model illustrating mechanisms for regulation of transmitter synthesis in noradrenergic neurons. TH, tyrosine hydroxylase; NE, norepinephrine; DA, dopamine; DOPA, 3,4-dihydroxyphenylalanine.

ceruleus appears to be accomplished primarily by increasing TH gene expression. In contrast, the same conditions and treatments that increase TH gene expression in brain stem noradrenergic neurons fail to increase TH mRNA levels in DA-containing neurons of the midbrain. In these dopaminergic neurons, it appears that synthesis is regulated primarily by altering the activity of TH, i.e., by phosphorylation (see Chapter 9).

Storage

A great conceptual advance made in the study of catecholamines more than 40 years ago was the recognition that in almost all tissues a large percentage of the NE present is located within highly specialized subcellular particles (later shown to be synaptic vesicles but colloquially referred to as "granules") in sympathetic nerve endings and chromaffin cells. Much of the NE in the CNS is also located within similar vesicles. These granules contain adenosine triphosphate (ATP) in a molar ratio of catecholamine to ATP of about 4:1. Some complex of the amines with ATP and protein is probable since the intravesicular concentration of amines, at least in the adrenal chromaffin granules and probably in the splenic nerve granules (0.3–1.1 M), would be hypertonic if present in free solution and might be expected to lead to osmotic lysis of the vesicles.

Two vesicular monoamine transporters (VMATs) localized to the membranes of synaptic vesicles and chromaffin granules have been cloned (VMAT-1 and VMAT-2). VMAT-1 is found in the adrenal medulla , in the adrenal chromaffin cells that synthesize and release monoamines. VMAT-2 is present in catecholamine and serotonin neurons in the CNS. VMAT-2, the isoform found in the brain, shows modest substrate specificity and can transport catecholamines and indoleamines, as well as histamine, into vesicles. These VMATs define another new family of transporter proteins that display the common motif of 12 hydrophobic, putative transmembrane domains and move transmitters into acidic intracellular compartments such as neurotransmitter vesicles. These transporters display no significant sequence homology with the plasma membrane Na^+/Cl^--coupled family (see below), but they do resemble bacterial drug-resistance transporters. VMATs can also be distinguished from plasma membrane transporters by their use of transmembrane H^+ gradients instead of Na^+ (see below). All of the amine storage vesicles studied in brain and adrenal chromaffin cells contain a vascular-type H^+-pumping ATPase similar to that found in lysosomes and Golgi membranes. The significant homology of these vesicular transporters to a group of bacterial drug-resistance transporters suggests that VMATs may play a role in detoxification. VMAT enables vesicles to sequester toxins and reduce their cellular toxicity. For example, mice heterozygous for one knockout copy of VMAT-2 show increased toxicity of DA neurons to the DA neurotoxin 1-methyl-4-phenyl-1,2,3,6-tetrahydropyridine. Loss of VMAT means less sequestration of the toxin , which can then exert its toxic action by targeting mitochondrial respiration (see Chapter 9).

Release

The release of NE, similar to the release of acetylcholine (ACh), is Ca^{2+}-dependent and appears to occur by the same Ca^{2+}-dependent process (exocytosis) that has been described for other transmitters (see Chapter 2). Exocytosis requires that the entire content of the granular vesicle be released (i.e., catecholamine, ATP, and soluble protein). Since the nerve terminal region, as far as we know, cannot synthesize any protein, high rates of axonal flow from the nerve cell body would be required to replenish the protein lost during exocytosis. Alternatively, one might propose a "protein reuptake" mechanism, to recapture that protein released during synaptic transmission. In peripheral nerves, release of NE is frequency-dependent within a physiological range of frequencies. Some evidence has also been presented that newly synthesized NE may be released preferentially. This preferential release is additional evidence to support the contention that NE exists in more than one pool within the sympathetic neuron.

It has been a great deal more difficult to demonstrate NE release from its nerve endings in the CNS. However, with the independent development of the push–pull cannula, microdialysis techniques, and electrochemical detectors, catecholamines can be detected extracellularly from certain deep nuclear masses of the CNS or measured directly by in vivo voltammetry. NE release detemined by the above techniques is Ca^{2+}-dependent and influenced by the functional activity of the noradrenergic neuron.

Regulation of Release

The major homeostatic mechanism for regulation of catecholamine release in both central and peripheral catecholamine neurons involves interaction of the released catecholamine with specific presynaptic receptors (autoreceptors), which are located on the nerve terminals.

In most catecholamine-containing systems, administration of catecholamine agonists attenuates stimulus-induced release while administration of catecholamine receptor blockers augments release. These pharmacological studies have established the concept that presynaptic receptors modulate release by responding to the concentration of catecholamine in the synapse (high concentrations inhibiting release and low concentrations augmenting release). Presynaptic autoreceptors have also been implicated in the regulation of DA synthesis (see Chapter 9).

Several types of presynaptic receptor are involved in the inhibition of transmitter release from adrenergic nerves. These include α_2-adrenergic autoreceptors as well as muscarinic, opiate, and DA receptors. Different presynaptic receptors are linked to facilitation of NE release, including β_2-adrenergic adrenoceptors, nicotinic cholinergic receptors, and angiotensin II receptors. The precise mechanism by which autoreceptors influence neurotransmitter release from adrenergic neurons varies depending on their location on the neuron. Activation of α_2-adrenoceptors inhibits NE release by several possible mechanisms, including *(1)* attenuation of the rate of Ca^{2+} entry through inhibition of voltage-gated Ca^{2+} channels; *(2)* opening of K channels, leading to hyperpolarization of the neuron terminals; and *(3)* inhibition of adenylate cyclase, resulting in a decrease of intracellular cAMP and Ca^{2+} concentration.

Presynaptic β-adrenergic receptors are also present in some sympathetic tissues. These autoreceptors are usually believed to be of the β_2 subtype. The β_2-adrenoceptor-mediated facilitation of NE release may be due to stimulation of adenylate cyclase, leading to an increase in cAMP and a subsequent increase in intracellular Ca^{2+} concentrations. Alternatively, it has been suggested that the facilitation may be due to a transsynaptic signal involving a local renin–angiotensin response. This hypothesis posits that β-adrenoceptor agonists activate postsynaptic β_2-adrenoceptors in the vascular wall. This activation results in the synthesis of angiotensin II, which diffuses across the synaptic cleft to activate presynaptic angiotensin II receptors that facilitate NE release.

The presence of presynaptic facilitatory and inhibitory adrenergic receptors on the same nerve terminals may provide for a fine-tuning control of stimulus-evoked NE release. Since the affinity of E for β_2-adrenoceptors is much greater than that of NE, this subset of adrenoceptors is more likely to be activated by endogenous E than NE. Thus, low synaptic concentrations of E could activate presynaptic β_2-adrenoceptors preferentially, leading to an increase in NE release. At higher concentrations of E or NE, activation of α_2-adrenoceptors would predominate and NE release would be diminished rather than enhanced.

Prostaglandins of the E series are also potent inhibitors of neurally induced release of NE in a great number of tissues, and their action appears to be dissociated from any interaction with presynaptic receptors. These substances are released from sympathetically innervated tissues, and most evidence indicates that inhibition of local prostaglandin production is associated with an increase in the release of NE and subsequent effector responses induced by neuronal activity. The control of NE release by this prostaglandin-mediated feedback mechanism appears to operate through restriction of calcium availability for the NE release process and to be most efficient within the physiological frequency range of nerve impulses.

Metabolism

The metabolism of exogenously administered or endogenous catecholamines is markedly slower than that of ACh by the ACh–ACh-esterase system. The major mammalian enzymes of importance in the metabolic degradation of catecholamines are monoamine oxidase (MAO) and catechol-*O*-methyltransferase (COMT) (Fig. 8–5). MAO is an enzyme that converts catecholamines to their corresponding aldehydes. This aldehyde intermediate is rapidly metabolized, usually by oxidation by the enzyme aldehyde dehydrogenase to the corresponding acid. In some circumstances, the aldehyde is reduced to the alcohol or glycol by aldehyde reductase. In the case of brain NE, reduction of the aldehyde metabolite appears to be the favored route of metabolism. MAO is a particle-bound protein, localized largely in the outer membrane of mitochondria, although partial microsomal localization cannot be excluded. There is also some evidence for a riboflavin-like material in MAO isolated from liver mitochondria. MAO is usually considered to be an intraneuronal enzyme, but it occurs in abundance extraneuronally. In fact, most experiments indicate that chronic denervation of a sympathetic end organ leads only to a relatively small reduction in MAO, suggesting that the greater proportion of this enzyme is extraneuronal. However, it is the intraneuronal enzyme that seems to be important in catecholamine metabolism. MAO present in human and rat brain exists in at least two different forms, designated MAO-A and MAO-B based on substrate specificity and sensitivity to inhibition by selected inhibitors.

Clorgyline is a specific inhibitor of MAOA, which has a substrate preference for NE and serotonin. Deprenyl is a selective inhibitor of MAOB, which has a substrate preference for β-phenylethylamine and benzylamine as substrates. DA, tyramine, and tryptamine appear to be equally good substrates for both forms of the enzyme. MAO-A and -B arise from distinct genes on chromosome X and are expressed in different regions of the nervous system. The differential expresssion of the two forms of MAO in biogenic amine neurons is noteworthy. MAO-A mRNA appears to be the main form of the enzyme expressed in peripheral and central noradrenergic neurons, whereas MAO-B mRNA is found primarily in serotonergic neurons in the midbrain raphe and histaminergic neurons in the hypothalamus. Neither mRNA appears be expressed in midbrain dopaminergic neurons, although this remains a matter of controversy.

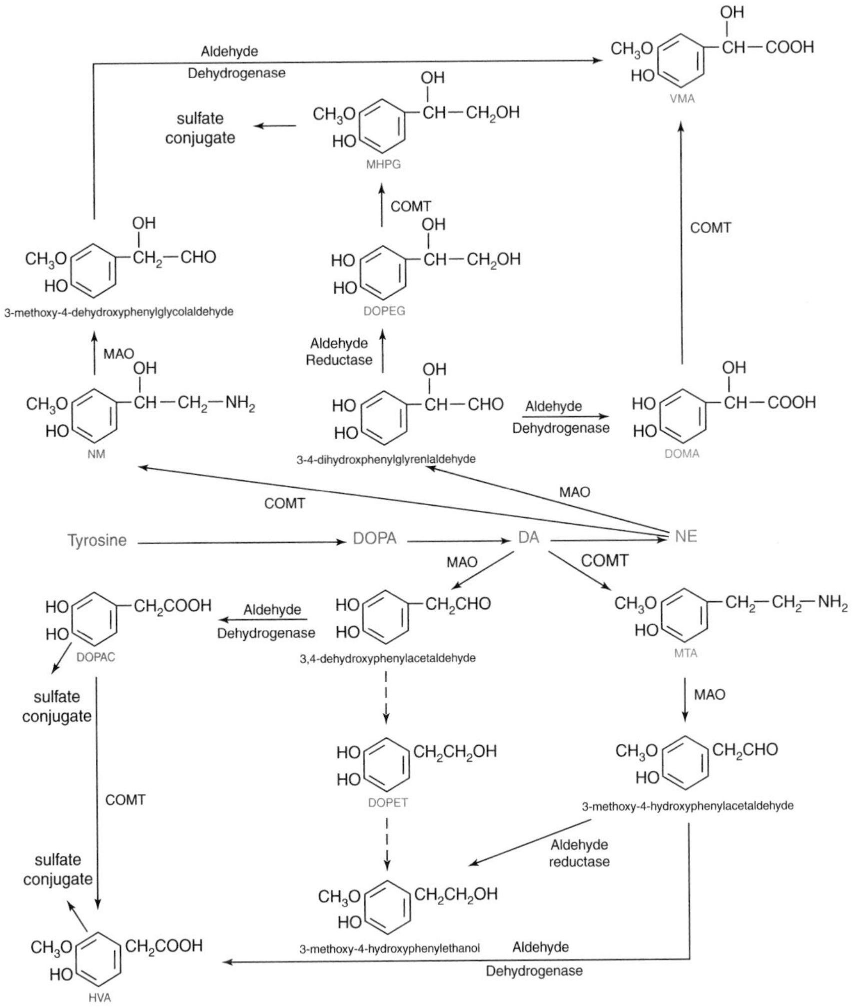

FIGURE 8–5. Dopamine and norepinephrine metabolism. DOPA, 3,4-dihydroxyphenylalanine; DA, dopamine; NE, norepinephrine; DOMA, 3,4-dihydroxymandelic acid; DOPAC, 3,4-dihydroxyphenylacetic acid; DOPEG, 3,4-dihydroxyphenylglycol; DOPET, 3,4-dihydroxyphenylethanol; MHPG, 3-methoxy-4-hydroxyphenylglycol; HVA, homovanillic acid; VMA, 3-methoxy-4-hydroxymandelic acid; NM, normetanephrine; MTA, 3-methoxytyramine; MAO, monoamine oxidase; COMT, catechol-O-methyltransferase. Dashed arrows indicate steps that have not been firmly established.

The role and relative importance of these two types of MAO in physiological and pathological states is currently unknown, but this is an important area for further research. It has been speculated that under certain circumstances MAO could regulate NE biosynthesis by controlling the amount of substrate DA available to the enzyme dopamine-β-hydroxylase. MAO is not an exclusive catabolic enzyme for the catecholamines, since it also oxidatively deaminates other biogenic amines, such as 5-hydroxytryptamine, tryptamine, and tyramine. The intraneuronal localization of MAO in mitochondria or other structures suggests that this would limit its action to amines that are present in a free (unbound) form in the axoplasm. Here, MAO can act on amines that have been taken up by the axon before they are sequestered by VMAT or even on amines that are released from the VMAT before they pass out through the axonal membrane. Interestingly, the latter possibility seems of minor physiological importance since MAO inhibition does not potentiate the effects of peripheral sympathetic nerve stimulation.

The second enzyme of importance in the catabolism of catecholamines is COMT, discovered by Axelrod in 1957. This enzyme is a relatively nonspecific enzyme that catalyzes the transfer of methyl groups from *S*-adenosylmethionine to the *m*-hydroxyl group of catecholamines and various other catechol compounds. COMT is found in the cytoplasm of most animal tissue, being particularly abundant in kidney and liver. A substantial amount is also found in the CNS and in various sympathetically innervated organs. The precise cellular localization of COMT has not been determined, although it has been suggested that it functions extraneuronally. The purified enzyme requires *S*-adenosylmethionine and Mg^{2+} ions for activity. As with MAO, inhibition of COMT activity does not markedly potentiate the effects of sympathetic nerve stimulation, although in some tissues it tends to prolong the duration of the response to stimulation. Therefore, neither MAO nor COMT seems to be the primary mechanism for terminating the action of NE liberated at sympathetic nerve terminals. It may be, however, that these enzymes play a more important role in terminating transmitter action and regulating catecholamine function in the CNS. In fact, recent data have suggested that COMT may play an important role in the regulation of synaptic DA in the prefrontal cortex (see Chapter 9) in contrast to its minimal action in the mesoaccumbens and mesostriatal DA systems.

Reuptake

When sympathetic postganglionic nerves are stimulated at frequencies low enough to be comparable to those encountered physiologically, very little intact NE overflows into the circulation, suggesting that local inactivation is very efficient. This local inactivation is not significantly blocked when COMT, MAO, or both are inhibited; and it is believed to involve mainly reuptake of the transmitter by sympathetic neurons.

Much attention has been focused on the role of tissue-uptake mechanisms in the physiological inactivation of catecholamines, but only in recent years has this concept received direct experimental support via isolation of the NE transporter and the cloning of its gene.

This uptake process is a saturable membrane transport process dependent on temperature and requiring energy. The stereochemically preferred substrate is L-NE; furthermore, NE is taken up more efficiently than its *N*-substituted derivatives. It is now appreciated that neuronal reuptake of catecholamines is a major means of inactivation of the released transmitter in the brain and peripheral nervous system. The reuptake process conserves transmitter and allows intracellular enzymes that degrade catecholamines to act, thus bolstering the actions of extracellular enzymes. Neuronal reuptake of catecholamines, and indeed of all transmitters for which a reuptake process has been identified, has several characteristics. The reuptake process is energy-dependent, is saturable, and depends on Na^+ cotransport. In addition, extracellular Cl^- is necessary for transport. Because reuptake depends on coupling to the Na^+ gradient across the neuronal membrane, toxins that inhibit Na^+,K^+-ATPase also inhibit reuptake. These plasma membrane transporters, in contrast to the monoamine transporters associated with neuronal vesicles, are not Mg^{2+}-dependent and not inhibited by reserpine. A DNA clone encoding a human NE transporter has been isolated. The isolated cDNA sequence predicts a protein of 617 amino acids with the typical transporter motif of 12 highly hydrophobic regions, probable membrane-spanning domains. Expression of the cDNA clone in transfected HeLa cells indicates that the NE transport activity is sodium-dependent and sensitive to NE-transport inhibitors. Striking sequence homology is notable between the NE transporter and the rat and human γ-aminobutyric acid (GABA) transporters, suggesting a new transporter gene family.

The plasma membrane transporters have been intensively studied at the biochemical, pharmacological, and molecular levels. It has become clear that these Na^+- and Cl^--coupled transporters represent a group of integral membrane proteins encoded by a closely related family of genes that includes the transporters of monoamines, GABA, glycine, and choline. A different class of plasma membrane transporters is represented by the glutamate transporter (see Chapter 6). Plasma membrane transporters can be classified into families and subfamilies based on ion dependence, topology, and sequence relatedness. The NE transporters (NETs) are members of the Na^+/Cl^--dependent neurotransmitter transporter family, which includes the DA transporter

(DAT) and the serotonin transporter (SERT). Expression of NET, SERT, and DAT in nonneuronal cells has established model systems for analysis of the structural basis of transporter specificity for transmitters and transporter-specific antagonists. The catecholamine transporters, NET and DAT are not very specific, with each accumulating both DA and NE. In fact, the NET has a higher affinity for DA than for NE and, under certain conditions, can play a role in the modulation of DA transmission (see Chapter 9). No specific transporter for E-containing neurons has been found in mammalian species, but one has been identified in the frog. The availability of pure transporter proteins has facilitated the development of transporter-specific antibodies and nucleic acid probes and stimulated renewed interest in the endogenous control mechanisms acutely regulating monoamine transport in vivo. Also, the availability of human cDNA encoding the NET offers the opportunity to determine whether alterations in transporter genes could have important etiological implications for major depression or affective disorder.

Since many aspects of NE neurotransmission, including synthesis and release, are tightly regulated, it would not be unexpected to find that the NET is also subject to regulatory control. In fact, almost 40 years ago, Schneider and Gillis noted that, following stimulation of the sympathetic input to the heart, a rapid increase in the retention of NE occurred in the cat atrium. A number of similar observations have been made over the past four decades, suggesting that NE reuptake increases in parallel with an increased rate of sympathetic discharge and NE release. Although the mechanism responsible for the increase in NET activity remains to be determined, the presence of serine/threonine phosphorylation sites on human NET raises the possibility that this effect might be mediated by protein phosphorylation. The role of protein phosphorylation or other posttranslational modifications in regulating transporter function is only beginning to be evaluated. However, the fact that NET proteins can now be visualized suggests that this should be a fertile area for future investigations. Little is currently known about promotor regions and other regulatory elements involved in the transcriptional regulation of the NET and SERT genes. However, rats treated chronically with antidepressant drugs exhibit a decrease in SERT mRNA levels and a reduction in NET using [^{3}H]nisoxetine autoradiography.

The availability of cDNAs encoding transporter proteins may facilitate the development of sensitive screening techniques to help develop new and selective agents that target specific neurotransmitter transporters. Although this screening technology has not yet led to the development of uniquely selective inhibitors of the monoamine transporters, a number of selective inhibitors of NE uptake are available, several of which are used clinically in the treatment of affective disorders. Several of these NE-uptake blockers are illustrated in Figure 8–6.

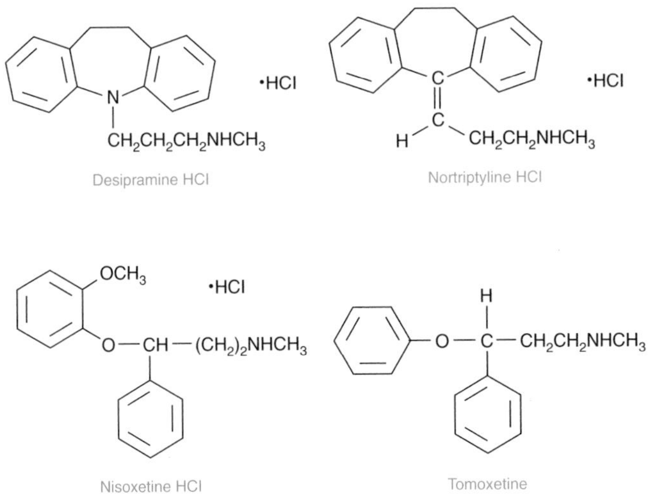

FIGURE 8–6. Representative compounds that selectively inhibit norepinephrine reuptake.

Adrenergic Receptors

When catecholamines are released from either noradrenergic nerve terminals or the adrenal medulla, they are recognized by and bind to specific receptor molecules on the plasma membrane of the neuroeffector cells that transduce catecholamine interactions with the cell into a physiological response. Depending on the nature of the receptor, this interaction sets off a cascade of membrane and intracellular events, which cause the cell to carry out its specialized function. Classically, peripheral adrenergic receptors have been divided into two distinct classes, called α and β receptors. The development of synthetic compounds active at adrenergic receptors has allowed further differentiation of adrenergic receptors into α_1, α_2, β_1, and β_2 subtypes (Fig. 8–7). Genes for these subtypes of adrenergic receptor have been cloned; this has shown that these receptors are members of a larger family of hormone receptors that mediate their activity through interaction with one of a series of guanine nucleotide–binding regulatory proteins (G proteins). In addition to traditional classification based on their pharmacological profile, adrenergic receptors can be divided into three major classes by their differential coupling to G proteins (Table 8–1). β-Adrenergic receptors activate G_s to stimulate adenylate cyclase. α_2-Adrenergic receptors decrease adenylyl cyclase activity through coupling to G_i. α_1-Adrenergic receptors stimulate phospholipase C action through a still ill-defined G_q. As schematically illustrated in Figure 8–7, β-adrenoceptor signals are transmitted

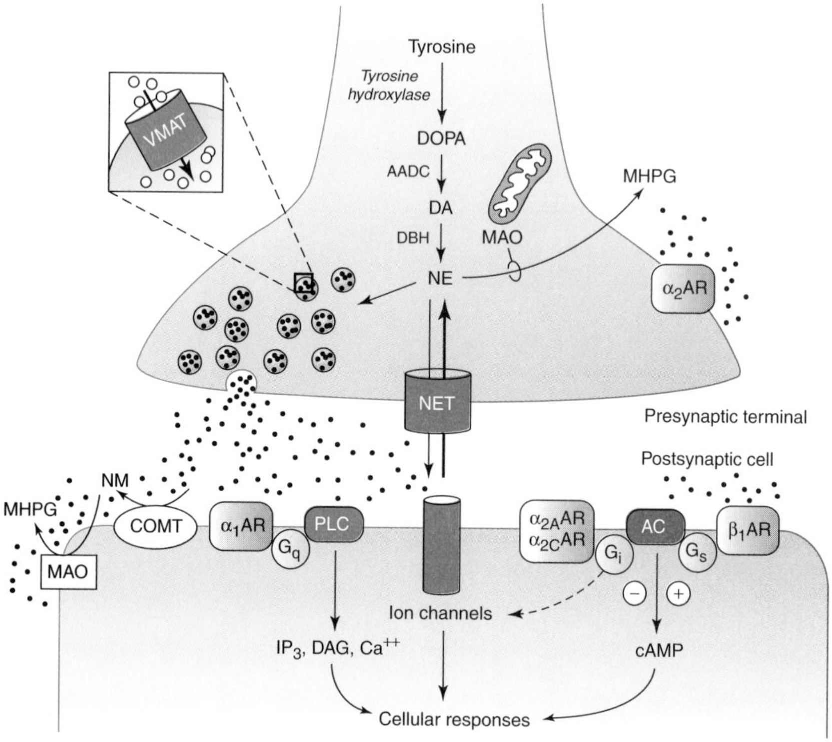

FIGURE 8–7. Model of a norepinephrine synapse illustrating the presynaptic and postsynaptic molecular entities involved in the synthesis, storage, release, reuptake, and signaling of norepinephrine. Tyrosine is transported into the presynaptic terminal by an active uptake mechanism and converted to NE by a series of enzymatic steps. NE is taken up from the axoplasm and stored by the VMAT. Once released, NE can interact with two categories of adrenergic receptors, alpha and beta. Alpha 2 autoreceptors localized in the nerve terminals modulate the synthesis and release of NE, while cell body and dendritic α_2 autoreceptors modulate impulse flow. The majority of G protein–coupled adrenergic receptors are localized postsynaptically where they mediate the cellular responses of the postsynaptic neuron. NE is metabolized by MAO and COMT giving rise to MHPG, a major brain NE metabolite. The synaptic action of NE is terminated by reuptake into the presynaptic terminal by NET. AC, adenylyl cyclase; AR, adrenergic receptor; DAG, diacylglycerol; IP3, inositol triphosphate; MHPG, 3-methoxy-4-hydroxy phenethylene glychol; NM, normetanephrine; PLC, phospholipase C; NET, plasma membrane norepinephrine transporter, VMAT, vesicular monoamine transporter. (Modified from Nesther et al., *Molecular Neuropharmacology*)

Table 8–1. Characterization of Adrenergic Receptor Subtypes

Receptor Subtype	*Receptor-Specific Agents*		*Tissue Distribution*	*Major G Protein*	*Effector System*
	Agonist	*Antagonist*			
α_1A	Cirazoline	Corynanthine	Vas deferens, brain	G_q	↑ Phospholipase C
α_1B	Methoxamine		Liver, brain		↑ Ca^{2+} channel
α_1C	Phenylephrine	Indoramin	Olfactory bulb		↑ Phospholipase A_2
α_1D		Prazosin	Vas deferens, brain		↑ Phospholipase D
					↓ Adenylylate cyclase
α_2A	Guanabenz	RX 821002	Aorta, brain	G_i	↑ K^+ ↓ Ca^{2+}
$\alpha_2\beta$	*p*-Aminoclonidine	Yohimbine	Liver, kidney		↑ Na^+-H^+ exchange
α_2C	BHT-920	SKF 86426	Brain		↑ Phospholipases C, A_1
β_1	Isoproterenol	Alprenolol	Heart, brain, pineal	G_s	↑ Adenylylate cyclase
β_2		Propranolol	Lung, prostate		
β_3		Pindolol	Adipose tissue		↑ Ca^{2+} channel

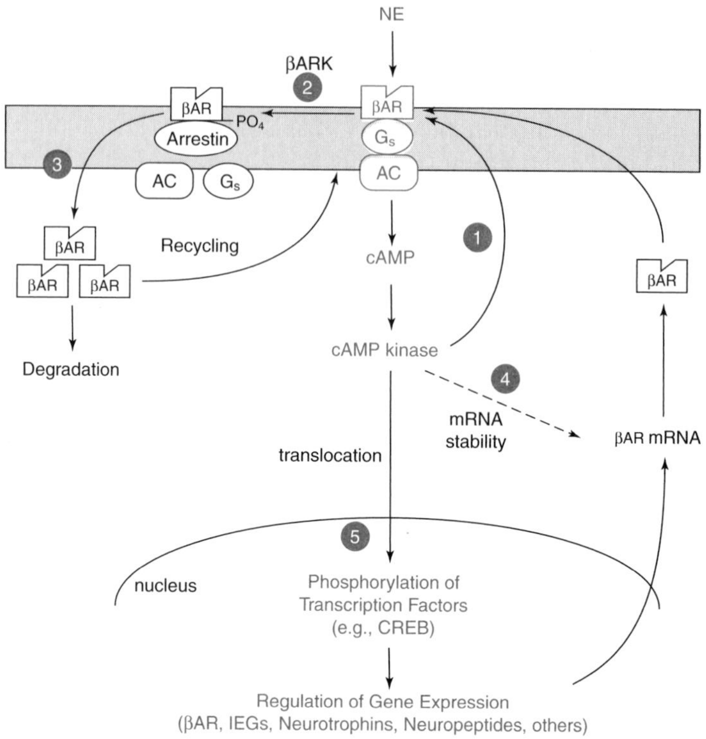

FIGURE 8–8. Schematic illustration of multiple mechanisms underlying regulation of the β-adrenergic receptor (βAR). *(1)* βAR stimulation of the cyclic adenosine monophosphate (cAMP) system results in phosphorylation of the receptor by cAMP-dependent protein kinase, which leads to uncoupling of the receptor from G_s. The activated protein kinase would also phosphorylate many other proteins not shown in the figure, which would then mediate many of the actions of βAR activation. *(2)* Prolonged activation of the receptor leads to phosphorylation by another kinase, βAR kinase (βARK), which phosphorylates only the agonist-activated form of the receptor. This results in binding of the receptor to β-arrestin, which competes with G_s and thereby inhibits βAR stimulation of adenylyl cyclase (AC). *(3)* Loss of βARs from the membrane occurs when receptors are internalized and sequestered into intracellular vesicles. This pool of receptors is then available for either recycling back to the membrane or degradation. Such sequestration, internalization, degradation, and membrane reinsertion may be mediated via receptor phosphorylation and dephosphorylation involving the cAMP and/or βARK pathways. Another mechanism by which receptor activation leads to downregulation of the βAR is via regulation of receptor mRNA levels, which may occur by two primary mechanisms. *(4)* The level of receptor mRNA is regulated by the stability or half-life of the mRNA. Although the mechanisms responsible for regulation of mRNA stability have not been identified, they may also involve cAMP-dependent protein kinase. *(5)* The level of receptor mRNA is also regulated via changes in βAR gene transcription. This effect is mediated by the cAMP pathway and appears to involve translocation of the cAMP-dependent protein kinase catalytic subunit into the nucleus and

through the effector cell membrane via the adenylyl cyclase system. Occupation of the β adrenoceptor by the catecholamines stimulates adenylyl cyclase to generate cAMP by a series of intramembrane events (see Chapter 5). Initially, the receptor interacts with a guanosine triphosphate (GTP)–dependent protein, G_s. G_s is linked to a catalytic unit, which generates cAMP upon activation by G_s; the catalytic unit then converts adenosine monophosphate (AMP) to cAMP. The latter compound triggers a series of intracellular events involving protein kinases. The protein kinases activate further unknown biochemical changes to generate the final biological response to the transmitter.

The order of potency for stimulation of β_1-adrenergic receptors by catecholamines is isoproterenol > NE ≥ E. For the β_2 adrenoceptor, the order of potency is isoproterenol > E > NE. Adrenergic blocking agents such as propranolol and alprenolol prevent the activation of β receptors. The interaction of adrenergic agents with their receptors is saturable, stereospecific, and reversible. Prolonged exposure of β adrenoceptors to endogenous or exogenous agonists often reduces the responsiveness of these receptors (desensitization). Desensitization can also be caused by uncoupling of β receptors from the adenylyl cyclase and by a decrease in the number of receptors (downregulation) (Fig. 8–8). Depriving the β adrenoceptors of catecholamine (by chemical or surgical denervation) increases their responsiveness.

The recently identified β_3 adrenoceptor appears to be responsible for lipolysis in white adipose tissue and thermogenesis in brown adipose tissue found in rodents. The β_3 adrenoceptor shows lower affinity for agonists, with a rank order of potency of NE > isoproterenol > E and a lower affinity for known β-adrenoceptor antagonists. All three β-adrenoceptor subtypes appear to be linked to adenylyl cyclase activation through a stimulatory G protein, and at present there is no evidence for subtype-related differences in receptor-mediated cyclase interactions. However, the β_3 adrenoceptor is insensitive to blockade by most β-adrenoceptor antagonists, making pharmacological evaluation of its function difficult. Several selective β-adrenocep-

---

the phosphorylation of constitutively expressed transcription factors (e.g., cAMP response element–binding protein [CREB]). It might also depend on the subsequent induction of other transcription factors (e.g., immediate early gene [IEG] products such as c-Fos). In addition to regulation of receptor gene transcription, such regulation of transcription factors would mediate the effects of βAR activation on the expression of many other genes, for example, those for G proteins, cAMP-dependent protein kinase, neurotrophins, and neuropeptides. This, in turn, would mediate many of the more long-term effects of βAR activation on brain function. (From Duman and Nestler, 1995.)

tor antagonists have been developed that exhibit high specificity for the β_3 adrenoceptor. These new agents have facilitated further studies on the functions subserved by this receptor subtype. In contrast to the β_1 and β_2 adrenoceptors, the β_3 adrenoceptor is resistant to both agonist-mediated short-term desensitization and, on a prolonged time scale, agonist-induced receptor downregulation. The biological importance of this differential response to the regulation by agonists is uncertain but suggests that during prolonged activation of the sympathetic nervous system β_3 receptor–mediated effects might be preserved while β_1- and β_2-mediated responses are diminished. The α_1 adrenoceptor appears to be associated with calcium mobilization (Fig. 8–7), possibly via stimulation of phospholipase C and phospholipase. The α_1 receptor is distinguished from the α_2 adrenoceptor by its inhibition by the selective α_1-blocking agent prazosin. The α_2-adrenergic receptor is negatively linked to the adenylyl cyclase complex via an inhibitory G protein (G_i). Like G_s, G_i is activated by GTP, but in this case it inhibits the generation of AMP by the catalytic unit. In the peripheral sympathetic nervous system, the α_2 receptor is localized mainly on presynaptic nerve terminals, where it modulates NE release. Stimulation of this receptor by catecholamines inhibits impulse-dependent release of transmitter; blockade of this receptor facilitates release.

Although it is clear that activation of α_2 adrenoceptors inhibits adenylyl cyclase activity (Fig. 8–7) mediated through an inhibitory G protein, this may not in all cases represent the signal-transduction mechanism responsible for the effects associated with receptor activation. For example, the α_2-mediated inhibition of neurotransmitter release is generally insensitive to inactivation of inhibitory G proteins by pertussis toxin, suggesting the involvement of another as yet undetected signal-transduction mechanism.

Dynamics of Adrenergic Receptors

Adrenergic receptors do not seem to be static entities but change in number and affinity in response to altered synaptic activity. In both brain and the peripheral sympathetic nervous system, destruction of adrenergic neurons is associated with functional supersensitivity of postsynaptic sites. Conversely, increasing synaptic NE by administering uptake blockers (tricyclic antidepressants) or MAO inhibitors leads to functional subsensitivity. These changes appear to be a compensatory response to altered levels of synaptic transmitter. The number of α_1 and α_2 receptors also increases after noradrenergic neurons in the brain, and sympathetically innervated tissues have been destroyed by administration of 6-hydroxydopamine. After lesions in

NE-containing neurons are made in the cerebral cortex, the number of β_1 receptors increases markedly but no change occurs in the number of β_2 receptors. This may be because β_2 receptors have a low affinity for NE and the concentration of E in the cerebral cortex is relatively low. Likewise, chronic administration of tricyclic antidepressants leads to a selective decrease in the density of β_1-adrenergic receptors in the cerebral cortex. This finding implies that the β_1-adrenergic receptors in the cortex are functionally innervated by adrenergic neurons.

Desensitization

The waning of a stimulated response in the face of continuous agonist exposure is termed *desensitization*. Desensitization has been demonstrated in many hormone- and neurotransmitter-receptor systems, including those that activate different G proteins and effector systems. Desensitization of the cAMP response elicited by β-adrenergic agonists is a useful model system for studying this phenomenon. Desensitization has been viewed historically as two separate processes: heterologous and homologous. *Heterologous* desensitization occurs when exposure of cells to a desensitizing agent leads to diminished responsiveness to a number of different stimuli. In contrast, *homologous* desensitization is much more specific and involves only the loss of responsiveness to the specific desensitizing agent. A number of molecular mechanisms operating at the receptor level have been implicated in the process of desensitization. The heterologous pattern of desensitization is thought to be mediated by phosphorylation of receptors coupled to G_s by protein kinase A, whereas the homologous pattern of desensitization is thought to involve phosphorylation of receptors by a novel, cAMP-independent kinase, such as β-adrenergic receptor kinase. Recent studies on the β-adrenergic receptor have suggested that the principal mechanisms underlying rapid, agonist-induced desensitization of β-adrenergic receptor function in intact mammalian cells are a combination of these two processes and involve phosphorylation of the receptor by both types of kinase. Phosphorylation of the receptor is thought to lead to uncoupling of the β-adrenergic receptor from the stimulatory G protein (G_s). It is not yet clear exactly how phosphorylation leads to uncoupling of β-adrenergic receptor from G_s; for example, phosphorylation of one specific residue may directly impair interaction of the receptor with G_s, whereas phosphorylation of another residue may allow cytosolic factors such as β-arrestin to bind to the phorphorylated β-adrenergic receptor and decrease or prevent the coupling of the β-adrenergic receptor with G_s, leading to desensitization.

CNS Catecholamine Neurons

The cellular organization of the brain and spinal cord has been studied for many decades by classic histological and silver impregnation techniques. With the development in the 1960s of completely different histological techniques based on the presence of a given type of transmitter substance or on specific synthetic enzymes involved in the formation of a given transmitter, it became possible to map chemically defined neuronal systems in the CNS of many species. By the time such techniques had been employed for several years, it became clear that the distribution of these chemically defined monoamine neuronal systems did not necessarily correspond to that of systems described earlier with the classic techniques.

However, it was really not until the mid-1960s that the histochemical fluorescent technique of Falck and Hillarp was applied to brain tissue and the anatomy of the monoamine-containing neuronal systems was described. By a variety of pharmacological and chemical methods, it has subsequently been possible to discriminate between NE-, E-, and DA-containing neurons and to describe in detail the distribution of these catecholamine-containing neurons in the mammalian CNS. Several recent and thorough surveys of these systems are available using multiple histochemical measures including mapping of mRNAs, immunodetection of synthetic proteins, and ligand binding to receptors and transporters.

Figure 8–9 provides a schematic model of a central monoamine-containing neuron as observed by fluorescence and electron microscopy. The cell bodies contain relatively low concentrations of amine (about 10–100 mg/g), while the terminal varicosities contain a very high concentration (about 1000–3000 mg/g). The axons, however, consist of highly branched, largely unmyelinated fibers that have such a low concentration of amine that they are barely visible in untreated adult animals. With the electron microscope, depending on the type of fixation, it is possible to observe small granular vesicles that are thought to represent the subcellular storage sites containing the catecholamines. These granular vesicles are concentrated in the terminal varicosities of the central noradrenergic neuron, just as they are observed in the peripheral sympathetic neuron.

Systems of Catecholamine Pathways in the CNS

Detailed analysis of catecholamine pathways in the CNS has been greatly accelerated by improvements in the application of fluorescence histochemistry, such as the use of glyoxylic acid as the fluorogen and by the application of numerous auxiliary mapping methods mentioned above. Extensive progress in the functional analysis of these systems has also been made possible by evaluation with microelectrodes of the effects produced by selective electrical stimulation of the catecholamine (especially NE) cell-body groups. From such studies, it is clear that the systems can be characterized in simple terms only by ignoring large amounts of detailed cytological and functional data and that the earlier catecholamine wiring diagrams are no longer tenable. Furthermore, anatomic details for monoamine systems in rodents seem to

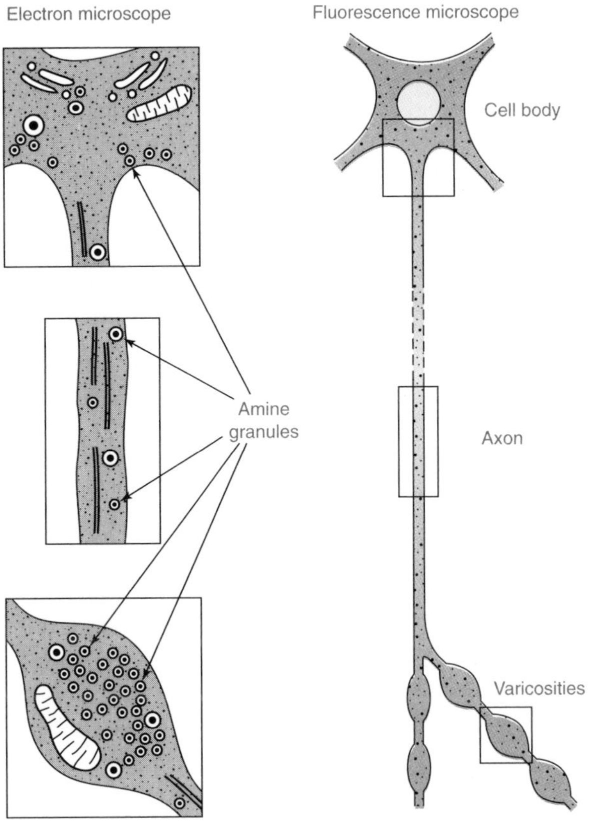

FIGURE 8–9. Schematic illustration of a central monoamine-containing neuron. *Right side* depicts the general appearance and intraneuronal distribution of monoamines based on fluorescence microscopy. Cell bodies contain a relatively low concentration of catecholamine (about 10–100 μg/g), while the terminal varicosities contain a very high concentration (1000–3000 μg/g). Preterminal axons contain very low concentrations of amine. At the electron microscopic level (*left side*), dense core granules can be observed in the cell bodies and axons but appear to be highly concentrated in the terminal varicosities. (Modified from Fuxe and Hökfelt, 1971.)

bear only rudimentary homology to their detailed selective anatomic configurations in human and nonhuman primates. There are two major clusterings of NE cell bodies from which axonal systems arise to innervate targets throughout the entire neuraxis.

Locus Ceruleus

This compact cell group within the caudal pontine central gray is named for the pigment the cells bear in humans and some higher primates; in the rat, the nucleus contains about 1500 neurons on each side of the brain (Fig. 8–10). In humans, the locus ceruleus is composed of about 12,000 large neurons on each side of the brain. The axons of these neurons form extensive collateral branches, which project widely along well-defined tracts. At the electron microscope level, terminals of these axons exhibit, under appropriate fixation

Figure 8–10. Formaldehyde-induced fluorescence of the rat nucleus locus ceruleus. In this frontal section through the principal portion of the nucleus, intensely fluorescent neurons can be seen clustered closely together. Within the neurons, the cell nucleus, which is not fluorescent after this treatment, appears dark except for the nucleolus. (Bloom, unpublished) (×650).

methods, the same type of small granular vesicles seen in the peripheral sympathetic nerves (Fig. 8–2).

Fibers from the locus ceruleus form five major noradrenergic tracts (Fig. 8–11): the central tegmental tract (or dorsal bundle described by Ungerstedt), a central gray dorsal longitudinal facsiculus tract, and a ventral tegmental-medial forebrain bundle tract. These tracts remain largely ipsilateral, although there is a crossing over in some species of up to 25% of the fibers. These three ascending tracts then follow other major vascular and fascicular routes to innervate all cortices, specific thalamic and hypothalamic nuclei, and the olfactory bulb. Another major fascicle ascends in the superior cerebellar peduncle to innervate the cerebellar cortex. The fifth major tract descends into the mesencephalon and spinal cord, where the fibers course in the ventral-lateral column. At their terminals, the locus ceruleus fibers form a plexiform network in which the incoming fibers gain access to a cortical region by passing through the major myelinated tracts, turning vertically toward the outer cortical surface and then forming characteristic T-shaped branches, which run parallel to the surface in the molecular layer; this pattern is seen in the cerebellar, hippocampal, and cerebral cortices.

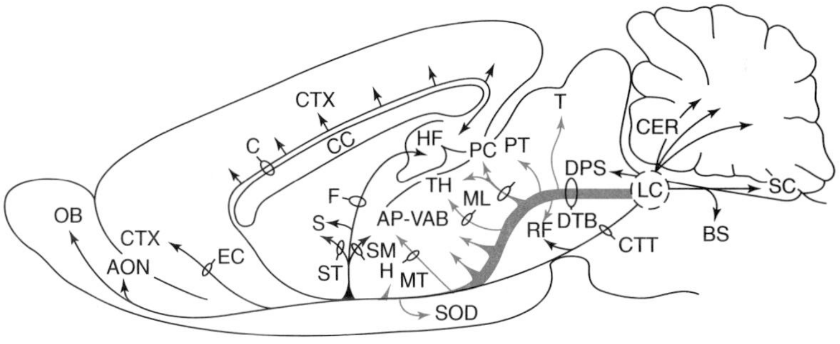

FIGURE 8–11. Diagram of the projections of the locus ceruleus viewed in the sagittal plane. See text for description. AON, anterior olfactory nucleus; AP–VAB, ansa peduncularis–ventral amygdaloid bundle system; BS, brain stem nuclei; C, cingulum; CC, corpus callosum; CER, cerebellum; CTT, central tegmental tract; CTX, cerebral neocortex; DPS, dorsal periventricular system; DTB, dorsal catecholamine bundle; EC, external capsule; F, fornix; H, hypothalamus; HF, hippocampal formation; LC, locus ceruleus; ML, medial lemniscus; MT, mammillothalamic tract; OB, olfactory bulb; PC, posterior commissure; PT, pretectal area; RF, reticular formation; S, septal area; SC, spinal cord; SM, stria medullaris; SOD, supraoptic decussations; ST, stria terminalis; T, tectum; TH, thalamus. (Diagram compiled by R. Y. Moore, from the observations of Lindvall and Björklund, 1974; Jones and Moore, 1977.)

Virtually all of the noradrenergic pathways that have been studied physiologically so far are efferent pathways of the locus ceruleus neurons; in cerebellum, hippocampus, and cerebral cortex, the major effect of activating this pathway is to inhibit spontaneous discharge. This effect has been associated with the slow type of synaptic transmission, in which the hyperpolarizing response of the target cell is accompanied by increased membrane resistance. The mechanism of this action has been experimentally related to the second-messenger scheme in which the noradrenergic receptor elicits its characteristic action on the target cells by activating the synthesis of cAMP in or on the postsynaptic membrane. Pharmacologically and cytochemically, target cells responding to NE or the locus ceruleus projection in these cortical areas exhibit β-adrenergic receptors.

Lateral Tegmental Noradrenergic Neurons

A large number of noradrenergic neurons lie outside of the locus ceruleus, where they are more loosely scattered throughout the lateral ventral tegmental fields. In general, the fibers from these neurons intermingle with those arising from the locus ceruleus, those from the more posterior tegmental levels contributing mainly descending fibers within the mesencephalon and spinal cord and those from the more anterior tegmental levels contributing to the innervation of the forebrain and diencephalon. Because of the complex intermingling of the fibers from the various noradrenergic cell body sources, the physiological and pharmacological analysis of the effects of brain lesions becomes extremely difficult. The neurons of the lateral tegmental system may be the primary source of the noradrenergic fibers observed in the basal forebrain, such as amygdala and septum. No specific analysis of the physiology of these projections has yet been reported; therefore, it remains to be established whether the β-receptive cAMP mechanism associated with the cortical projections of the locus ceruleus group also apply to the synapses of the lateral tegmental neurons.

Epinephrine Neurons

In the sympathetic nervous system and the adrenal medulla, E shares with NE the role of final neurotransmitter, the proportion of this sharing being a species-dependent, hormonally modified arrangement. Until method refinement, however, little evidence could be gathered to document the existence of E in the CNS because the chemical methods for analyzing E levels or for detecting activity attributable to the synthesizing enzyme PNMT were unable to provide unequivocal data. With the development of sensitive immunoassays for PNMT and their application in immunohistochemistry and with the application of gas chromatography–mass fragmentography and high-performance liquid chromatography with electrochemical detection to brain neurotransmitter measurements, the necessary data were rapidly acquired and the existence of E-containing neurons in the CNS confirmed. By immunohistochemistry, E-containing neurons are defined as those that are positively stained (in serial sections) with antibodies to TH, dopamine-β-hydroxylase, and PNMT. These cells are found largely in two groups: one, called C-1 (see Hokfelt et al., 1984), is intermingled with noradrenergic cells of the lateral tegmental system; the other, called C-2, is found in the regions in which the noradrenergic cells of the dorsal medulla are also found. The axons of these two E systems ascend to the hypothalamus with the central tegmental tract, then via the ventral periventricular system into the hypothalamus. A third group of cells (C-3) in the midline (medial longitudinal fascicle) has also been described (Fig. 8–12). Within the mesencephalon, the E-containing fibers innervate the nuclei of visceral efferent and afferent systems, es-

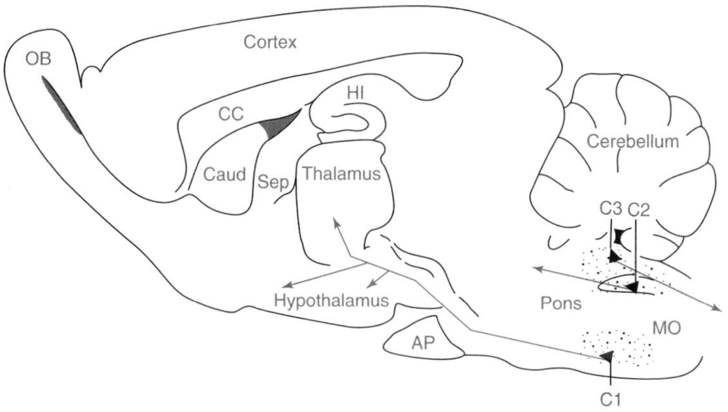

Figure 8–12. Schematic illustration of the distribution of the main central neuronal pathways containing epinephrine. The C1 group represents a rostral continuation of the noradrenergic cell group in the ventrolateral medulla oblongata. The dorsal C2 group of cells is located mainly in the dorsal vagal complex, with the vast majority of cell bodies in the solitary tract nucleus. A C3 group of cells in the midline (medial longitudinal fascicle) has also been described. The ventral group of epinephrine-containing neurons gives rise to both ascending and descending pathways, innervating largely periventricular regions such as the periaqueductal central gray and various hypothalamic nuclei and the lateral sympathetic column in the spinal cord. OB, olfactory bulb; CC, corpus callosum; CAUD, caudate; SEP, septum; AP, ansa peduncularis; HI, hippocampus; MO, Medulla Oblongata. (Data from Hokfelt et al., 1984.)

pecially the dorsal motor nucleus of the vagus nerve. In addition, E fibers innervate the locus ceruleus, the intermediolateral cell columns of the spinal cord, and the periventricular regions of the fourth ventricle.

Although there are considerably fewer E than DA and NE neurons in the brain, they appear to have a discrete anatomic distribution and are believed to subserve physiological functions discrete from other catecholamines. Selective plasma membrane transporters have been described for DA and NE, and these transporters have been extensively studied both in vivo and in vitro. However, little attention has been given to the hypothetical E transporter, and at the present time a specific transporter for E has not been isolated from mammalian brain nor have drugs that selectively block E reuptake been described. All of the pharmacological agents that have been shown to block the NET (see Fig. 8–7) selectively are also effective inhibitors of E uptake. Thus, it remains uncertain whether a selective transporter for E actually exists in the mammalian CNS or peripheral nervous system. The cloning and expression of the E transporter might permit a search for selective inhibitors, which, if discovered, could provide valuable pharmacological tools for elucidating the functional role of central E-containing neurons.

Except for tests of E in the locus ceruleus (where it inhibits firing), no other tests of the cellular physiology of this system have thus far been reported. Our understanding concerning the function of E-containing neurons in the brain is still very limited, but based on the distribution of E in specific brain regions, attention has been directed to their possible role in neuroendocrine mechanisms and blood pressure control.

Coexistence of Classic Transmitters and Peptides

The finding of numerous peptides in the CNS has raised the question of the relationship of the neurons containing these substances to neurons containing classic neurotransmitters. It was originally believed that the peptide-containing nerves represented separate identifiable systems. However, when studies in the peripheral nervous system clearly indicated the coexistence of peptides with the sympathetic and parasympathetic transmitters NE and ACh, respectively, this initiated a series of systematic studies in the CNS to determine if neurotransmitters in the brain coexisted with one or more peptides. These studies revealed that the coexistence of classic transmitters and peptides may represent a rule rather than an exception. The functional significance of the occurrence of a catecholamine and a peptide in the same neuron and their possible release from the same nerve ending is still unclear in the CNS. However, some insight may be obtained from studies in the periphery where two distinct types of interaction have been noted.

In the salivary gland, the parasympathetic cholinergic neurons contain a vasoactive intestinal polypeptide (VIP)-like peptide while the sympathetic NE neurons contain a neuropeptide Y-like peptide. In this system, the peptides seem to augment the action of the classic transmitters. Thus, VIP induces vasodilation and enhances the secretory effects of ACh while neuropeptide Y causes vasoconstriction like NE. In contrast, in the vas deferens, where the NE neurons innervating this tissue also contain neuropeptide Y, the peptide seems to inhibit the release of NE via a presynaptic action. There is also some indication of preferential storage and release of NE and neuropeptide Y, with release of the peptide occurring preferentially at higher frequencies or during burst firing.

The physiological significance of multiple messengers at the synapse in the CNS is still unclear, but an appreciation of coexistence phenomena may influence our view on interneuronal communication and in a larger perspective may be of importance in our efforts to treat or prevent various disease states or abnormalities of the nervous system.

CNS Catecholamine Metabolism

Catecholamine metabolism was discussed earlier, and only a few aspects pertinent to central catecholamine metabolism will be covered here.

In the peripheral sympathetic nervous system, the aldehyde intermediate produced by the action of MAO on NE and normetanephrine can be oxidized to the corresponding acid or reduced to the corresponding glycol. In contrast to the CNS, peripheral findings are that oxidation usually exceeds reduction and, quantitatively, vanilylmandelic acid (VMA) is the major metabolite of NE and is readily detectable in the urine. In fact, urinary levels of VMA are routinely measured in clinical laboratories to provide an index of peripheral sympathetic nerve function as well as to diagnose the presence of catecholamine-secreting tumors such as pheochromocytomas and neuroblastomas. In the CNS, however, reduction of the intermediate aldehyde formed by the action of MAO on catecholamines or catecholamine metabolites predominates, and a major metabolite of NE found in the brain is the glycol derivative 3-methoxy-4-hydroxy-phenethyleneglycol (MHPG). Very little, if any, VMA is found in the brain.

In many species, a large fraction of the MHPG formed in the brain is sulfate-conjugated. However, in primates, MHPG exists primarily in an unconjugated "free" form in the brain. Some normetanephrine is also found in the brain and spinal cord. Destruction of noradrenergic neurons in the brain or spinal cord causes a reduction in the endogenous levels of these metabolites although not as marked as the corresponding reduction in endogenous

NE. Direct electrical stimulation of the locus ceruleus or severe stress produces an increase in the turnover of NE as well as an increase in the accumulation of the sulfate conjugate of MHPG in the rat cerebral cortex. These effects are completely abolished by ablation of the locus ceruleus or by transection of the dorsal pathway, suggesting that the accumulation of MHPG-sulfate in the brain may reflect the functional activity of central noradrenergic neurons. Since MHPG-sulfate readily diffuses from the brain into the cerebrospinal fluid (CSF) or general circulation, estimates of its concentration in the CSF or in urine are thought to reflect the activity of noradrenergic neurons in the brain. Even though MHPG is proportionately a minor metabolite of NE in the peripheral sympathetic nervous system, a fairly large portion of the MHPG in the urine still derives from the periphery. In rodents, it has been estimated that the brain provides a minor contribution (25%-30%) to urinary MHPG, while in primates, the brain contribution is quite large (60%). Thus, at least in rodents, it is quite probable that relatively large changes in the formation of MHPG by the brain are necessary to produce detectable changes in urinary MHPG. For example, in rats, destruction of the majority of NE-containing neurons in the brain by treatment with 6-hydroxydopamine leads to only about a 25% decrease in urinary MHPG levels. Nevertheless, measurement of urinary changes in MHPG is still a reasonable strategy for obtaining some insight into possible alterations of brain NE metabolism in patients with psychiatric illnesses. Measurement of CSF levels of MHPG also provides another possible approach to assessing central adrenergic function in human subjects. More recent studies have suggested that plasma levels of MHPG might provide a reflection of central noradrenergic activity. In these studies, stimulation of the locus ceruleus in the rat was shown to result in a significant increase in the levels of plasma MHPG; and administration of drugs that are known to alter noradrenergic activity in rodents has predictably changed plasma levels and venous–arterial differences in MHPG in nonhuman primates. Changes in urinary, plasma, and CSF levels of MHPG and their relationship to central noradrenergic function have to be interpreted with considerable caution.

Pharmacology of Central Catecholamine-Containing Neurons

The psychotropic drugs (drugs that alter behavior, mood, and perception in humans and behavior in animals) can be divided into two main categories: the psychotherapeutic drugs and the psychotomimetic agents. Psychotherapeutic drugs can be further divided, on the basis of their activity in humans, into at least four general classes: antipsychotics, antianxiety drugs, antidepressants, and stimulants.

A great deal of information has become available concerning the anatomy, biochemistry, and functional organization of catecholamine systems. This has fostered knowledge concerning the mechanisms and sites of action of many psychotropic drugs. It is now appreciated that drugs can influence the output of monoamine systems by interacting at several distinct sites. For example, drugs can influence the output of catecholamine systems by *(1)* acting presynaptically to alter the life cycle of the transmitter (i.e., synthesis, storage, and release, etc. [see Figs. 8–13 and 9–8]); *(2)* acting postsynaptically to mimic or block the action of the transmitter at the level of postsynaptic receptors; and *(3)* acting at the level of cell body autoreceptors to influence the physiological activity of catecholamine neurons. The activity of catecholamine neurons can also be influenced by postsynaptic receptors via negative neuronal feedback loops. In the latter two actions, drugs appear to exert their influence by interacting with catecholamine receptors. Figure 8–14 illustrates the various neuronal pathways and local mechanisms regulating synaptic homeostasis in central noradrenergic neurons. In general, catecholamine receptors can be subdivided into two broad categories: those that are localized directly on catecholamine neurons (often referred to as *presynaptic receptors*) and those on other cell types, often termed simply *postsynaptic receptors* since they are postsynaptic to the catecholamine neurons, the source of the endogenous ligand. When it became appreciated that catecholamine neurons, in addition to possessing receptors on the nerve terminals, appear to have receptors distributed over all parts of the neurons (i.e., soma, dendrites, and preterminal axons), the term *presynaptic receptors* really became inappropriate as a description for all of these receptors. Carlsson, in 1975, suggested that *autoreceptor* was a more appropriate term to describe these receptors since the sensitivity of these catecholamine receptors to the neuron's own transmitter seemed more significant than their location relative to the synapse. The term *autoreceptor* achieved rapid acceptance, and pharmacological research in the succeeding years resulted in the detection and description of autoreceptors on neurons in almost all chemically defined neuronal systems.

The presence of autoreceptors on some neurons may turn out to be only of pharmacological interest since they may never encounter effective concentrations of the appropriate endogenous agonist in vivo. Others, however, in addition to their pharmacological responsiveness, may play a very important physiological role in the regulation of presynaptic events. This certainly appears to be the case for DA autoreceptors (see Chapter 9).

The pharmacology of central and peripheral NE neurons is quite similar (Fig. 8–13). The main differences seem to be related primarily to drug delivery and the greater complexity of the neuronal pathways regulating synaptic homeostasis of the central NE systems (Fig. 8–14).

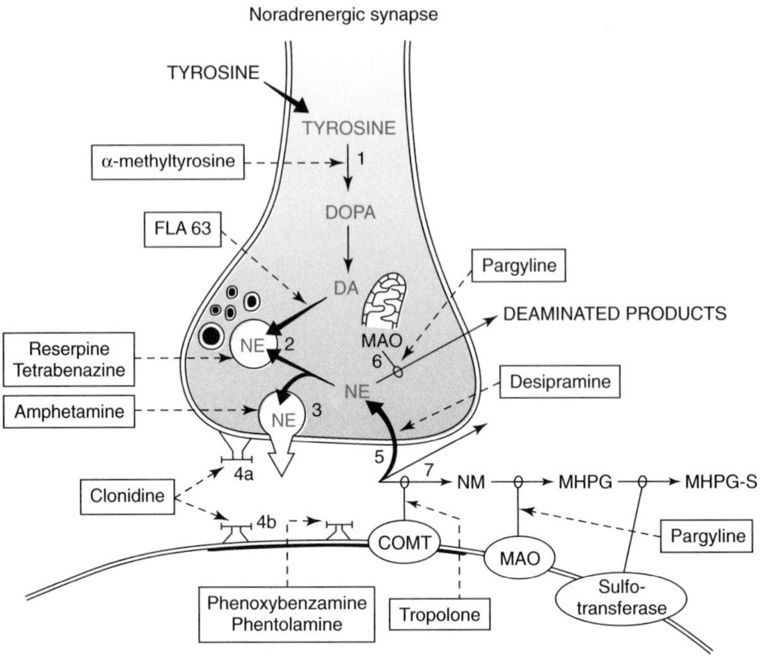

FIGURE 8–13. Schematic model of central noradrenergic neuron indicating possible sites of drug action.

Site 1: *Enzymatic synthesis:*

a. Tyrosine hydroxylase reaction blocked by the competitive inhibitor, α-methyltyrosine.

b. Dopamine-β-hydroxylase reaction blocked by a dithiocarbamate derivative, Fla-63-bis-(1-methyl-4-homopip erazinyl-thiocarbonyl)-disulfide (FLA 63).

Site 2: *Storage:* Reserpine and tetrabenazine interfere with the uptake–storage mechanism of the amine granules. Depletion of norepinephrine (NE) produced by reserpine is long-lasting, and the storage granules are irreversibly damaged. Tetrabenazine also interferes with the uptake–storage mechanism of the granules, except the effects of this drug are of a shorter duration and do not appear to be irreversible.

Site 3: *Release:* Am chanism by which amphetamine causes release is by its ability to block effectively the reuptake mechanism.

Site 4: *Receptor interaction:*

a. Presynaptic α_2 autoreceptors

b. Postsynaptic α receptors

Clonidine appears to be a very potent autoreceptor-stimulating drug. At higher doses, clonidine will also stimulate postsynaptic receptors. Phenoxybenzamine and phenotolamine are effective α receptor–blocking agents. These drugs may also have some presynaptic α_2 receptor–blocking action. However, yohimbine and piperoxane are more selective as α_2 receptor–blocking agents.

Site 5: *Reuptake:* NE has its action terminated by being taken up into the presynaptic terminal. The tricyclic drug desipramine is an example of a potent inhibitor of this uptake mechanism.

Site 6: *Monoamine Oxidase (MAO):* NE or dopamine (DA) present in a free state within the presynaptic terminal can be degraded by the enzyme MAO, which appears to be located in the outer membrane of mitochondria. Pargyline is an effective inhibitor of MAO.

Site 7: *Catechol-O-methyltransferase (COMT):* NE can be inactivated by the enzyme COMT, which is believed to be localized outside the presynaptic neuron. Tropolone is an inhibitor of COMT. The normetanephrine (NM) formed by the action of COMT on NE can be further metabolized by MAO and aldehyde reductase to 3-methoxy-4-hydroxyphenylglycol (MHPG). The MHPG formed can be further metabolized to MHPG-sulfate (MHPG-S) by the action of a sulfotransferase found in the brain. Although MHPG-S is the predominant form of this metabolite found in rodent brain, the free unconjugated MHPG is the major form found in primate brain.

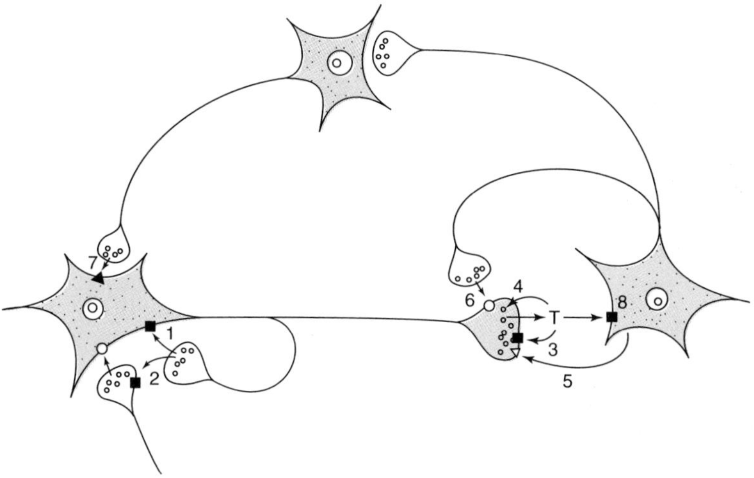

FIGURE 8–14. Neural pathways regulating synaptic homeostasis of locus ceruleus neurons. Influence of locus ceruleus neuron (shaded) on its target cells can be regulated both by modulation of transmitter release *(1–7)* and by amplification of signal provided by that release *(8)*. This involves a large variety of cell surface receptors, including receptors that respond to norepinephrine itself (■) as well as receptors responding to other chemical signals (○, □, ▲, △). Principal pathways for regulating norepinephrine release: *1,* direct action of recurrent collaterals onto soma; *2,* indirect action of recurrent collaterals, mediated via influence on presynaptic afferents; *3,* direct action of norepinephrine on presynaptic terminal; *4,* alterations in rate of norepinephrine reuptake; *5,* humoral signals generated by target; *6,* neural signals providing short-loop negative feedback from target; *7,* neural signals providing long-loop negative feedback from target; *8,* extent to which signal is amplified can be modulated by short-term modification of the sensitivity of the target, by long-term changes in number of receptors, and by other means such as release of cotransmitter. (Modified from Stricker and Zigmond, 1986.)

Effect of Drugs on the Electrophysiological Activity of Noradrenergic Neurons

A number of drugs influence the activity of the noradrenergic neurons in the locus ceruleus. Table 8–2 lists the drugs that have been studied and their influence on locus cell firing. Amphetamine inhibits locus cell firing, apparently by activating a neuronal feedback loop. The effects of amphetamine on locus firing are partially blocked by chlorpromazine. L-Amphetamine is equipotent with D-amphetamine in its inhibitory effects. α-Receptor agonists and antagonists also influence locus cell firing. This is due in part to the in-

TABLE 8–2. Correlation Between Brain Levels of 3-Methoxy-4-hydroxyphenylglycol (MHPG) and Electrophysiological Activity of Noradrenergic Neurons in the Locus Ceruleus

Drug or Experimental Condition	*Change in Locus Ceruleus Unit Activity*	*Change in Brain MHPG*
α_2-agonists	Decrease	Decrease
Clonidine		
Guanfacine		
α_2-antagonists	Increase	Increase
Yohimbine		
Piperoxane		
Idazoxane		
Tricyclic antidepressants	Decrease	Decrease
Desipramine		
Imipramine		
Amitriptyline		
Amphetamine	Decrease	Decrease, increase (dose-dependent)
Morphine	Decrease	Increase
Methylxanthines	Increase	Increase
Isobutylmethylxanthine		
Naloxone	No change	No change
Naloxone-precipitated morphine withdrawal	Increase	Increase
Stress	Increase	Increase
Electrical stimulation of locus ceruleus	Increase	Increase
Transection of dorsal norepinephrine bundle	Decrease	Decrease

teraction of these adrenergic agents with autoreceptors on the NE cell bodies or dendrites in the locus. α_2-Agonists, such as clonidine and guanfacine, suppress locus cell firing, and these inhibitory effects are reversed by α_2 antagonists. Piperoxane, yohimbine, and idazoxane (α_2 antagonists) cause a marked increase in single-cell activity. Morphine and morphine-like peptides, such as enkephalins and endorphins, exert an inhibitory influence on locus cell firing. The inhibitory effects of morphine or enkephalin are reversed by the opiate antagonist naloxone but are uninfluenced by α_2 antagonists. Methylxanthines, such as isobutyl-methylxanthines, which induce a quasi-opiate withdrawal syndrome, cause an increase in locus cell firing; this increase is antagonized by α_2 agonists, such as clonidine. Locus ceruleus cells in chronic morphine-treated rats dramatically increase locus cell firing during naloxone-induced withdrawal, and this increase in firing is suppressed by clonidine. These results, viewed in conjunction with the clinical data demonstrating the effectiveness of treating opiate withdrawal with clonidine, have suggested that NE hyperactivity may be an important component of the opiate withdrawal syndrome in humans. These studies have thus provided new insight into the rational design of a new class of drugs (α_2 agonists) to treat opiate withdrawal, although the role of the locus ceruleus in this treatment effect has not been established.

Numerous pharmacological studies, conducted mostly in rodents, have demonstrated a good correlation between drug-induced changes in the firing rate of locus ceruleus neurons monitored by extracellular single-unit recording and alterations in brain levels of MHPG. For example, drug-induced suppression of central noradrenergic activity produced by administration of clonidine or tricyclic antidepressants is accompanied by a reduction in MHPG. The α_2 antagonists (e.g., idazoxane, piperoxane, or yohimbine) or experimental conditions (e.g., stress or naloxone-precipitated withdrawal) that cause an increase in noradrenergic activity produce an increase in brain levels of MHPG. The magnitude of the increase in MHPG produced by naloxone-precipitated morphine withdrawal exceeds that produced by administration of supramaximal doses of α_2 antagonists such as piperoxane or yohimbine. The biochemical observation is consistent with electrophysiological studies indicating that the increase in locus ceruleus cell activity produced during naloxone-precipitated withdrawal exceeds that achieved by administration of α_2 antagonists such as piperoxane and yohimbine. Acute severe stress increases the turnover of NE in the CNS, apparently as a result of increased impulse flow in noradrenergic neurons. If impulse flow in noradrenergic neurons projecting to the cortex is acutely interrupted by destruction of the locus ceruleus, the increase in NE turnover and accumulation of MHPG-sulfate induced by stress are completely blocked. Also, microdialysis experiments have demonstrated stress-induced increases in NE release in hipocampus and frontal cortex. The stress induced activation of NE release and turnover can be blocked or attenuated with α_2 agonists such as clonidine or guanfacine.

Since MHPG levels in brain measured under controlled experimental conditions provide an index of physiological activity in the locus ceruleus, measures of this metabolite in accessible body fluids may be useful for the assessment of changes in central NE function in intact animals or patients. Clinical researchers are currently monitoring this metabolite in CSF, plasma, and urine in the hope that this will provide insight concerning alterations in central noradrenergic function.

Physiological Functions of Central Noradrenergic Neurons

Many functions have been proposed for the central NE neurons and their several sets of synaptic connections. Among the hypotheses that have the most supportive data are those concerning their role in affective psychoses (described below), learning and memory, reinforcement, sleep–wake cycle regulation, and the anxiety–nociception hypothesis. It has also been suggested that a major function of central noradrenergic neurons is not on neuronal activity and related behavioral phenomena at all but, rather, a more general role in cerebral blood flow and metabolism. However, available data fit better into a more general proposal: the main function of the locus ceruleus and its projections is to determine the brain's global orientation concerning events in the external world and within the viscera. Such a hypothesis of central noradrenergic neuron function has been generated by observations of locus ceruleus unit discharge in untreated, awake, behaviorally responsive rats and monkeys. These studies reveal the locus ceruleus units to be highly responsive to a variety of nonnoxious sensory stimuli and that the responsiveness of these units varies as a function of the animal's level of behavioral vigilance. Increased neuronal activity in the locus ceruleus is associated with unexpected sensory events in the subject's external environment, while decreased noradrenergic activity is associated with behaviors that mediate tonic vegetative behaviors. Such a global-orienting function can also incorporate other aspects of presumptive function expressed by earlier data, but none of those more discrete functions can be documented as necessary or sufficient explanations of locus ceruleus function.

Pharmacology of Adrenergic Neurons

Limited experiments have suggested that the pharmacology of central adrenergic neurons is similar to that of central noradrenergic neurons. Agents that block TH, DOPA-decarboxylase, and dopamine-β-hydroxylase lead to a reduction of both NE and E in the brain. Depleting agents such as reserpine, which cause release of NE and DA, also are effective at releasing E. MAO inhibitors cause an elevation of NE, DA, and E; but inhibitors of MAO-A are much more effective at elevating E. In fact, most of the pharmacological data are consistent with the hypothesis that, at least in the rat hypothalamus, oxidative deamination is an important metabolic process by which E is degraded and that MAO-A is predominantly involved in this degradation. Similar to observations made in central noradrenergic neurons, α_2 agonists such as clonidine decrease E formation and α_2 antagonists increase E turnover. These data are also consistent with the possibility that α_2 receptors are involved in the regulation of synthesis and release of E and perhaps in the control of the functional activity of adrenergic neurons.

Selected References

Abell, C. W. and S. W. Kwan (2000). Molecular characterization of monoamine oxidases A and B. *Prog. Nucleic Acid Res. Mol. Biol.* 65, 129–156.

Amara, S. G. (1995). Monoamine transporters: basic biology with clinical implications. *Neuroscience* 1, 259–268.

Aston-Jones, G., J. Rajkowski, and J. Cohen (1999). Role of locus coeruleus in attention and behavioral flexibility. *Biol. Psychiatry* 46, 1309–1320.

Blakely, R. D. and A. L. Bauman (2000). Biogenic amine transporters: regulation in flux. *Curr. Opin. Neurobiol.* 10, 328–336.

Bonisch, H. and M. Bruss (1993). The noradrenaline transporter of the neuronal plasma membrane. Ann. N.Y. Acad. Sci. 733, 193–202.

Bylund, D. B., D. C. Eikenberg, J. P. Hieble, S. Z. Langer, R. J. Lefkowitz, K. P. Minneman, P. B. Molinoff, R. R. Ruffolo, Jr., and U. Trendelenburg (1994). International Union of Pharmacology Nomenclature of Adrenoceptors. *Pharmacol. Rev.* 46, 121–136.

Charney, D. S., J. D. Bremner, and D. E. Redmond, Jr. (1995). Noradrenergic neuronal substrates for anxiety and fear: clinical associations based on preclinical research. In *Psychopharmacology: The Fourth Generation of Progress* (F. E. Bloom and D. J. Kupfer, eds.). Raven Press, New York, pp. 387–396.

Dahlström, A. and A. Carlsson (1986). Making visible the invisible. In *Discoveries in Pharmacology. Pharmacology Methods, Receptors and Chemotherapy*, vol. 3 (M. J. Parnham and J. Bruinvels, eds.). Elsevier, Amsterdam, pp. 97–125.

Duman, R. S. and E. J. Nestler (1995). Signal transduction pathways for catecholamine receptors. In *Psychopharmacology: The Fourth Generation of Progress* (F. E. Bloom and D. J. Kupfer, eds.). Raven Press, New York, pp. 303–320.

Foote, S. L. and G. S. Aston-Jones (1995). Pharmacology and physiology of central noradrenergic systems. In *Psychopharmacology: The Fourth Generation of Progress* (F. E. Bloom and D. J. Kupfer, eds.). Raven Press, New York, pp. 335–346.

Fuller, R. W. (1982) Pharmacology of brain epinephrine neurons. *Annu. Rev. Pharmacol. Toxicol.* 22, 31–55.

Goldstein, M. (1995). Long- and short-term regulation of tyrosine hydroxylase. In *Psychopharmacology: The Fourth Generation of Progress* (F. E. Bloom and D. J. Kupfer, eds.). Raven Press, New York, pp. 189–196.

Hökfelt, T., O. Johansson, and M. Goldstein (1984). Central catecholamine neurons as revealed by immunohistochemistry with special reference to adrenaline neurons. In *Handbook of Chemical Neuroanatomy. Classical Transmitters in the CNS*, vol. 2, part I (A. Bjorklund, and T. Hökfelt, eds.), Elsevier, Amsterdam, pp. 157–259.

Holmes, P. V. and J. N. Crawley (1995). Coexisting neurotransmitters in central noradrenergic neurons. In *Psychopharmacology: The Fourth Generation of Progress* (F. E. Bloom and D. J. Kupfer, eds.). Raven Press, New York, pp. 347–354.

Kumer, S. C. and K. E. Brana (1996). Intricate regulation of tyrosine hydroxylase activity and gene expression. *J. Neurochem.* 67, 443–462.

Lindvall, O. and A. Bjorklund (1974). The organization of the ascending catecholamine neuron systems in the rat brain as revealed by the glyoxylic acid fluorescence method. *Acta Physiol Scand.* 412, 1–47.

Lindvall, O. and A. Bjorklund (1983). Dopamine and norepinephrine containing neuron systems: their anatomy in the rat brain. In *Chemical Neuroanatomy* (P. C. Emson, ed.). Raven Press, New York.

Maas, J. W., ed. (1983). *MHPG: Basic Mechanisms and Psychopathology.* Academic Press, New York.

Mefford, I. N. (1988). Epinephrine in mammalian brain. *Prog. Neuropsychopharmacol. Biol. Psychiatry* 12, 365–388.

Melikian, H. E., J. K. McDonald, H. Gu, G. Rudnick, K. R. Moore, and R. D. Blakely (1994). Human norepinephrine transporter: biosynthetic studies using a site-directed polyclonal antibody. *J. Biol. Chem.* 269, 12290–12297.

Milligan, G., P. Svoboda, and C. M. Brown (1994). Why are there so many adrenoceptor subtypes? *Biochem. Pharmacol.* 48, 1059–1071.

Moore, R. Y. and F. E. Bloom (1979). Central catecholamine neuron systems: anatomy and physiology of the norepinephrine and epinephrine systems. *Annu. Rev. Neurosci.* 2, 113.

Nagatomo, T. and K. Koike, (2000). Recent advances in structure, binding sites with ligands and pharmacological function of beta-adrenoceptors obtained by molecular biology and molecular modeling. *Life Sci.* 66, 2419–2426.

Nagatsu, T. and H. Ichinose (1999). Regulation of pteridine-requiring enzymes by the cofactor tetrahydrobiopterin. *Mol. Neurobiol.* 19, 79–96.

Redmond, D. E., Jr. (1987). Studies of the nucleus locus coeruleus in monkeys and hypothesis for neuropsychopharmacology. In *Psychopharmacology: The Third Generation of Progress* (H. Y. Meltzer, ed.). Raven Press, New York, pp. 967–975.

Reimer, R. J., E. A. Fon, and R. H. Edward (1998). Vesicular neurotransmitter transport and the presynaptic regulation of quantal size. *Curr. Opin. Neurobiol.* 8, 405–412.

Robbins, T. W. and B. J. Everitt (1995). Central norepinephrine neurons and behavior. In *Psychopharmacology: The Fourth Generation of Progress* (F. E. Bloom and D. J. Kupfer, eds.). Raven Press, New York, pp. 363–372.

Schuldiner, S., A. Shirvan, and M. Linial (1995). Vesicular neurotransmitter transporters: from bacteria to humans. *Physiol. Rev.* 75, 369–392.

Tam, S.-Y. and R. H. Roth (1997). Commentary: Mesprefrontal dopaminergic neurons: Can tyrosine availability influence the function of the mesoprefrontal dopaminegic neurons? *Biochem. Pharmacol.* 53, 441–453.

Thony, B., G. Auerbach, and N. Blau (2000). Tetrahydrobiopterin biosynthesis, regeneration and functions. *Biochem. J.* 347, 1–16.

Valentino, R. J. and G. S. Aston-Jones (1995). Physiological and anatomical determinants of locus coeruleus discharge: behavioral and clinical implications. In *Psychopharmacology: The Fourth Generation of Progress* (F. E. Bloom and D. J. Kupfer, eds.). Raven Press, New York, pp. 373–386.

von Euler, U. S. (1956). *Noradrenaline.* Charles C. Thomas, Springfield, IL.

Yamamoto, B. K. and S. Novotney (1998). Regulation of extracellular dopamine by the norepinephrine transporter. *J. Neurochem.* 71, 274–280.

9

Dopamine

Dopamine is the most recently discovered catecholamine transmitter in the mammalian brain. Until the mid-1950s it was exclusively considered to be an intermediate in the biosynthesis of the catecholamines norepinephrine and epinephrine. Significant tissue levels of dopamine were first demonstrated in peripheral organs of ruminant species. A short time later, Montagu, Carlsson, and co-workers found that dopamine was also present in the brain in about equal concentrations to those of norepinephrine but with a quite different distribution. The very marked differences in regional distribution of the two catecholamines dopamine and norepinephrine, both within the central nervous system (CNS) and in bovine peripheral tissue, led Swedish investigators to propose a biological role for dopamine independent of its function as a precursor for norepinephrine biosynthesis. Studies demonstrating that most brain dopamine is confined to the basal ganglia led to the hypothesis that it might be involved in motor control and that decreased striatal dopamine could be the cause of extrapyramidal symptoms in Parkinson's disease. The discovery of profound depletions of dopamine in the striatum of parkinsonian patients and the demonstration that L-dihydroxyphenylalanine (L-DOPA) has beneficial effects in these patients substantiated the clinical relevance of this theory. These and other largely pharmacological studies were the impetus for the almost explosive developments in dopamine research during the past four decades. With the development of histochemical methods for the visualization of dopamine or its main synthetic enzyme in brain tissue, the anatomy of brain dopamine systems could also be described, paving the way for more direct studies on this neurotransmitter. This research ultimately culminated in the award of the Nobel Prize in 2000 to three scientists, Carlsson, Greengard, and Kandel for their seminal contributions to this field.

Dopaminergic Systems

The central dopamine-containing systems are considerably more complex in their organization than the noradrenergic and adrenergic systems. Not only are there many more dopamine cells (the number of mesencephalic dopamine cells has been estimated at about 15,000–20,000 on each side, while the number of noradrenergic neurons in the entire brain stem is reported to be about 5000 on each side) but there are also several major dopamine-containing nuclei. From anatomical studies of the dopamine systems in the 1970s (Fig. 9–1), these systems have been divided into three major categories based on the length of the efferent dopamine fibers.

1. *Ultrashort systems.* Among the ultrashort systems are the *interplexiform amacrine-like neurons*, which link the inner and outer plexiform layers of the retina, and the *periglomerular dopamine cells* of the olfactory bulb, which link together mitral cell dendrites in separated adjacent glomeruli. These neurons make extremely localized connections.
2. *Intermediate-length systems.* The intermediate-length systems include *(1)* the *tuberohypophysial dopamine cells*, which project from arcuate and periventricular nuclei into the intermediate lobe of the pituitary and into the median eminence (often referred to as the *tuberoinfundibular system*); *(2)* the *incertohypothalamic neurons*, which link the dorsal and posterior hypothalamus with the dorsal anterior hypothalamus and lateral septal nuclei; and *(3)* the *medullary periventricular group*, which includes those dopamine cells in the perimeter of the dorsal motor nucleus of the vagus nerve, the nucleus tractus solitarius, and the cells scattered in the tegmental radiation of the periaqueductal gray matter.
3. *Long systems.* The long systems are the long projections linking the ventral tegmental (A8, A10) and substantia nigra (A9) dopamine cells with three principal sets of targets: the neostriatum (principally the caudate and putamen); the limbic cortex (medial prefrontal, cingulate, and entorhinal areas); and other limbic structures (the regions of the septum, olfactory tubercle, nucleus accumbens septi, amygdaloid complex, and piriform cortex). These latter two groups have frequently been termed the *mesocortical* and *mesolimbic dopamine pro-*

jections, respectively. Under certain conditions, these limbic target systems, when compared to the nigrostriatal system, exhibit some unique pharmacological properties, which are discussed in detail in the latter portion of this chapter. When dopamine systems were first visualized in the CNS, it was thought that all dopamine cells within the zona compacta of the substantia nigra projected to the caudate putamen nucleus. Dopamine cells in the ventral tegmental area sur-

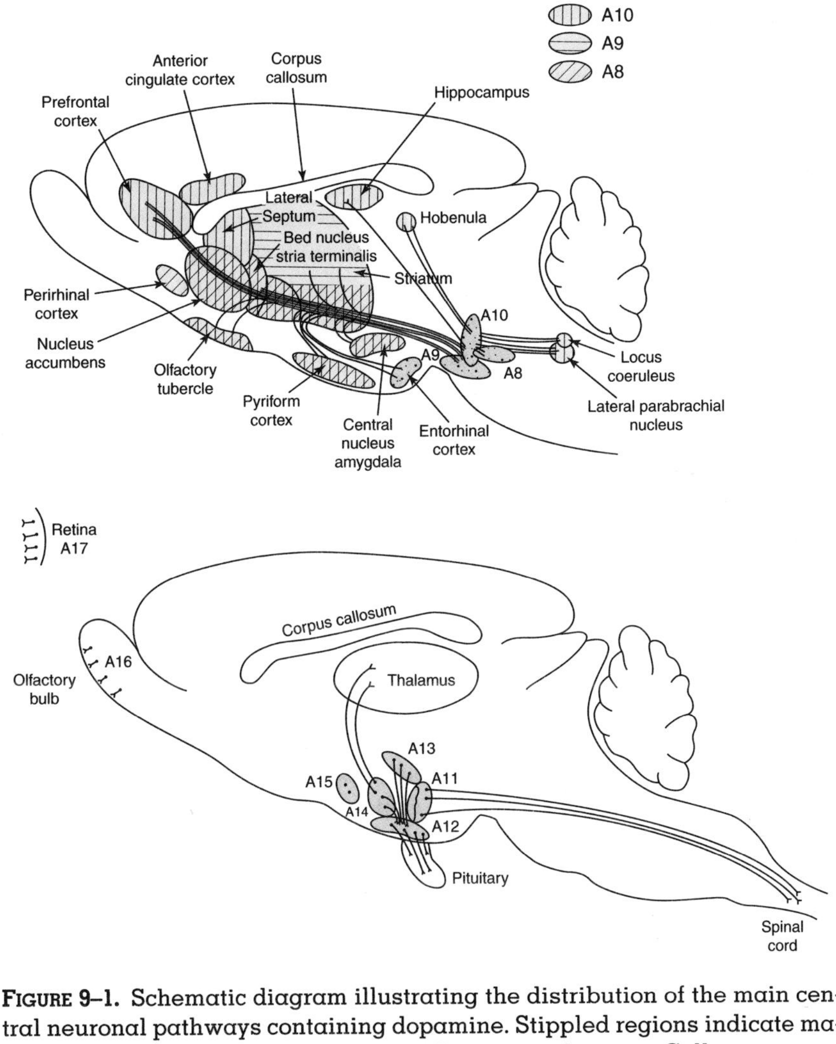

FIGURE 9–1. Schematic diagram illustrating the distribution of the main central neuronal pathways containing dopamine. Stippled regions indicate major nerve terminal areas and their cell groups of origin. Cell groups are named according to the nomenclature of Dahlström and Fuxe (1965).

rounding the interpeduncular nucleus were believed to project exclusively to parts of the limbic system. This beautiful simplicity lasted a relatively short time. Soon, dopamine inputs to the cortex were discovered; and within a few years, primarily through the use of retrograde tracing techniques, it became apparent that dopamine cells within the A8, A9, and A10 areas form an anatomically heterogeneous population in terms of their projection areas (Björklund and Lindvall, 1984). Retrograde tracing techniques also led to another important discovery, that dopamine cells project topographically to the areas that they innervate. Thus, although there is some overlap, dopamine cells that are near each other in a given area are more likely to innervate a common region than are dopamine cells distant from each other.

With regard to cellular analysis of function, the majority of reported data are from studies of the nigrostriatal dopamine projection. Here, most studies using the iontophoretic administration of dopamine indicate that the predominant qualitative response is inhibition. This effect, like that of apomorphine and cyclic adenosine monophosphate (cAMP), is potentiated by phosphodiesterase inhibitors, providing the physiological counterpart to the second-messenger hypothesis suggested by biochemical studies. However, the effects on the properties of caudate neurons when electrical stimulation is applied to the ventral tegmentum and substantia nigra are considerably less homogeneous, ranging from excitation to inhibition with considerable variations in latency and sensitivity to dopamine antagonism. In one study, neither the excitations nor the inhibitions elicited by nigral stimulation were prevented by 6-hydroxydopamine–induced destruction of the dopamine cell bodies. Unambiguous pharmacological and electrophysiological analyses of the dopamine pathway thus remain to be accomplished. Although at present the bulk of the evidence favors an inhibitory role for dopamine, inquisitive students will want to examine the arguments raised by both sides and make their own evaluations of the effects reported (Grenhoff and Johnson, 1997). Important basic information is needed to rule out the spread of current to nearby nondopamine tracts and to determine accurately the latency of conduction reported for these extremely fine unmyelinated fibers. This may be a mute point since the actions of dopamine can be better described not in terms of inhibition or excitation, but rather as related to the gating of inputs and modulation of the states of neuronal activity of postsynaptic follower cells. The modulation of the integration of information can then be further influenced at the network level via the actions of dopamine on interneurons or cell coupling. This arrangement is consistent with the behavioral actions of dopamine. Dopamine does not directly produce reward or motor activity but instead modulates inputs and adjusts the state of the organism in order to redirect the stimulus response output to achieve the most effective behavioral outcome.

It has become apparent that midbrain dopamine neurons are quite heterogeneous in terms of their biochemistry, physiology, pharmacology, and regulatory properties when compared to the prototypic nigrostriatal system, in which most of the earlier studies were performed. While midbrain dopamine neurons differ in a number of important ways, their functional organization generally reflects features of transmitter dynamics that are shared by all dopamine neurons. These features have been most thoroughly studied in the nigrostriatal pathway and are summarized below (see Fig. 9–2).

Dopamine Synthesis

Dopamine synthesis originates from tyrosine, and its rate-limiting step is the conversion of L-tyrosine to L-DOPA by the enzyme tyrosine hydroxylase (TH). DOPA is subsequently converted to dopamine by L-aromatic amino acid decarboxylase at rates so rapid that DOPA levels in the brain are negligible under normal conditions. Because endogenous levels of DOPA are normally low in the brain, the formation of dopamine can be enhanced dramatically by providing this enzyme with increased amounts of L-DOPA. Since the levels of tyrosine in the brain are relatively high and above the K_m for TH, under normal conditions it is not feasible to augment dopamine synthesis significantly by increasing brain levels of this amino acid. Endogenous mechanisms for regulating the rate of dopamine synthesis in dopamine neurons primarily involve modulation of TH activity through four major regulatory influences:

1. Dopamine and other catecholamines function as end-product inhibitors of TH by competing with a tetrahydrobiopterin (BH_4) cofactor for a binding site on the enzyme.
2. The availability of BH_4 may also play a role in regulating TH activity. Since endogenous levels of BH_4 are controlled by guanosine triphosphate (GTP) cyclohydrolase activity, this rate-limiting enzyme in BH_4 synthesis can indirectly influence tyrosine hydroxylation. TH can exist in two kinetic forms, which exhibit different affinities for BH_4. Conversion from low- to high-affinity forms is thought to involve direct phosphorylation of the enzyme, and the proportion of TH molecules in the high-affinity state appears to be a function of the state of neuronal firing.

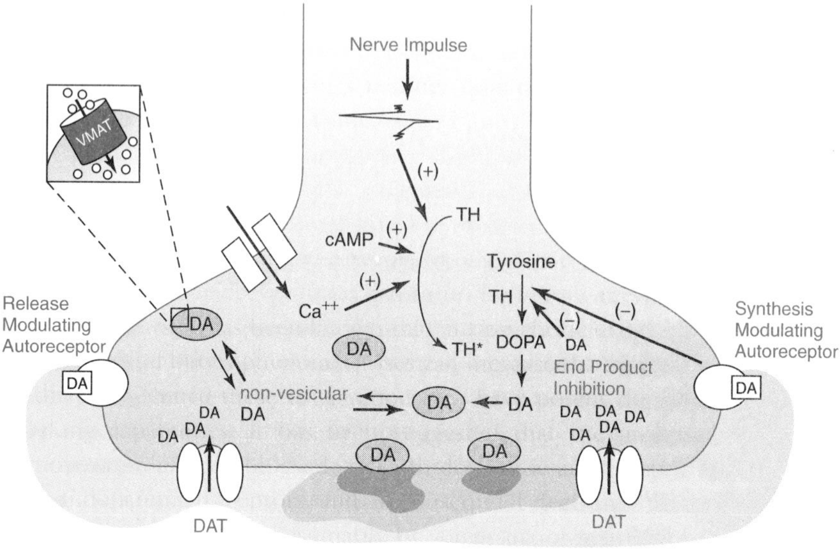

FIGURE 9–2. Schematic model of a prototypic dopaminergic nerve terminal illustrating the life cycle of dopamine (DA) and the mechanisms that modulate its synthesis, release, and storage. Invasion of the terminal by a nerve impulse results in the Ca^{2+}-dependent release of dopamine. This release process is attenuated by release-modulating autoreceptors. Increased impulse flow also stimulates tyrosine hydroxylation. This appears to involve the phosphorylation of tyrosine hydroxylase (TH), resulting in conversion to an activated form with greater affinity for tetrahydrobiopterin cofactor and reduced affinity for the end-product inhibitor dopamine. The rate of tyrosine hydroxylation can be attenuated by *(1)* activation of synthesis-modulating autoreceptors, which may function by reversing the kinetic activation of TH, and *(2)* end-product inhibition by intraneuronal dopamine, which competes with cofactor for a binding site on TH. Release- and synthesis-modulating autoreceptors may represent distinct receptor sites. Alternatively, one site may regulate both functions through distinct transduction mechanisms. The plasma membrane dopamine transporter is a unique component of the dopamine terminal that serves an important physiological role in the inactivation and recycling of dopamine release into the synaptic cleft. The vesicular monoamine transporter (VAT) transports cytoplasmic dopamine into storage vesicles, decreasing the cytoplasmic concentration of dopamine and preventing metabolism by monoamine oxidase. VAT modulates the concentration of free dopamine in the nerve terminals.

3. Presynaptic dopamine receptors also modulate the rate of tyrosine hydroxylation. These receptors are activated by dopamine released from the nerve terminal, resulting in feedback inhibition of dopamine synthesis. Autoreceptors can modulate both synthesis and release of dopamine and represent important sites for the pharmaco-

logical manipulation of dopaminergic function by dopamine agonists and antagonists.

4. Dopamine synthesis in the striatum also depends on the rate of impulse flow in the nigrostriatal pathway. During periods of increased impulse flow, the rate of tyrosine hydroxylation is increased primarily through kinetic activation of TH, which increases its affinity for BH_4 and decreases its affinity for the normal end-product inhibitor dopamine. Under conditions of increased impulse flow, tyrosine hydroxylation is also more susceptible to precursor regulation by tyrosine availability.

Calcium-dependent release of dopamine from the nerve terminal is thought to occur in response to invasion of the terminal by an action potential. The extent of dopamine release appears to be a function of the rate and pattern of firing. The burst-firing mode leads to enhanced release of dopamine. Dopamine release is also modulated by presynaptic release-modulating autoreceptors. In general, dopamine agonists inhibit while dopamine antagonists enhance the evoked release of dopamine.

Dopamine Uptake and the Dopamine Transporter

Dopamine nerve terminals possess high-affinity dopamine-uptake sites, which are important in terminating transmitter action and maintaining transmitter homeostasis. Uptake is accomplished by a membrane carrier, the dopamine transporter (DAT), which can transport dopamine into and out of the terminal depending on the existing concentration gradient. Substantial progress in the 1980s led to the development of a new class of very potent and selective dopamine-uptake inhibitors, the GBR series (Fig. 9–3). With these drugs, the stage was set for the isolation and molecular characterization of the DAT, successfully achieved in 1991 by several groups.

The DNA encoding the rat DAT exhibits high sequence similarity with the previously cloned norepinephrine and γ-aminobutyric acid transporters. The DAT is a 619–amino acid protein with 12 putative hydrophobic membrane-spanning domains and is a member of the family of Na^+/Cl^--dependent plasma membrane transporters. Using the energy provided by the Na^+ gradient generated by the Na^+/K^--transporting adenosine triphosphatase (ATPase), the DAT recaptures dopamine soon after its release, thereby modulating its concentration in the synapse and its time-dependent interaction with both pre- and postsynaptic receptors. Molecular characterization and cloning of rat, bovine, and human DATs have shown that these proteins are highly conserved between species with similar orientation in the plasma membrane and potential sites of glycosylation and phosphorylation (Fig. 9–4).

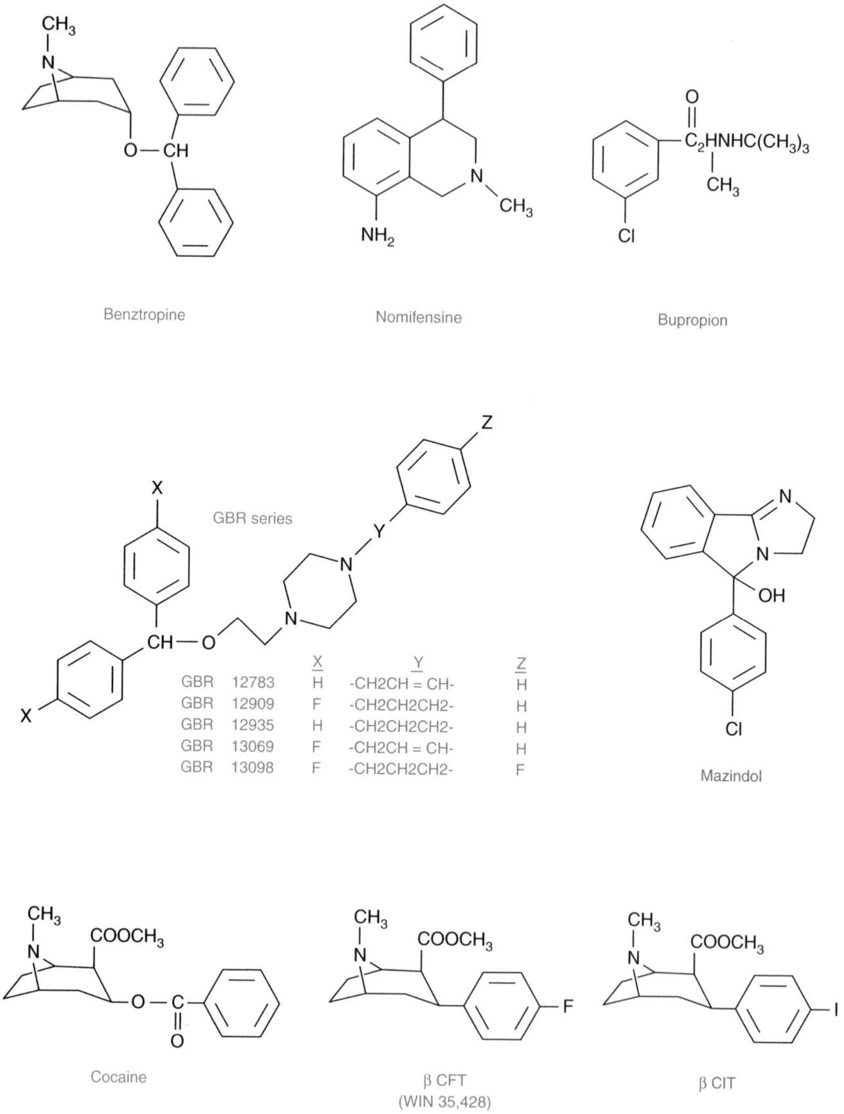

FIGURE 9–3. Chemical structures of some dopamine-uptake inhibitors.

A number of studies have suggested that DAT is a useful phenotypic marker for dopamine neurons and their nerve terminals and perhaps even better in some cases than TH. Nevertheless, caution should be exercised in the use of this DAT marker since its expression varies significantly among dopamine cell groups. The tuberoinfundibular dopamine neurons (A12, see Fig. 9–1), which release dopamine into the pituitary portal blood system, lack demonstrable DAT mRNA and protein. Because dopamine released from tuberoinfundibular neurons is carried away rapidly in the vascular system, the existence of a transporter protein on these dopamine neurons seems superfluous.

During development, embryonic midbrain dopamine neurons express dopamine and TH well before they express DAT. Although the catecholamine transporters have highly similar molecular features, they exhibit important differences in their selectivity for their catecholamines and for neurotoxins like 1-methyl-4-phenylpyridinium (MPP^+) and very distinct pharmacologies (cf. Figs. 9–3 and 8–6).

Immunohistochemical studies of the subcellular localization of these transporters led to an unexpected finding. The use of antibodies generated against DAT revealed that the transporter is typically expressed outside of the

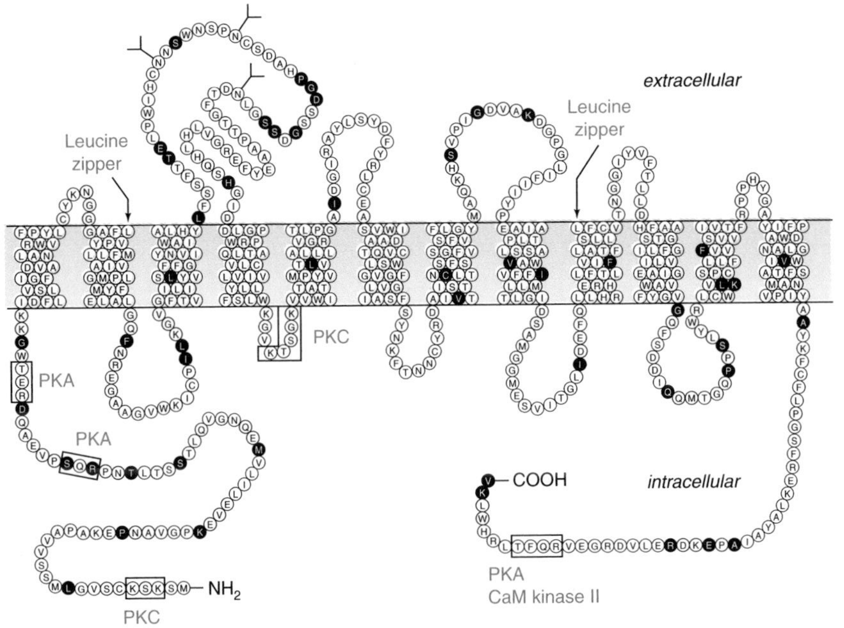

FIGURE 9–4. Schematic diagram of the primary amino acid sequence of the human dopamine transporter. Amino acid residues in black are those that differ between rat and human sequences. Y-shaped symbols represent potential sites for *N*-glycosylation. Boxed residues represent consensus phosphorylation sites for protein kinase A (PKA), protein kinase C (PKC), or Ca^{2+}–calmodulin-dependent kinase II (CaM kinase II). Disposition of the transporter protein with respect to the plasma membrane is putative. Two leucine zipper motifs are indicated. (From Giros and Caron, 1993.)

synapse, in the extrasynaptic region of the axon terminal. This suggests that the transporter may be used to inactivate (accumulate) dopamine that has escaped from the synaptic cleft and, thus, that diffusion is the initial process by which dopamine is removed from the synapse. This is consistent with recent in situ studies indicating that perisynaptic concentrations of dopamine can reach approximately 1.0 mM. Receptors for dopamine and many other transmitters are also found extrasynaptically (indeed, along the length of axons); this observation coupled with the presence of catecholamine transporters in extrasynaptic regions suggests that extrasynaptic ("paracrine" or "volume") neurotransmission may be of considerable importance for catecholaminergic signaling.

Mesolimbic dopamine neurons are implicated in the reinforcing properties of a variety of drugs of abuse, including psychomotor stimulants such as cocaine and amphetamine. Cocaine and related drugs bind to DAT and prevent dopamine transport in a fashion that correlates well with their behavioral reinforcing and psychomotor-stimulating properties. In fact, DATs have often been referred to as the brain's principal "cocaine receptors."

Receptor-binding studies have demonstrated that compounds that bind to the DAT also inhibit dopamine uptake, with a rank order of potency proportional to the affinity demonstrated in binding studies. This relationship of uptake inhibitory potency and binding potency suggests that the two processes may be intimately linked and that any compound that binds to DAT will also block dopamine transport. However, point mutation studies of the cloned DAT indicate that reuptake inhibition and binding potency may be distinct processes that, under certain conditions, are separable, not inextricably linked. These and other studies on chimeric DAT proteins have revealed that the cocaine-binding site on DAT is distinct from the substrate-recognition site. These observations suggest that it may be possible to develop agents that can prevent binding of stimulants like cocaine to the DAT while still allowing normal dopamine transport to ensue, thus supporting the feasibility of developing cocaine antagonists for the treatment of drug overdose, withdrawal, or addiction.

The DAT has also assumed importance in the study of 1-methyl-4-phenyl-1,2,3,6-tetrahydropyridine (MPTP)–induced and idiopathic Parkinson's disease. The selectivity of dopamine neurotoxins like MPTP seems to depend on their high affinity for the DAT. In primates, MPTP toxicity can be prevented by pretreatment with DAT inhibitors, but once transported into the neuron, the toxin destroys the dopamine neurons, ultimately producing parkinsonism (see Chapter 13). Expression of the cloned DAT in COS cells confers sensitivity to MPP^+ toxicity, while expression of a vesicular transporter clone confers resistance to MPP^+ in sensitive CHO cells. Thus, the levels of vesicular transporter and DAT expression in combination could conceivably dictate the response to exogenous or endogenously generated neurotoxins. Interestingly, regional differences in the levels of DAT expression appear to correlate with the extent of dopamine cell loss after MPTP treatment or in Parkinson's disease.

Dopamine Metabolism

Released dopamine is converted to dihydroxyphenylacetic acid (DOPAC) by intraneuronal monoamine oxidase (MAO) after reuptake by the nerve terminal. Released dopamine is also converted to homovanillic acid (HVA), probably at an extraneuronal site, through the sequential action of catechol-*O*-methyltransferase (COMT) and MAO. In rat brain, DOPAC is the major metabolite and considerable amounts of DOPAC and HVA are present in sulfate-conjugated as well as free forms. In humans and other primates, however, the major brain metabolite is HVA, of which only a small amount is found in the conjugated form.

The primary metabolites of dopamine in the CNS are HVA, DOPAC, and a small amount of 3-methoxytyramine (3-MT). In dopamine systems, in contrast to norepinephrine systems (see Chapter 8), the acidic rather than the neutral metabolites appear to predominate. Accumulation of HVA in the brain or cerebrospinal fluid (CSF) has often been used as an index of the functional activity of dopaminergic neurons in the brain. 3-MT is also a useful index, provided precautions are taken to minimize postmortem increases in this metabolite. Antipsychotic drugs, which increase the turnover of dopamine (in part because of their ability to increase the activity of dopaminergic neurons and to augment dopamine release), also increase the amount of HVA and 3-MT in the brain and CSF. In addition, electrical stimulation of the nigrostriatal pathway increases brain levels of HVA and 3-MT (normally quite low) as well as the release of HVA into ventricular perfusates. In Parkinson's disease, where substantia nigra dopamine neurons die, reduced HVA is observed in the CSF. Similar changes are observed in MPTP-induced parkinsonism in humans and nonhuman primates.

In rat brain, short-term accumulation of DOPAC in the striatum can be taken as an accurate reflection of activity in nigrostriatal dopaminergic neurons. Cessation of impulse flow after the placement of acute lesions in the nigrostriatal pathway leads to a rapid decrease in striatal DOPAC. Conversely, electrical stimulation of the nigrostriatal pathway results in a frequency-dependent increase in DOPAC within the striatum. Drugs that increase impulse flow in the nigrostriatal pathway, such as the antipsychotic phenothiazines and butyrophenones, anesthetics, and hypnotics, also increase striatal DOPAC. Drugs that block or decrease impulse flow, such as γ-hydroxybutyrate, (−)3-amino-1-hydroxypyrrolid-2-one [(−)HA-966], apomorphine, and amphetamine, reduce DOPAC levels. Thus, there appears to be an excellent correlation between changes in impulse flow in dopaminergic neurons, which are induced either pharmacologically or mechanically, and changes in the steady-state levels of DOPAC. In primates, DOPAC is a minor brain metabolite. Not only is it difficult to measure DOPAC in CSF but in primate brain this metabolite, in contrast to HVA, is unresponsive to drug treatments that cause large changes in dopamine metabolism.

By means of the sensitive and specific technique of gas chromatography–mass fragmentography, it has been possible to measure DOPAC, HVA, and their conjugates accurately in rat and human plasma. Studies in rats have demonstrated that stimulation of the nigrostriatal pathway and administration of antipsychotic drugs increase plasma levels of DOPAC and HVA. Several studies in humans have also indicated that dopaminergic drugs can influence plasma levels of HVA, although the effect is quite modest in comparison to that observed in rodents, limiting their clinical usefulness.

Functional Regulation

The synthesis and release of dopamine are clearly influenced by the activity of dopaminergic neurons, but these neurons behave differently from peripheral or central noradrenergic neurons; indeed, differences exist between dopamine neurons. Increased impulse flow in the nigrostriatal or mesolimbic dopamine system does lead to both an increase in dopamine synthesis and turnover and a frequency-dependent increase in the accumulation of dopamine metabolites in the striatum and olfactory tubercle. This parallels other monoamine systems, where an increase in impulse flow causes an increase in the synthesis and turnover of transmitter.

Short-term stimulation of central dopaminergic neurons increases tyrosine hydroxylation by kinetic alterations in TH, with an increased affinity for pteridine cofactor and a decreased affinity for the natural end-product inhibitor dopamine. As in central noradrenergic systems, it seems that a finite period of time is necessary for this activation to occur and that, once activated, the enzyme remains in this altered physical state for a short period after the stimulation ends. The activation appears to involve TH phosphorylation.

However, if impulse flow is interrupted in the nigrostriatal or mesolimbic dopamine system, either mechanically or pharmacologically by treatment with γ-hydroxybutyrate, the neurons respond in a rather peculiar fashion, by rapidly increasing the concentration of dopamine in the nerve terminals of the respective dopamine systems. Not only do the terminals accumulate dopamine by reducing release but there is also a dramatic increase in the rate of dopamine synthesis. This increase occurs despite the steadily increasing concentration of endogenous dopamine within the nerve terminal.

The actual mechanisms whereby a cessation of impulse flow initiates changes in the properties of TH are unclear, although they may involve a decrease in the availability of intracellular calcium. Similar changes in the activity or properties of TH are not observed in central noradrenergic neurons or in other dopamine neurons (e.g., the mesoprefrontal dopamine neurons) lacking synthesis-modulating autoreceptors. At present, the physiological significance of this paradoxical response to a cessation of impulse flow is unclear. However, it is conceivable that these neurons achieve some operational advantage by increasing their supply of transmitter rapidly during periods of quiescence.

Potential Sites of Drug Action on Dopaminergic Neurons

There are many sites at which drugs can influence the function of dopamine neurons. The potential sites for modulation are illustrated in Figure 9–5 and summarized in Table 9–1. For the purpose of this discussion, drug effects can be divided into three broad categories:

1. Nonreceptor-mediated effects on presynaptic function
2. Dopamine receptor–mediated effects
3. Effects mediated indirectly as a result of drug interaction with other neurotransmitter systems that interact with dopamine neurons

The relative importance of each of these potential sites of drug action will vary among different dopamine systems, depending on factors such as the presence or absence of autoreceptors, the efficiency of postsynaptic receptor-mediated neuronal feedback pathways, and the nature of the afferent inputs impinging on the dopamine neurons in question.

Nonreceptor-Mediated Effects

There are several stages in the life cycle of dopamine (synthesis, storage, and release) where drugs can influence transmitter dynamics, as illustrated in Figure 9–5. There are many useful pharmacological tools for modifying dopaminergic activity and manipulating dopaminergic function at these sites, but most of these agents are not very selective for dopaminergic synapses and will interact with other catecholamine (norepinephrine and epinephrine) systems (in some cases, with other monoamine [5-hydroxytryptamine] systems as well). Some drugs which interact with the plasma membrane transporter

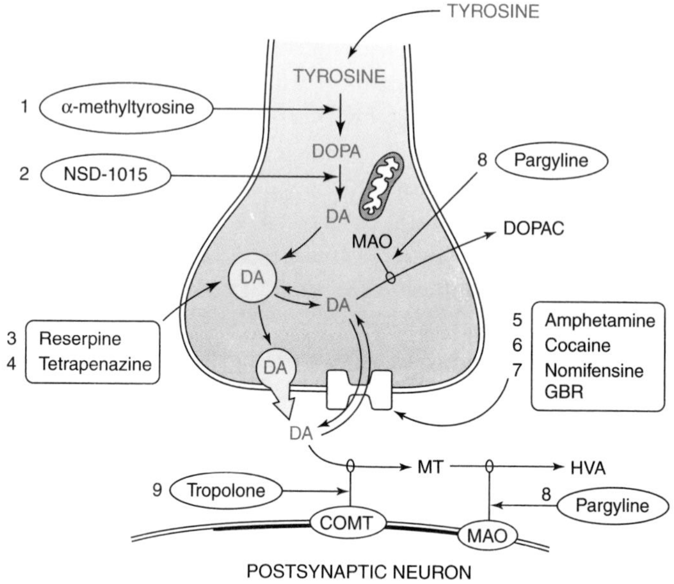

FIGURE 9–5. Schematic model of striatal dopaminergic nerve terminal. Drugs which alter the dopamine (DA) life cycle include *(1)* α-methyltyrosine, a competitive inhibitor of tyrosine hydroxylase; *(2)* NSD-1015, an inhibitor of DOPA decarboxylase; *(3)* reserpine, which irreversibly damages DA uptake/storage mechanisms and produces long-lasting depletion of DA; *(4)* tetrabenazine, which also interferes with DA uptake/storage but the effects are of shorter duration than those of reserpine and do not appear to be irreversible; *(5)* amphetamine, which increases synaptic DA through a number of mechanisms, including induction of release of DA and blockade of DA reuptake; *(6)* cocaine, which also blocks DA reuptake and induces DA release; *(7)* nomifensine and GBR, which also block DA reuptake but lack DA-releasing ability; *(8)* pargyline, an inhibitor of monoamine oxidase (MAO); *(9)* tropolone, an inhibitor of catechol-O-methyltransferase (COMT). HVA, homovanillic acid; MT, 3-methoxytyramine. DOPA, dihydroxyphenylalanine; DOPAC dihydrophenylacetic acid.

do not have a high degreee of specificity. For example, amphetamine, cocaine, benztropine, and nomifensine interact with the plasma membrane transporter that normally functions in the reuptake of released dopamine. However, these drugs also have an appreciable affinity for noradrenergic (and in some cases serotonergic) uptake sites.

Nevertheless, drugs that are highly selective for the dopamine transport complex have been developed and employed as valuable experimental tools for visualization of the integrity of dopamine systems in vivo. In fact, striking results have been obtained with several new cocaine derivatives, such as

TABLE 9–1. Potential Sites for Modulating Dopaminergic Function

Site	*Consequences*
Modulatory effects at dopamine (DA) receptors	
1. Stimulate postsynaptic DA receptors	A. Enhance dopaminergic transmission B. Enhance function of neuronal feedback loops
Block postsynaptic DA receptors	A. Block dopaminergic transmission B. Interfere with function of neuronal feedback loops
2. Stimulate presynaptic DA autoreceptors	A. Decrease DA synthesis and release B. Decrease firing rate and diminish DA output from nerve terminal
Block presynaptic DA autoreceptors	A. Increase DA synthesis and release B. Decrease somatodendritic DA turnover(?)
3. Stimulate somatodendritic DA autoreceptors	A. Interfere with feedback regulation of firing rate and DA output from terminal
Block somatodendritic DA autoreceptors	B. Interfere with feedback regulation of somatodendritic DA turnover
Modulatory effects at non-DA receptors on DA neurons	
4. Modify afferent input to cell body (i.e., block or mimic effects of transmitter released by afferent terminal)	A. Alter firing rate of DA cell and thus alter DA output from nerve terminal B. Alter somatodendritic turnover of DA (and co-localized peptides?)
5. Modify afferent input to nerve terminal	A. Alter release from nerve terminal of DA (and co-localized peptides?)

3β-(4-fluorophenyl)tropane-2β-carboxylate (CFT) and 3β-(4-iodo-phenyl) tropane-2β-carboxylate (β-CIT), which exhibit high affinity for the DAT (see Fig. 9–3). These agents have been used in autoradiographic experiments and in positron emission tomography (PET) and single-photon emission computed tomography (SPECT) studies to image the striatal DAT in both normal and parkinsonian monkeys and humans. These studies have demonstrated *(1)* loss of striatal DAT in both experimental and idiopathic Parkinson's disease and *(2)* restoration of DAT density and improvement of behavioral functions following nigral grafts in the caudate of transplanted MPTP monkeys. DAT ligands used for in vivo imaging in the future may be routinely employed in the diagnosis of certain diseases like parkinsonism and for following the progression of the disease and the response to treatment. Imaging of DAT may also prove very useful for monitoring the viability of dopamine grafts and their outgrowth once transplanted into parkinsonian recipients.

Dopamine Receptor–Mediated Effects

Originally, it was thought that all drugs that affect dopamine activity, including the neuroleptics, worked through nonreceptor-mediated mechanisms such as those described above. However, it is now clear that many therapeutically important drugs interact with dopamine receptors. Drugs that affect dopamine receptors can be classified into two groups (Table 9–2):

1. Receptors on nondopamine cell types, which are usually referred to as "postsynaptic" receptors since they are postsynaptic to a dopamine-releasing cell
2. Receptors on dopamine cells, which are referred to as "autoreceptors" to indicate their sensitivity to the neuron's own transmitter

Postsynaptic Dopamine Receptors • Postsynaptic dopamine receptors can be classified as either D_1 or D_2, based on the functional and pharmacological criteria described below. Both types of receptor have been found in the projection areas of midbrain dopamine neurons, although it is unclear whether they are located on distinct subsets of dopamine-receptive cells in various projection fields. In the striatum, postsynaptic dopamine receptors regulate the activity of neuronal feedback pathways by which striatal neurons can communicate with dopamine cell bodies in the substantia nigra. This enables dopamine-innervated cells in the striatum to modulate the physiological activity of nigrostriatal dopamine neurons. In general, increased postsynaptic receptor stimulation results in decreased nigrostriatal dopamine activity.

TABLE 9–2. Biochemistry, Physiology, and Pharmacology of D_1 and D_2 Dopamine (DA) Receptors

Biochemical Manifestations Induced by Receptor Stimulation			
D_1 Receptors (Increase in cAMP Formation, Phosphorylation of DARPP-32)		*D_2 Receptors (Decrease in cAMP Formation or No Change)*	
Location	*Function*	*Location*	*Function*
CNS: postsynaptic to DA neuron terminals dendrites (striatum, nuc. acc., olf. tub., SN, etc.)	Enabling effect on behavioral and electrophysiological effect elicited by stimulation of D_2 receptors. Function uncertain	Striatum and nuc. acc., DA nerve terminals	Autoreceptors inhibit DA synthesis and release and modulate turnover
		Retina	Mediate light-adaptive response of photoreceptors ($\uparrow$ blink)
Bovine parathyroid gland	Increases parathyroid hormone release	SN and VTA: soma dendrites	Inhibits DA cell firing
Vascular smooth muscle (canine renal and mesenteric bed most used model system)	Vascular relaxation	Striatum: cholinergic interneurons	Inhibits acetylcholine release
Vertebrate retina: in teleost, localized specifically to horizontal cells	Mediate responses of horizontal cells	Pituitary gland: anterior lobe	Inhibits cAMP and prolactin release. May also regulate Ca^{2+} channels

(*continued*)

TABLE 9–2. Biochemistry, Physiology, and Pharmacology of D_1 and D_2 Dopamine (DA) Receptors (*Continued*)

Biochemical Manifestations Induced by Receptor Stimulation			
D_1 Receptors (Increase in cAMP Formation, Phosphorylation of DARPP-32)		*D_2 Receptors (Decrease in cAMP Formation or No Change)*	
Location	*Function*	*Location*	*Function*
		Pituitary gland: intermediate lobe melanotrophs	Inhibits cAMP and αMSH release
		Chemosensitive trigger zone	Emesis
		Carotid body	Depresses spontaneous chemosensory discharge
		Sympathetic nerve terminals (numerous tissues)	Inhibits norepinephrine release
PHARMACOLOGY			
Selective agonists	SKF-38393 (partial agonist), dihydrexidine (full agonist), SKF-82526 (fenoldopam)	LY-171555 (quinpirole), RU-24926, [+]PHNO EMD-23-448 (autoreceptor-selective)	
Selective antagonists	SCH-23390, SKF-83566, SCH-39166	(−)-Sulpiride, YM-09151-2, domperidone, raclopride	

cAMP, cyclic adenosine monophosphate; DARPP-32, dopamine and cAMP–regulated phosphoprotein of 32 kDa; CNS, central nervous system; nuc. acc., nucleus accumbens; olf. tub., olfactory tubercle; SN, substantia nigra; VTA, ventral tegmental area; MSH, melanocyte-stimulating hormone.

Following chronic exposure to dopamine agonists or antagonists, postsynaptic dopamine receptors exhibit adaptive changes. For example, chronic exposure to dopamine antagonists or chemical denervation with 6-hydroxydopamine (6-OHDA) produces an increase in the number of dopamine-binding sites measured in receptor-binding assays. This may be related to the behavioral supersensitivity to dopamine agonists that also develops as a result of chronic antagonist administration or denervation. Conversely, repeated administration of dopamine agonists decreases the number of dopamine-binding sites and produces subsensitivity to subsequent administration of dopamine agonists in behavioral as well as biochemical and electrophysiological models. Changes such as these may be relevant to understanding the state of dopamine receptors in diseases believed to involve chronic dopaminergic hyper- or hypoactivity.

Autoreceptors • Autoreceptors can exist on most portions of dopamine cells, including the soma, dendrites, and nerve terminals. Stimulation of dopamine autoreceptors in the somatodendritic region slows the firing rate of dopamine neurons, while stimulation of autoreceptors on the nerve terminals inhibits dopamine synthesis and release. Thus, somatodendritic and nerve terminal autoreceptors work in concept to exert feedback on dopaminergic transmission. Both somatodendritic and nerve terminal autoreceptors can be classified as D_2 receptors and exhibit similar pharmacological properties. Like postsynaptic receptors, somatodendritic and nerve terminal autoreceptors develop supersensitivity after chronic antagonist treatment or prolonged decreases in dopamine release and desensitize in response to repeated administration of dopamine agonists. Interestingly, the autoreceptors are more readily desensitized than postsynaptic dopamine receptors. This has been suggested to play a role in the "on–off" effects observed during chronic L-DOPA therapy in Parkinson's disease.

Dopamine autoreceptors can be defined functionally in terms of the events they regulate and are therefore divided into three categories: synthesis-modulating, release-modulating, and impulse-modulating autoreceptors. However, it is not yet known whether distinct receptor proteins modulate each of these functions or whether the same receptor protein is coupled to each function through distinct transduction mechanisms. It is clear, however, that autoreceptor-mediated pathways for the regulation of dopamine release from the nerve terminal are distinct from autoreceptor-mediated pathways for the regulation of dopamine synthesis since dopamine terminals in the prefrontal and cingulate cortices possess autoreceptors that regulate release but lack functional synthesis-modulating autoreceptors.

Autoreceptors Versus Postsynaptic Receptors: Pharmacological and Functional Considerations • Autoreceptors and postsynaptic dopamine receptors differ in several ways. The most clear-cut difference is that autoreceptors are 5–10 times more sensitive to the effects of dopamine and apomorphine than postsynaptic dopamine receptors in behavioral, biochemical, and electrophysiological models. In the low-dose range, therefore, autoreceptor-mediated effects of dopamine agonists predominate, resulting in diminished dopaminergic function, while higher doses also stimulate postsynaptic receptors, leading to enhanced dopaminergic function.

Autoreceptors also differ from postsynaptic receptors in their pharmacological profile. Dopamine agonists that are relatively selective for autoreceptors have been synthesized. As would be predicted, autoreceptor-selective agonists inhibit dopamine release, synthesis, and impulse flow in dopamine neurons and elicit behavioral responses associated with diminished dopaminergic function. These agonists are very useful experimental probes for studying dopamine receptor function and may prove useful in diseases thought to involve excessive dopaminergic activity. Dopamine antagonists that selectively block autoreceptors have also been synthesized. By blocking dopamine autoreceptors, they enhance dopamine function. Some of these agents appear to have a built-in ceiling on their response since as the dose is increased,

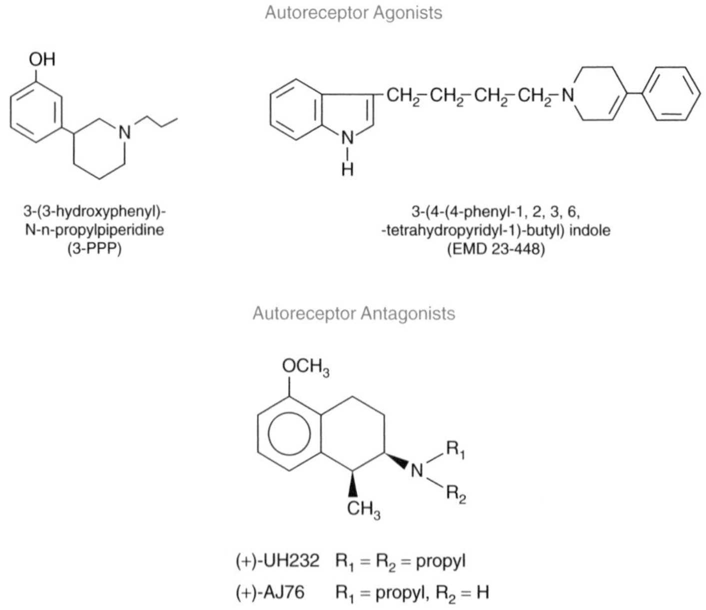

FIGURE 9–6. Structures of some selective dopamine autoreceptor ligands.

they exert an antagonistic action on postsynaptic dopamine receptors. The structures of several of these autoreceptor-selective agents are illustrated in Figure 9–6.

Dopamine agonists and antagonists may act on several types of dopamine receptor to elicit biochemical changes in dopamine metabolism and alter the functional output of dopaminergic systems (Table 9–2). A drug's net effect on dopaminergic activity will depend on both its pre- and postsynaptic effects and the selectivity with which it acts at these different sites.

Drug Interactions at D_1 and D_2 Receptors

In the preceding discussion, dopamine receptors were broadly divided into presynaptic and postsynaptic categories. A second popular dopamine receptor classification that received increasing attention in the 1980s is based on the presence or absence of positive coupling between the receptor and adenylate cyclase activity. On the basis of biochemical, physiological, and pharmacological studies, it is now well established that dopamine can act on at least two types of brain receptor, termed D_1 and D_2 receptors (Table 9–3). These two classes of receptors are clearly distinguished by their biochemical characteristics. D_1 receptors mediate the dopamine-stimulated increase of adenylate cyclase activity. D_2 receptors are thought to mediate effects that are independent of D_1-mediated effects and to exert an opposing influence on adenylate cyclase activity. The D_2 site is further characterized by picomolar affinity for antagonist, while the D_1 site is characterized by millimolar affinity for antagonist. The arrangements of D_1 and D_2 receptors and the nerve terminal autoreceptor are diagrammed in Figure 9–7.

With the development of D_1- and D_2-selective agonists and antagonists, however, it has become common to rely on pharmacological characteristics

Table 9–3. Signal-Transduction Mechanisms Associated with Dopamine Receptors

D_2 Receptors	*D_1 Receptors*
Inhibition of adenylate cyclase	Stimulation of adenylate cyclase
Inhibition of Ca^{2+} entry through voltage-sensitive Ca^{2+} channels	Stimulation of phosphoinositide turnover
Enhancement of K^+ conductance	
Modulation of phosphoinositide metabolism	

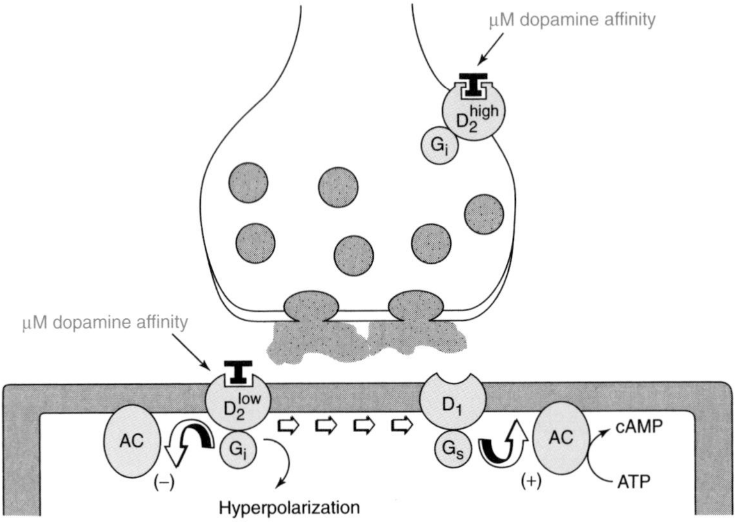

FIGURE 9–7. Schematic diagram depicting the anatomical arrangement of D_1 and D_2 receptors and the nerve terminal autoreceptor. The D_1 site is positively coupled with adenylate cyclase via a G_s protein. The D_2 postsynaptic site is negatively coupled with adenylate cyclase via a G_i protein, stimulation of which leads to hyperpolarization. The autoreceptors appear to exert their dopamine synthesis- and release-modulating effects via a G_i regulatory protein. Dopamine autoreceptors are sensitive to very low (nanomolar) concentrations of dopamine. Although it is not known whether the D_1 and D_2 receptors are situated on the same neuron, they appear to be functionally linked, as depicted by the arrows. Innervated postjunctional dopamine receptors operate in the low-affinity state (D_1 low and D_2 low) in contrast to the D_2 autoreceptor, which operates in the high-affinity state (D_2 high).

when determining whether an effect is mediated by D_1 or D_2 receptors. The distinction between pharmacological and functional definitions is important because it is becoming clear that dopamine receptors with the same pharmacological characteristics do not necessarily have the same functional characteristics. For example, dopamine receptors with D_2 pharmacology are present in both the striatum and nucleus accumbens but are coupled to inhibition of adenylate cyclase only in the striatum. Regional differences in coupling between dopamine receptors and GTP-binding proteins have also been reported. Furthermore, dopamine receptors can influence cellular function through mechanisms other than stimulating or inhibiting adenylate cyclase (Table 9–3). These may include direct effects on potassium and calcium channels as well as modulation of inositol phosphate production. It therefore seems unrealistic to expect equivalence between pharmacological and biochemical classifications of dopamine receptor subtypes.

Dopamine Signaling

Although D_1 and D_2 receptors can have opposite effects on adenylate cyclase activity, it is apparent that the physiological significance of their interaction is more complex. While D_1 and D_2 agonists can have opposite effects on oral movements, they also produce a synergistic increase in locomotor activity and behavioral stereotypes under certain circumstances. Electrophysiological experiments have suggested that D_1 receptor activation is required for full postsynaptic expression of D_2 effects. The interaction of the D_1 receptor with other neurotransmitter systems is still being explored. Despite widespread interest in dopamine signaling, little was known until recently about the molecular and cellular basis for the action of dopamine on its target cells. A phosphoprotein named DARPP-32 (dopamine and cAMP-regulated phosphoprotein of 32 kDa) plays a key role in the biology of dopaminoceptive neurons (Fig. 9–8). By acting on the D_1 receptors, dopamine stimulates adenylyl cyclase via a G protein to increase cAMP formation and the activity of cAMP-dependent protein kinase (protein kinase A, PKA), leading to phosphorylation of DARPP-32 on a single threonine residue. Phosphorylation converts this phosphoprotein into a potent inhibitor of protein phosphatase-1. At least two intracellular pathways that decrease DARPP-32 phosphorylation are involved in the modulation of dopamine signaling via D_2 receptors. One mechanism involves inhibition of adenylyl cyclase, a decrease in cAMP, a decrease in the activity of PKA, and a decrease in DARPP-32 phosphorylation. The other D_2-mediated effect involves an increase in intracellular calcium and activation of calcineurin. One of the actions of calcineurin is to dephosphorylate DARPP-32 and thus relieve the inhibition of protein phosphatase-1. Striatonigral neurons receive glutamate input from the cerebral cortex as well as a rich dopamine innervation from the substantia nigra. In these neurons, glutamate acting on *N*-methyl-D-aspartate (NMDA) receptors also gives rise to a large influx of calcium (Fig. 9–8). Thus, in this system, stimulation of NMDA receptors also results in activation of calcineurin and enhanced dephosphorylation of DARPP-32, producing an effect very similar to the stimulation of D_2 receptors.

Dynamics of Dopamine Receptors

Destruction of the nigrostriatal dopamine systems has clear and reproducible behavioral consequences. Unilateral lesions of this system produce rotational behavior. Behavioral studies in lesioned rats indicate that dopamine receptors in the denervated striatum are supersensitive. Administration of dopamine agonists (e.g., apomorphine) that selectively stimulate dopamine receptors produces rotational behavior in rats with unilateral lesions of the

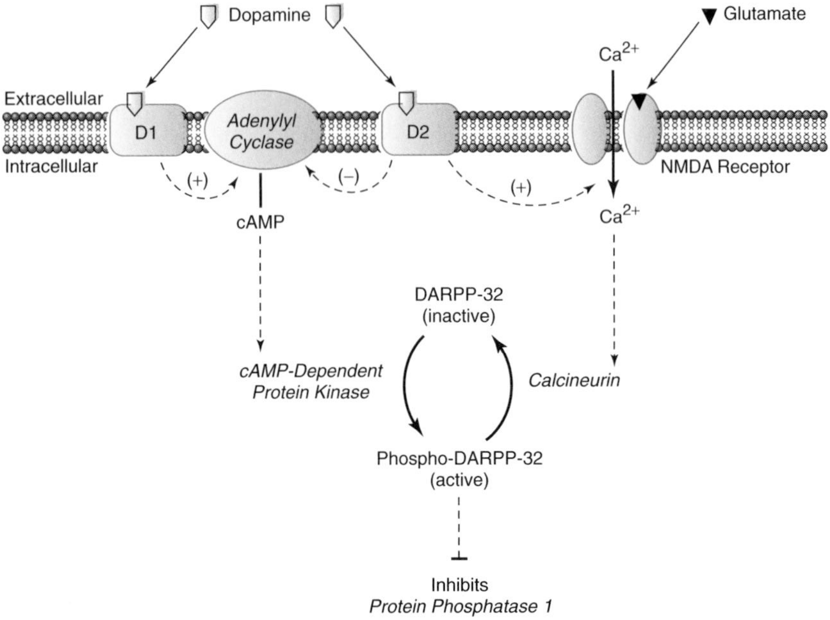

FIGURE 9–8. Postulated pathways by which dopamine and glutamate may regulate dopamine and cAMP–regulated phosphoprotein of 32 kDa (DARPP-32) phosphorylation. Dopamine, by acting on the D_1 class of receptors, stimulates adenylyl cyclase via a G protein to increase cyclic adenosine monophosphate (cAMP) formation and the activity of cAMP-dependent protein kinase (protein kinase A, PKA), leading to phosphorylation of DARPP-32. Phosphorylation converts this phosphoprotein into a potent inhibitor of protein phosphatase-1. Inhibition of protein phosphatase-1 increases the phosphorylation state of numerous phosphoproteins involved in the regulation of important physiological processes. Dopamine, via D_2 receptors, decreases DARPP-32 phosphorylation by two intracellular signaling pathways. One mechanism involves inhibition of adenylyl cyclase, a decrease in cAMP, a decrease in the activity of PKA, and decreased phosphorylation of DARPP-32. The other D_2-mediated effect involves an increase in intracellular calcium, activation of calcineurin, and increased dephosphorylation of phospho-DARPP-32. Glutamate acting on *N*-methyl-D-aspartate (NMDA) receptors also gives rise to a large influx of calcium. Thus, stimulation of NMDA receptors can also lead to activation of calcineurin and enhanced dephosphorylation of DARPP-32, producing an effect very similar to stimulation of D_2 receptors.

nigrostriatal dopamine neurons. The degree of receptor sensitivity can be quantified by measuring the amount of rotational behavior. The number of dopamine receptors in the striatum ipsilateral to the lesion increases markedly, and this increase appears to correlate with the extent of the behavioral supersensitivity reflected by the rotational behavior. Thus, an increase in dopamine receptor density appears to be related to the behavioral supersensitivity observed following unilateral destruction of the nigrostriatal dopamine system.

Changes in the number of dopamine receptors are also observed in the striatum following chronic administration of dopamine antagonists. This led to the speculation that serious side effects, such as tardive dyskinesia, following chronic treatment with a neuroleptic drug might be due to supersensitivity of dopamine receptors that have been chronically blocked.

Dopamine receptors have been observed to change in disease states. In schizophrenia, the density of the DAT and of the D_1 dopamine receptor is normal. However, the D_2 receptor density is consistently elevated in postmortem studies of brain regions such as caudate and putamen, even in tissue obtained from neuroleptic-free individuals. Some preliminary evidence indicating abnormal D_2 structure as well as reduced linkage between D_1 and D_2 receptors is available, warranting a detailed study of the genes for these two receptors in schizophrenia. Loss of midbrain dopamine in Parkinson's disease is accompanied by a matching loss of the DAT and a rise in density of both D_1 and D_2 receptors. These alterations are found in the caudate nucleus and putamen tissues from unmedicated patients. Long-term treatment with L-DOPA appears to revert the receptor densities back toward normal levels. D_1 and D_2 receptors are decreased in the striatum of patients with Huntington's chorea, and there appears to be reduced or absent linkage between them.

Molecular Biology of Dopamine Receptors

D_1 and D_2 receptors are distinct molecular entities, utilize different transducing units (Table 9–3), and have a different distribution in the brain (Table 9–2).

Developments in molecular biology, including cloning of the cDNA and/or genes for several members of the large family of G protein–coupled receptors, have revealed that heterogeneity in the biochemical characteristics or pharmacology of individual receptors often indicates the presence of previously unsuspected molecular subtypes (see Chapter 11). For dopamine systems, even though the D_1/D_2 receptor classification is widely accepted, biochemical, pharmacological, and behavioral approaches have produced data that are increasingly difficult to reconcile with the existence of only two dopamine receptor subtypes and suggested the presence of several novel subtypes of both receptor types. Cloning studies have already identified four subtypes of the D_2 receptor and two subtypes of the D_1 receptor (Table 9–4). Two forms of the D_2 receptor, $D_{2(short)}$ and $D_{2(long)}$, were identified by gene cloning and shown to be derived from alternative splicing of a common gene. These two subtypes appear to have an identical pharmacology. A third sub-

TABLE 9–4. Comparison of Dopamine Receptor Subtypes

Receptor Isoforms	D_1	$D_{2(short)}$	$D_{2(long)}$	D_3	D_4	D_5
Chromosome	5q35.1	11q22–23		3q13.3	11p15.5	4p15–16
Brain regions enriched	C/P OT NA	C/P OT NA	C/P OT NA	OT NA IC	FCX Midbrain AMG	TH Hi Hyp
Posterior pituitary	Absent	Present	Present	Absent	?	Absent
Nigral dopamine cells	No	Yes	Yes	Yes	No	No
GTP regulation	Yes	Yes	Yes	Yes	Yes	No
Adenylylate cyclase	Stimulates	Inhibits	Inhibits	Inhibits	Inhibits	Stimulates
Affinity for dopamine	Micromolar	Micromolar	Micromolar	Nanomolar	Submicromolar	Submicromolar
Characteristic agonist	SKF-38393	Bromocriptine	Bromocriptine	7-OH-DPAT	CP-226, 269 PD-106, 077	SKF-38393
Characteristic antagonist	SCH-23390	Sulpiride	Sulpiride	UH-232	Clozapine, NGD-94-1	SCH-23390
Amino acids						
Rat	446	415	444	446	385	475
Human	446	414	443	400	387	477
Amino acid sequence homology in transmembrane vs. $D_{2(long)}$	44%	100%	100%	75%	53%	47%

C/P, caudate/putamen; OT, olfactory tubercle; NA, nucleus accumbens; IC, islands of Calleja; FCX, frontal cortex; AMG, amygdala; TH, thalamus; Hi, hippocampus; Hyp, hypothalamus; GTP, guanosine triphosphate.

type of the D_2 receptor, the D_3 receptor, was isolated by screening rat brain cDNA and genomic libraries by reverse transcription-polymerase chain reaction. This new D_3 receptor exhibits several novel characteristics. It has a different anatomical distribution, with the highest levels found in limbic brain structures, and its pharmacological profile, although similar to the $D_{2(short)}$ and $D_{2(long)}$ forms, shows some distinct differences; the D_3 receptor exhibits about a 100-fold increase in affinity for the dopamine agonist quinpirol.

The fourth subtype of the D_2 receptor, cloned in 1991, the D_4 receptor gene, has high homology to the human D_2 and D_3 receptor genes. The pharmacological profile of this receptor resembles that of the D_2 and D_3 receptors, but its affinity for the atypical antipsychotic drug clozapine is an order of magnitude higher. The D_4 RNA has an interesting regional distribution in monkey brain, with high levels observed in the frontal cortex, midbrain, amygdala, and medulla and lower levels detected in the basal ganglia. The function of these D_2 receptor subtypes is presently unknown. All known varieties of the D_2 receptor have seven membrane-spanning domains, similar to the structure originally proposed for β-adrenergic receptors. Differences in ligand-binding and transduction mechanisms are presumably related to variations in the sequence of the receptor. The D_1 receptor of humans and rats has also been cloned, expressed, and characterized by several laboratories; this work in conjunction with other studies is consistent with the idea that other D_1 receptor subtypes may also exist. In fact, a gene encoding a 477–amino acid protein has been cloned that has a striking homology to the cloned D_1 receptor. This D_1 receptor subtype, called D_5, has a pharmacological profile similar to that of the cloned D_1 receptor but displays a 10-fold higher affinity for the endogenous agonist dopamine. Similar to the D_1 receptor, the D_5 receptor stimulates adenylate cyclase activity. This receptor is neuron-specific and located primarily in the limbic areas of the brain but is absent from the parathyroid, kidney, and adrenal gland.

Distribution of Subtype-Specific Dopamine Receptor mRNA in Brain

Advances in molecular biology have made it feasible to determine which specific cells express a given gene, thus allowing the anatomical determination not only of what population of cells expresses a given gene but also of how these genes may be regulated in normal and pathological conditions. At least five genes encoding dopamine receptors have been discovered. Currently, there are few pharmacological or immunological tools for accurately measuring the distribution of the receptor proteins of the new dopamine receptors. Thus, our knowledge concerning the tissue distribution of these receptors, especially in cases where very specific ligands or selective antibodies are just beginning to be developed, has come primarily from in situ hybridization experiments. The tissue distribution and characteristics of the mRNAs of the five different dopamine receptors are illustrated in Tables 9–4 and 9–5. These dopamine receptors appear to have overlapping as well as some unique anatomical distributions and, in some cases, distinct pharmacological profiles. In general, the distribution patterns found in rodents parallel those observed in primates, with several notable exceptions alluded to below. The D_1 and D_2 receptor mRNAs are present in all dopaminoceptive regions of the rat brain. In brain regions such as the substantia nigra and ventral tegmental area, high levels of D_2, but not D_1, mRNA are detected. The absence of D_1 and D_5 receptor mRNA in the substantia nigra and ventral tegmental area argues against these receptors playing a role as autoreceptors. Receptor mRNAs for D_3, D_4, and D_5 are largely present in tissues where D_1 and/or D_2 receptor mRNAs are also expressed. However, in most cases, the relative abundance of mRNA for these receptors is one or several orders of magnitude lower than that found for D_1 and/or D_2 receptors. While the primate substantia

TABLE 9–5. Neuroanatomical Distribution of Dopamine Receptor mRNAs[a]

	D_1	D_2	D_3	D_4	D_5
Dopamine cell body regions (autoreceptors)					
Substantia nigra	−	+++	+	?+	−
Ventral tegmental area	−	+++	+	?+	−
Zona incerta	−	+++	−	−	−
Dopaminoceptive regions (postsynaptic receptors)					
Caudate putamen	+++	+++	+	−	−
Nucleus accumbens	+++	+++	+++	+	−
Septum	+	+	+	−	−
Olfactory tubercle	+++	+++	+	−	−
Amygdala	+++	+	+	+	−
Hippocampus	+	+	+	+	++
Cortex	+	+	+	+	−
Hypothalamus	+	+	+	+	+
Thalamus	+	+	+	+	++
Cerebellum	+	+	+	−	−

[a]Relative abundance of given mRNA in rat brain: +++, abundant expression; ++, moderate level of expression; +, low level of expression; −, no mRNA observed. (Data summary from Meador-Woodruff, 1994).

nigra contains high levels of D_2 receptor mRNA, the ventral tegmental area in the primate brain does not contain readily measurable levels of D_2 or D_3 receptor mRNA, suggesting that the primate ventral tegmental area may not contain appreciable numbers of dopamine autoreceptors.

Anatomically, D_5 receptor mRNA has a rather discrete distribution in rat brain: it is found only in the hypothalamus, hippocampus, and parafascicular nucleus of the thalamus. In the primate, this distribution extends to other temporal lobe structures as well. The consensus from a number of studies is that the only brain region that expresses all five dopamine receptors is the hippocampus. Dopamine receptor mRNA has also been found outside the CNS. D_2 receptor mRNA is abundant in the pituitary and adrenal glands and in retina. Northern blot analyses have shown that neither D_1 nor D_3 receptor mRNA is detected outside the CNS, although in kidney and heart D_1- and D_2-like activities have been described. Since low levels of D_5 receptor mRNA are expressed in the kidney, this could account for the D_1-like activity. The D_4 receptor mRNA found in rat heart may account for the previously described D_2-like activity. The level of D_4 receptor mRNA found in rat brain is about 20-fold lower than the level in heart. In peripherally innervated tissue, heart seems to be the exception since no D_4 mRNA has been found in adrenal, kidney, or liver.

Pharmacology of Dopamine Receptor Subtypes

At present, no selective ligands have been developed that can distinguish between D_1 and D_5 receptors. Pharmacologically, the only characteristic that distinguishes D_1 from D_5 receptors is the increased affinity of the D_5 receptor for dopamine. Most neuroleptic drugs exhibit a higher affinity for D_2 receptors than for D_3 or D_4 receptors. The affinities of the five dopamine receptor subtypes for selected clinically relevant dopamine antagonists are summarized in Table 9–6.

Uniquely among the subtypes, D_4 receptors respond to low concentrations of norepinephrine and epinephrine as well as to dopamine. The most interesting feature of the human D_4 receptor is its apparent high affinity for clozapine (an atypical neuroleptic) and its unique distribution in primate brain (frontal cortex > midbrain > amygdala > striatum), differing markedly from D_2 and D_3 receptor mRNA. This interesting pharmacology and unique distribution in brain has generated a great deal of excitement, particularly from a clinical standpoint. The possibility that clozapine exerts its therapeutic effects via a D_4 receptor mechanism was seen immediately as offering a new and rational target for drug development. Seeman and colleagues provided tantalizing evidence that there may be an increase in the number of D_4 re-

TABLE 9–6. Affinities of Clinically Relevant Antagonists for the Five Dopamine Receptors

	K_I values (nM)				
	D_1-like		D_2-like		
Ligand	D_1	D_5	D_2	D_3	D_4
Antagonists					
Chlorpromazine	90	130	3	4	35
Clozapine	170	330	230	170	21
Haloperidol	80	100	1.2	7	2.3
Nemonapride			0.06	0.3	0.15
Raclopride	18,000		1.8	3.5	2400
Remoxipride			300	1600	2800
Risperidone			5	6.7	7
SCH-23390	0.2	0.3	1100	800	3000
Spiperone	350	3500	0.06	0.6	0.08
S-Sulpiride	45,000	77,000	15	13	1000

Dissociative constants (K_i) for ligands at the various dopamine receptors were derived from Seemen and Van Tol (1994).

ceptors in schizophrenic patients, further fueling the impetus to develop D_4-selective antagonists as potential antipsychotic drugs; but this provocative, albeit indirect, study has not been replicated using more direct measures to assess D_4 receptor numbers in brains of normal and schizophrenic subjects. Also, the significance of this finding, even if replicated, would still be uncertain. The neuroleptics taken by patients throughout the course of their disease could modify dopamine receptor density, and the overabundance of D_4 receptors observed in the autopsied brains of schizophrenic patients could be a result of drug treatment rather than a cause of the disease. Future clinical research efforts might profitably be directed to the use of in vivo imaging techniques (i.e., SPECT and PET) to evaluate dopamine receptor subtypes in schizophrenia when appropriate D_4- and D_3-selective ligands become available. However, the low density of these receptors, especially D_4 receptors, may present an insurmountable obstacle.

The identification of novel dopamine receptor subtypes has already had a dramatic impact on our understanding of dopaminergic systems. Studies of the human D_4 receptor indicate that its DNA sequence is highly polymorphic at both the DNA and amino acid levels, exhibiting a least 25 alleles. A novel polymorphism of the D_4 receptor was observed within the putative third cytoplasmic loop of the protein, suggesting that some polymorphic variants may display different pharmacological properties. This high frequency of variation in the coding region of a functional receptor protein is unprecedented and could confer differences in efficacy of drug treatment and/or predispose an individual to the development of dopamine-dependent neuropsychiatric disorders. In fact, the D_4 receptor gene (*DRD4*) has been implicated in the pathophysiology of several common neuropsychiatric disorders, including mood disorders, attention-deficit/hyperactivity disorder, Parkinson's disease, and specific personality traits. The evidence is particularly strong for attention-deficit/hyperactivity disorder.

The availability of receptor clones, receptor antibodies, and expressed receptor proteins has permitted gene mapping as well as in-depth studies of the circuitry of the dopaminergic systems and the mechanisms regulating them at both the genomic and cytoplasmic levels. It has also allowed the physical structure of the receptors to be ascertained and should permit the design and development of highly specific ligands. It is hoped that these new selective agents will be helpful not only in studying the function of dopamine systems in normal and pathological states but also in the therapeutic management of disorders associated with malfunction of specific dopaminergic systems. Strides toward this goal had already begun with the successful development of selective D_4 antagonists by several pharmaceutical companies when the last edition of this text was completed. Some 5 years later, these D_4 antagonists have not yet provided the magic bullet for the treatment of schizophrenia. However, these agents have been useful in studying the localization and function of D_4 receptors in animals, including monkeys and humans, and may someday find a therapeutic use in the treatment of specific dopamine-dysregulated states.

Pharmacology of Dopaminergic Systems

Nigrostriatal and Mesolimbic Dopamine Systems

The nigrostriatal and mesolimbic dopamine neurons appear to respond in a similar manner to drug administration (Table 9–6). Acute administration of dopamine agonists (dopamine receptor stimulators) decrease dopamine cell activity, turnover, and catabolism. Acute administration of antipsychotic drugs (dopamine receptor blockers) increases dopaminergic cell activity, turnover, catabolism, and biosynthesis. The increase in dopamine biosynthesis occurs at the TH step and is in part a result of the ability of antipsychotic drugs to block postsynaptic receptors and to increase dopaminergic activity via a neuronal feedback mechanism (Table 9–1). Also, some of the observed effects are enhanced as a result of interaction with nerve terminal autoreceptors. Blockade of nerve terminal autoreceptors increases both the synthesis and the release of dopamine. These systems respond to MAO inhibitors (MAOIs) with an increase in dopamine and a decrease in dopamine synthesis, as do the other dopamine systems discussed below.

Long-term treatment with antipsychotic drugs produces a different spectrum of effects on central dopaminergic neurons. For example, following long-term treatment with haloperidol, nigrostriatal dopamine neurons in the rat become quiescent and dopamine metabolite levels and dopamine synthesis and turnover in the striatum return to normal limits. The kinetic activation of striatal TH, which occurs following an acute dose of an antipsychotic drug, also subsides following long-term treatment. These results are usually interpreted as indicative of the development of tolerance in the nigrostriatal dopamine system. In contrast (see below), tolerance to the biochemical effects observed following acute administration of antipsychotic drugs does not appear to develop in the mesoprefrontal and mesocingulate cortical dopamine pathways after chronic administration.

Mesocortical Dopamine System

The response of the mesocortical dopamine systems to dopaminergic drugs in most instances is qualitatively similar to that of the nigrostriatal and mesolimbic systems (Table 9–7), although some notable exceptions have been observed. The mesotelencephalic dopamine neurons, which were once believed to be three relatively simple and homogeneous systems, have been found to be an anatomically, biochemically, and electrophysiologically heterogeneous population of cells with differing pharmacological responsiveness. For example, although a great majority of midbrain dopamine neurons appear to possess autoreceptors on their cell bodies, dendrites, and nerve terminals, dopamine cells that project to the prefrontal and cingulate cortices appear either to have a greatly diminished number of these receptors or to lack them entirely. The absence (or insensitivity) of impulse-regulating somatodendritic as well as synthesis-modulating nerve terminal autoreceptors on the mesoprefrontal and mesocingulate cortical dopamine neurons may, in part, explain some of the unique biochemical, physiological, and pharmacological properties of these two subpopulations of midbrain dopamine neurons (Table 9–8). For example, the mesoprefrontal and mesocingulate dopamine neurons appear to have a faster firing rate and a more rapid turnover of transmitter than the nigrostriatal, mesolimbic, and mesopiriform dopamine neurons. Transmitter synthesis is also more readily influenced by altered availability of precursor tyrosine in midbrain dopamine neurons lack-

TABLE 9–7. Pharmacology of Central Dopaminergic Systems

Characteristics	*Nigro-striatal*	*Meso-accumbal*	*Meso-prefrontal*	*Meso-piriform*	*Tubero-infundibular*[a]	*Tubero-hypophyseal*[b]
Respond to DA antagonist (increase in synthesis, catabolism, and turnover)	Yes	Yes	Yes (small)	Yes	No	Yes (small)
Respond to DA agonists (decrease in synthesis, catabolism, and turnover)	Yes	Yes	Yes (small)	Yes	No	Yes
Respond to monoamine oxidase inhibitors (increase in DA, decrease in synthesis)	Yes	Yes	Yes	Yes	Yes	Yes
Presence of nerve terminal synthesis–modulating autoreceptors	Yes	Yes	No	Yes	No	Yes
High-affinity DA transport	Yes	Yes	Yes	Yes	No	No

(continued)

TABLE 9–7. Pharmacology of Central Dopaminergic Systems (*Continued*)

Characteristics	*Nigro-striatal*	*Meso-accumbal*	*Meso-prefrontal*	*Meso-piriform*	*Tubero-infundibular*[a]	*Tubero-hypophyseal*[b]
Respond to mild stress (increase in DA synthesis and catabolism, blocked by diazepam)	Yes	Yes? (small)	Yes	No	No	No
Respond to anxiogenic β-carbolines (increase in DA catabolism)	No	No	Yes	No	—	—

[a]Cell bodies of this group of neurons are located in the arcuate and periventricular nuclei, and their axons terminate in the external layer of the median eminence.

[b]Cell bodies of this group of neurons are located in the arcuate and periventricular nuclei, and their axons terminate in the neurointermediate lobe of the pituitary (posterior pituitary).

DA, dopamine.

TABLE 9–8. Unique Characteristics of Mesotelencephalic Dopamine Systems Lacking Autoreceptors (Mesoprefrontal and Mesocingulate) Compared to Those Possessing Autoreceptors (Mesopiriform, Mesolimbic, and Nigrostriatal)

1. A higher rate of physiological activity (firing) and a different pattern of activity (more bursting).
2. A higher turnover rate and metabolism of transmitter dopamine.
3. Greatly diminished biochemical and electrophysiological responsiveness to dopamine agonists and antagonists.
4. Lack of biochemical tolerance development following chronic antipsychotic drug administration.
5. Resistance to the development of depolarization-induced inactivation following chronic treatment with antipsychotic drugs.
6. Transmitter synthesis more readily influenced by altered availability of precursor tyrosine.

ing autoreceptors (mesoprefrontal and mesocingulate) than in those possessing autoreceptors. This may be related to the enhanced rate of physiological activity in this subpopulation of midbrain dopamine neurons, making them more responsive to precursor regulation. Mesoprefrontal and mesocingulate dopamine neurons also show diminished biochemical and electrophysiological responsiveness to dopamine agonists and antagonists. Low doses of apomorphine or autoreceptor-selective dopamine agonists, in contrast to their inhibitory effect on other midbrain dopamine neurons, are ineffective at decreasing the activity or dopamine metabolite levels in these two cortical dopamine projections. Dopamine receptor–blocking drugs, such as haloperidol, produce large increases in the synthesis and accumulation of dopamine metabolites in nigrostriatal, mesolimbic, and mesopiriform dopamine neurons but have only a modest effect in mesoprefrontal and mesocingulate dopamine neurons.

Heterogeneity among midbrain dopamine neurons is also found when one studies the effects of chronic antipsychotic drug administration. When classic antipsychotic drugs are administered repeatedly over time, the great majority of dopamine cells cease to fire due to the development of a state of depolarization inactivation. However, some midbrain dopamine cells appear to be unaffected by repeated antipsychotic drug administration. These dopamine cells are the neurons projecting to the prefrontal and cingulate cortices.

Parallel observations have been made biochemically. Following chronic administration of antipsychotic drugs, tolerance develops to the metabolite-elevating effects of these agents in the midbrain dopamine systems that possess autoreceptors but not in the systems that lack autoreceptors. When the atypical antipsychotic drug clozapine (which possesses therapeutic efficacy but lacks Parkinson-like side effects and an ability to produce tardive dyskinesia) is administered repeatedly, dopamine neurons in the ventral tegmental area develop depolarization inactivation but neurons in the substantia nigra do not. The reason for this differential effect is unknown. Foot shock, swim stress, and conditioned fear cause selective (benzodiazepine-reversible) metabolic activation of mesoprefrontal dopamine neurons without causing a marked or consistent effect on other midbrain dopamine neurons, including the mesocingulate dopamine neurons. Thus, this selective activation does not appear to be due solely to the absence of autoreceptors. The anxiogenic benzodiazepine receptor ligands, such as the β-carbolines, also produce a selective dose-dependent activation of mesoprefrontal dopamine neurons without increasing dopamine metabolism in other midbrain dopamine neurons.

In summary, certain mesotelencephalic dopamine systems, namely, the mesoprefrontal and mesocingulate dopamine neurons, possess many unique characteristics compared to the nigrostriatal, mesolimbic, and mesopiriform dopamine systems (Table 9–8). Many of these unique characteristics may be the consequence of a lack of impulse-regulating somatodendrite and synthesis-modulating nerve terminal dopamine autoreceptors. However, some, such as the response to stress and the anxiogenic β-carbolines, are clearly dependent on other regulatory influences and not solely related to the absence of autoreceptors. These findings suggest that dopamine action at autoreceptors may be one of the more critical ways that dopamine cells modulate their function. If valid, how do midbrain dopamine systems that lack autoreceptors regulate themselves? Perhaps it is through afferent systems by neuronal feedback. Some studies have suggested that a substance P/substance K innervation of the ventral tegmental area (A10) may influence the functional activity of mesocortical and mesolimbic dopamine neurons.

A number of studies have demonstrated the importance of NMDA receptors and of the glutamatergic input to the ventral tegmental area in the regulatory control of mesoprefrontal dopamine neurons. This input is believed to be at least partially responsible for converting pacemaker-like firing in dopamine cells into burst-firing patterns. NMDA receptors in the ventral tegmental area appear to modulate differentially the dopamine projections to the prefrontal cortex and nucleus accumbens. The NMDA receptor is selectively activated by NMDA and regulated at several pharmacologically distinct sites, including a high-affinity, strychnine-insensitive glycine-binding site (see Chapter 6). Competitive antagonists of this strychnine-insensitive glycine site, which cross the blood–brain barrier, have made possible the in vivo pharmacological modulation of the NMDA receptor via this site. In behavioral paradigms (restraint stress and conditioned fear) that cause metabolic activation of mesoprefrontal and mesaccumbens dopamine neurons, these agents (e.g., [+]-HA-966) selectively abolish the activation of mesoprefrontal dopamine neurons. The stress-induced activation of serotonin neurons in the prefrontal cortex and the dopaminergic activation of the nucleus accumbens are not altered by (+)-HA-966. Activation of mesoprefrontal dopamine neurons elicited by acute administration of phencyclidine and/or Δ-9-tetrahydrocannabinol, the active ingredient in marijuana, is also attenuated by HA-966. These data indicate that under certain perturbed states the NMDA receptor complex and the associated glycine-modulatory site play an important role in the afferent control of the dopamine neurons in the prefrontal cortex and provide a potential target for pharmacological regulation of this important dopamine projection.

The observation that central dopamine systems are quite heterogeneous from both a biochemical and a functional point of view holds promise that it will soon be possible to develop drugs targeted to modify or restore function to selective dopamine systems that are abnormal in various behavioral or pathological states. Some progress has already been achieved in developing agents that appear to act at selective dopamine receptor sites (Fig. 9–6). Whether these agents will be useful in selectively modifying the function of subsets of midbrain dopamine neurons remains to be determined.

Tuberoinfundibular and Tuberohypophysial Dopamine Systems

The tuberoinfundibular dopamine system responds to pharmacological and endocrinological manipulations in a manner that is qualitatively different from the other three dopamine systems (nigrostriatal, mesolimbic, and mesocortical) described above (Table 9–7). Tuberoinfundibular neurons appear to be regulated in part by circulating levels of prolactin. Prolactin increases the activity of these neurons by acting within the medial basal hypothalamus, possibly directly on the tuberoinfundibular neurons. These neurons in turn release dopamine, which then inhibits prolactin release from the anterior pituitary. Haloperidol and other antipsychotic drugs have no effect on dopamine turnover in the tuberoinfundibular dopamine system until about 16 hours after drug administration, whereas the biochemical effects in other systems are observed within minutes and are maximal in several hours. The absence of an acute response to dopamine antagonists and agonists may be related to the lack of autoreceptors in this system. While less is known about the pharmacology of the tuberohypophysial dopamine neurons, this system seems to respond to drugs in a manner qualitatively similar to the better-studied dopamine systems (Table 9–7).

Specific Drug Classes

Antipsychotic Drugs

For many years, it has been known that antipsychotic drugs of both the phenothiazine and butyrophenone classes can increase the turnover of dopamine in the CNS. Since these drugs appear to have potent dopamine receptor–blocking capabilities, it has been suggested that the increased dopamine turnover results from blockade of both dopamine autoreceptors and postsynaptic dopamine receptors and a consequent feedback activation of the dopaminergic neurons, presumably by some sort of neuronal feedback loop. This speculation has been verified in part by direct extracellular recording techniques.

The antagonism of central dopamine receptors by antipsychotic drugs has been postulated as a critical determinant of the therapeutic efficacy of this class of drugs. It is clear, however, that drugs with clinical antipsychotic effects influence dopamine transmission at several levels, including transmitter synthesis, release, and metabolism. Indeed, the "dopamine hypothesis" of schizophrenia was originally based on studies of alterations in brain dopamine metabolism produced by haloperidol and chlorpromazine in mice. The dopamine hypothesis later gained support from the finding that classical antipsychotic drugs such as haloperidol and chlorpromazine occupy dopamine D_2-like receptors in the brain.

After the superior efficacy of clozapine was demonstrated, a new generation of "atypical" antipsychotic drugs followed. The greater therapeutic efficacy of these drugs does not appear to be related to actions at the D_2 receptor alone (see Chapter 13). Moreover, the atypical antipsychotic drugs profoundly affect cortical dopamine metabolism and release after acute and chronic administration. Indeed, a variety of novel drugs have already been characterized as atypical antipsychotic drugs (clozapine, olanzapine, risperidone, amperozide, ziprasidone), and it has become customary to attribute their beneficial actions to a preferential increase in dopamine release in the frontal cortex relative to the striatum. Each of these drugs profoundly increases cortical, versus subcortical, dopamine release in a pattern unlike that found for typical antipsychotic drugs. Moreover, this increase in cortical dopamine efflux is directly related to increased impulse flow of mesocortical dopamine neurons. The argument that this cortical dopamine effect may be related to the superior efficacy of this class of drugs arises, in part, from the notion that the schizophrenic disease process may include a component of frontal cortical dopaminergic hypofunction, which these agents might reverse. The pharmacological mechanisms by which atypical antipsychotic drugs increase cortical dopamine transmission are currently unknown.

Compared to typical antipsychotic drugs, chronic administration of atypical antipsychotic drugs appears to affect dopaminergic systems in a distinct way. For example, basal cortical, but not subcortical, dopamine efflux is increased after chronic clozapine administration. Thus, after chronic clozapine administration, there may be a functional disconnection between impulse flow and release, possibly due to an emphasis on terminal level-dependent regulation of transmitter release. In any case, it appears that atypical antipsychotic drugs facilitate cortical dopamine transmission after acute or chronic exposure, and this relative activation of cortical versus subcortical dopamine release may underlie the strong therapeutic effects in the absence of observable extrapyramidal side effects.

Stimulants

The therapeutic use of this class of drugs is becoming less and less common as awareness of their abuse potential increases. At present, their use is largely restricted to the treatment of narcolepsy and of hyperkinetic children and as general anorectic agents. The principal drugs in this category are the various analogues and isomers of amphetamine and methylphenidate.

For many years it has been known that ingestion of large amounts of amphetamine often leads to a state of paranoid psychosis that may be hard to distinguish from the paranoia associated with schizophrenia. It now appears that this paranoid state can be readily and reproducibly induced in humans given large amounts of amphetamine, so the drug may provide a convenient "model psychosis" for experimental study. It is of interest in this regard that antipsychotic drugs such as chlorpromazine can readily reverse amphetamine-induced psychosis.

On the biochemical level, it was no surprise to learn that amphetamine and related compounds interact with catecholamine-containing neurons since amphetamine is a close structural analogue of the catecholamines. However, there was no clear evidence that amphetamine produced its CNS effects through a catecholamine mechanism until it was demonstrated that α-methyl-tyrosine (a potent inhibitor of TH) prevented most of the behavioral effects of D-amphetamine. The question as to which catecholamine, norepinephrine or dopamine, is involved in the behavioral effects of amphetamine remains

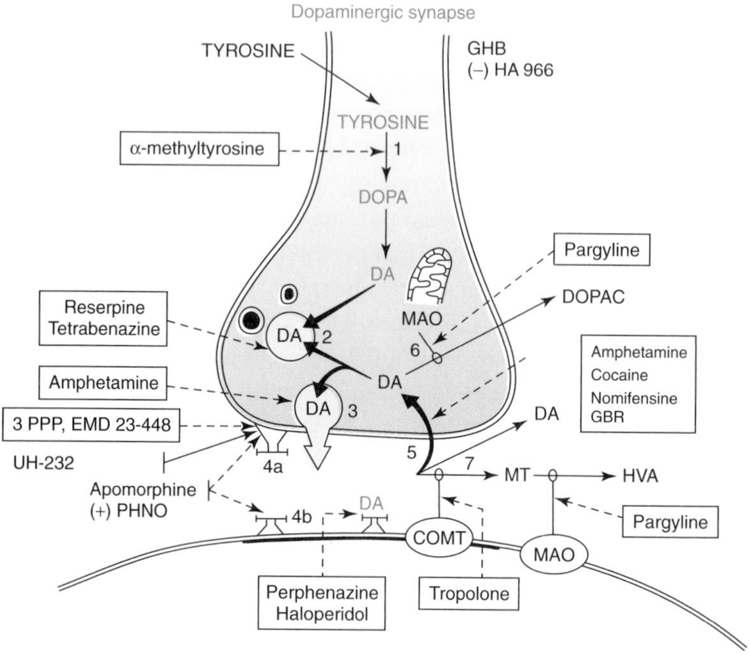

FIGURE 9–9. Schematic model of a central dopaminergic neuron indicating possible sites of drug action.

Site 1: *Enzymatic synthesis:* Tyrosine hydroxylase reaction blocked by the competitive inhibitor α-methyltyrosine and other tyrosine hydroxylase inhibitors.

Site 2: *Storage:* Reserpine and tetrabenazine interfere with the uptake–storage mechanism of the amine granules. The depletion of dopamine (DA) produced by reserpine is long-lasting, and the storage granules appear to be irreversibly damaged. Tetrabenazine also interferes with the uptake–storage mechanism of the granules, except that the effects of this drug do not appear to be irreversible.

Site 3: *Release:* γ-Hydroxybutyrate and HA966 effectively block the release of DA by blocking impulse flow in dopaminergic neurons. Amphetamine administered in high doses releases DA, but most of the releasing ability of amphetamine appears to be related to its ability to effectively block DA reuptake.

Site 4: *Receptor interaction:* Apomorphine is an effective DA receptor–stimulating drug, with both pre- and postsynaptic sites of action. Both 3-PPP and EMD-23-448 (an indolebutylamine) are autoreceptor agonists. Perphenazine and haloperidol are effective DA receptor–blocking drugs.

Site 5: *Reuptake:* DA has its action terminated by being taken up into the presynaptic terminal. Amphetamine as well as benztropine, the anticholinergic drug, are potent inhibitors of this reuptake mechanism.

Site 6: *Monoamine oxidase (MAO):* DA present in a free state within the presynaptic terminal can be degraded by the enzyme MAO, which appears unanswered, but it is generally believed that the so-called stereotypic behaviors in animals (i.e., compulsive gnawing, sniffing) induced by amphetamine are associated with a dopaminergic mechanism and that the increase in locomotor activity involves a noradrenergic mechanism or both.

Many classes of psychotropic drugs interact in one way or another with catecholamine-containing neurons. Figure 9–9 outlines the life cycle of the transmitters of dopaminergic and noradrenergic neurons in the CNS and indicates possible sites at which drugs may intervene in this cycle. This schematic model also provides examples of drugs or chemical agents that interfere at the various sites within the life cycle of the transmitter substances. These numerous sorts of interaction ultimately result in an increase, decrease, or no change in the functional activity of the catecholamine neuron in question.

Only recently did it become clear exactly how various pharmacological agents alter activity in defined catecholamine neuronal systems in the brain. In most cases, the turnover of monoamines depends essentially on impulse flow in the neuron. An increase in impulse flow usually causes an increase in turnover, and a decrease in impulse flow causes a reduction in turnover. As mentioned above, however, this is not always the case in the dopamine system, if synthesis is used as an index of turnover. Turnover measurement has been used to gain some insight into the activity of various types of monoamine-containing neuron during different behavioral states or after administration of different psychotropic drugs. As might be predicted, psychotropic drugs can have a variety of effects, and these effects can alter the turnover of a given transmitter substance without necessarily altering impulse flow in the neuronal system under study. For example, a drug can have a direct effect on the synthesis, degradation, uptake, or release of a given transmitter that will then alter the turnover of the transmitter in question but will not necessarily lead to an increase or decrease in the activity of the neuronal system that utilizes that substance as a transmitter. Thus, an alternation in turnover of a transmitter is not necessarily a clear indication that there has been a change in impulse flow in a given neuronal pathway. Therefore, the most direct way to determine if a drug alters impulse flow in a chemically

---

to be located in the outer membrane of the mitochondria. Dihydroxyphenylacetic acid (DOPAC) is a product of the action of MAO and aldehyde oxidase on DA. Pargyline is an effective inhibitor of MAO. Some MAO is also present outside the dopaminergic neuron.

Site 7: *Catechol-O-methyltransferase (COMT):* DA can be inactivated by the enzyme COMT, which is believed to be localized outside the presynaptic neuron. Tropolone is an inhibitor of COMT.

defined neuronal system is to measure the activity of that system while the animal is under the influence of the drug.

Drugs can alter impulse flow in several ways. For example, a drug can act directly on the nerve cell body; it can act on other neurons, which then influence impulse flow in the neuron under study; or it can act at the postsynaptic receptor to cause stimulation or blockade, which then results in some sort of feedback influence on the presynaptic neuron. This feedback information could be either neuronal or perhaps transsynaptically mediated by release of some local chemical from the postsynaptic membrane. The combined histochemical–neurophysiological identification of dopaminergic and noradrenergic neurons has made possible the direct study of the effects of various drugs on the firing of these chemically defined neurons.

Imaging Dopamine Transmission and Integrity in Neurological and Psychiatric Disorders

Advances in neuroimaging techniques have made it possible to visualize dopamine transmission in neuropsychiatric disorders. Radiotracer imaging with PET and SPECT can be used to measure pre-, post-, and intrasynaptic aspects of dopaminergic transmission. Presynaptic sites can be labeled with radiotracers for the DAT, VMAT or the synthetic enzyme aromatic L-amino acid decarboxylase. Postsynaptic sites can be labeled with radiotracers for the D_1 or D_2 receptor. Estimates of synaptic endogenous dopamine release can be made indirectly by measuring the displacement of receptor tracers by endogenous dopamine. Pharmacological agents that either release (i.e., amphetamine) or deplete (α-methyl-*p*-tyrosine) dopamine tissue stores are used to assess alterations in synaptic dopamine in normal and disease states.

Parkinson's Disease

Parkinson's disease is a progressive neurodegenerative disorder of the basal ganglia that is characterized by tremor, muscular rigidity, difficulty in initiating motor activity, and loss of postural reflexes. It is observed in approximately 1% of the population over age 55. It has been known for over 70 years that Parkinson's disease is characterized pathologically by loss of pigmented cells in the substantia nigra, but only since 1960 has it been appreciated that there is a substantial loss of dopamine in the striatum. It is now clear that Parkinson's disease can be defined in biochemical terms as primarily a dopamine-deficiency state resulting from degeneration or injury to dopamine neurons. The most striking degenerative loss of dopamine neurons is observed in the nigrostriatal system. Even in patients with mild symptoms, a striatal dopamine loss of 70%–80% is observed, while severely impaired subjects have striatal dopamine depletions in excess of 90%. Since the DAT is heavily expressed in the terminals of dopamine neurons that are lost in Parkinson's disease, it is not surprising that striatal binding of agents that label this site (cocaine, nomifensine, GBR, and mazindol) is lost in the parkinsonian striatum. This alteration corresponds well with the loss of functional dopamine uptake visualized in vivo by PET using (^{18}F)-L-DOPA uptake or nomifensine. Although striatal dopamine loss represents the primary neurochemical abnormality in the Parkinson's disease brain, typical parkinsonism is accompanied by loss of other dopamine systems and other monoamine neurons as well. Some degeneration of dopamine-containing neurons is also apparent in the mesolimbic, mesocortical, and hypothalamic systems; the norepinephrine-containing neurons in the locus ceruleus; and the serotonin neurons in the raphe nucleus. Nonmonoamine systems are also affected, with depletions observed in somatostatin, neurotensin, substance P, enkephalin, and cholecystokinin-8. Since many of these nondopamine systems indirectly interact with mesotelencephalic dopamine systems, changes in some of them are bound to influence in a complex way the function of dopamine neurons.

The strategy for treating Parkinson's disease has been to restore the dopamine deficit in the brain by pharmacological means or, more recently, by neural grafting of dopamine-containing cells. There are a number of theoretical strategies for drug therapy in Parkinson's disease, including substrate supplementation with direct and indirect dopamine agonists, metabolic inhibitors (MAOIs), and uptake inhibitors. The most successful treatment has been the use of L-DOPA. Direct dopamine agonists also have some benefit in patients whose responsiveness to L-DOPA is greatly reduced or erratic. So far, the only direct-acting dopamine agonist that has found extensive use is bromocriptine, primarily a D_2 agonist. Other agents belonging to this class will no doubt prove useful in the future as supplements or alternatives to L-DOPA.

In view of the behavioral and electrophysiological studies that suggest that D_1 receptor activation is necessary for the effects of D_2 receptor stimulation to be maximally expressed in normal animals as well as in animals with supersensitive dopamine receptors, the functional interaction between D_1 and D_2 receptors could have important implications in Parkinson's disease, where stimulation of postsynaptic dopamine receptors confers symptomatic benefit. Knowledge of the optimal ratio of relative drug activity at D_1 and D_2 receptors that is required to elicit effective stimulation of dopamine-mediated function may provide a basis for the design of new drugs. Also, as more knowledge accumulates concerning the distribution and function of various dopamine receptor subtypes, this should facilitate the development of new agents to treat dopamine-deficiency states.

The MPTP-treated parkinsonian primate has provided a very useful animal model in which to examine therapeutic strategies for treatment (see Chapter 13). In fact, this model has already been successfully exploited to design, refine, and evaluate neural transplantation and gene therapy techniques and to test new pharmacological strategies for the therapeutic management of Parkinson's disease.

Schizophrenia

The etiology of schizophrenia has also been linked to defective dopamine neurotransmission. The growing conviction that antipsychotic agents act therapeutically by decreasing central dopaminergic transmission led to the formulation of the dopamine theory of schizophrenia. This theory has been revised in recent years to take into account the heterogeneity of midbrain dopamine systems, their differential response to antipsychotic drugs, and data from neuroimaging of schizophrenic subjects and primate models of this disorder. In its simplest revised form, this hypothesis states that schizophrenia may be related to a relative excess of mesolimbic and a deficit of mesoprefrontal dopaminergic activity (see Chapter 13).

Selected References

Bannon, M. J., J. G. Granneman, and G. Kapatos (1995). The dopamine transporter: potential involvement in neuropsychiatric disorders. In *Psychopharmacology: The Fourth Generation of Progress* (F. E. Bloom and D. J. Kupfer, eds.). Raven Press, New York, pp. 179–188.

Bannon, M. J. and R. H. Roth (1983). Pharmacology of mesocortical dopamine neurons. *Pharmacol. Rev.* 35, 53.

Björklund, A. and O. Lindvall (1984). Dopamine-containing systems in the CNS. In *Handbook of Chemical Neuroanatomy. Classical Transmitters in the CNS*, vol. 2, part I (A. Björklund and T. Hökfelt, eds.). Elsevier, Amsterdam, pp. 55–122.

Chiodo, L. A. and A. S. Freeman (eds.) (1987). *Neurophysiology of Dopaminergic Systems: Current Status and Clinical Perspectives.* Lakeshore Publishing, Gross Pointe, MI.

Chiodo, L. A., A. S. Freeman, and B. S. Bunney (1995). Dopamine autoreceptor signal transduction and regulation. In *Psychopharmacology: The Fourth Generation of Progress* (F. E. Bloom and D. J. Kupfer, eds.). Raven Press, New York, pp. 221–226.

Civelli, O. (1995). Molecular biology of the dopamine receptor subtypes. In *Psychopharmacology: The Fourth Generation of Progress* (F. E. Bloom and D. J. Kupfer, eds.). Raven Press, New York, pp. 155–162.

Dahlström, A. and K. Fuxe (1964). Evidence for the existence of monoamine-containing neurons in the central nervous system. I: Demonstration of monoamines in the cell bodies of brain stem neurones. *Acta Physiol. Scand.*, 62. Suppl. 232.1–55.

Elsworth, J. D., M. Al-Tikriti, J. R. Sladek, Jr., J. R. Taylor, R. B. Innis, D. E. Redmond, Jr., and R. H. Roth (1994). Novel radioligands for the dopamine transporter demonstrate the presence of nigral grafts in caudate nucleus of MPTP treated monkey with improved behavioral function. *Exp. Neurol.* 126, 300–304.

Elsworth, J. D. and R. H. Roth (1995). Dopamine autoreceptor pharmacology and function: recent insights. In *The Dopamine Receptors* (K. A. Neve, ed.). Human Press, Totowa, NJ, pp. 223–265.

Fuxe, K. and T. Hökfelt (1971). Histochemical fluorescence detection of changes in central monoamine neurones provoked by drugs acting on the CNS. *Triangle*, 10(3): 73–84.

Gainetdinov, R. R., S. R. Jones, and M. G. Caron (1999). Functional hyperdopaminergia in dopamine transporter knock-out mice. *Biol. Psychiatry* 46, 303–311.

Gardner, E. L. and C. R. Ashby, Jr. (2000). Heterogeneity of the mesotelencephalic dopamine fibers: physiology and pharmacology. *Neurosci. Biobehav. Rev.* 24, 115–118.

Gingrich, J. A. and M. G. Caron (1993). Recent advances in the molecular biology of dopamine receptors. *Annu. Rev. Neurosci.* 16, 299–321.

Giros, B. and M. G. Caron (1993). Molecular characterization of the dopamine transporter. *Trends Pharmacol. Sci.* 14, 43–49.

Grace, A. (2002). Dopamine. In *Neuropsychopharmacology The Fifth Generation of Progress* (K. L. Davis, D. Charney, J. T. Cole and C. Nemeroff, eds.). Lippincott Williams & Wilkins, Philadelphia, PA, pp. 119–132.

Greengard, P., P. A. Allen, and A. C. Nairn (1999). Beyond the dopamine receptor: the DARPP-32/protein phosphatase-1 cascade. *Neuron* 23, 435–447.

Grenhoff, J., S. W. Johnson (1995). Electrophysiological Effects of Dopamine Receptor Stimulation. In *The Dopamine Receptors* (K. A. Neve, ed.) Human Press, Totowa, N.J., pp. 267–304.

Jentsch, J. D. and R. H. Roth (2000). Effects of antipsychotic drugs on dopamine release and metabolism in the central nervous system. In *Neurotransmitter Receptors in Actions of Antipsychotic Medications*, vol. 3 (M. Lidow, ed.). CRC Press, Boca Raton, FL, pp. 31–41.

Jentsch, J. D., J. R., Taylor, and R. H. Roth (2000). Phencyclidine model of frontal cortical dysfunction in non-human primates. *Neuroscientist* 6, 268–275.

Kitai, S. T., P. D. Shepard, J. C. Callaway, and R. Scroggs (1999). Afferent modulation of dopamine neuron firing patterns. *Curr. Opin. Neurobiol.* 9, 690–697.

Kim, D. S., M. S., Szczypka, and R. D. Palmiter (2000). Dopamine-deficient mice are hypersensitive to dopamine receptor argonists. *J. Neurosci.* 20, 4405–4413.

McCracken, J. T., S. L. Smalley, J. J. McGough, L. Crawford, M. Del'Homme, R. M. Cantor, A. Liu, and S. F. Nelson (2000). Evidence for linkage of a tandem duplication polymorphism upstream of the dopamine D_4 receptor gene (*DRD4*) with attention deficit hyperactivity disorder (ADHD). *Mol. Psychiatry* 5, 531–536.

Meador-Woodruff, J. H. (1994). Update on dopamine receptors. *Ann. Clin. Psychiatry* 6, 79–90.

Miller, G. W., R. R. Gainetdinov, A. I. Levey, and M. C. Caron (1999). Dopamine transporters and neuronal injury. Trends Pharmacol Sci. 20, 424–429.

Missale, C., S. R., Nash, S. W., Robinson, M. Jaber, and M. G. Caron (1998). Dopamine receptors: from structure to function. *Physiol. Rev.* 78, 189–225.

Moore, K. E. and K. J. Lookingland (1995). Dopaminergic neuronal systems in the hypothalamus. In *Psychopharmacology: The Fourth Generation of Progress* (F. E. Bloom and D. J. Kupfer, eds.). Raven Press, New York, pp. 245–256.

O'Malley, K. L., S. Haron, L. Tang, and R. D. Todd (1992). The rat dopamine D_4 receptor: sequence, gene structure and demonstration of expression in the cardiovascular system. *New Biol.* 4, 137–146.

Pani, L., A. Porcella, and G. L. Gessa (2000). The role of stress in the pathophysiology of the dopaminergic system. *Mol. Psychiatry* 5, 14–21.

Roth, R. H. and J. D. Elsworth (1995). Biochemical pharmacology of midbrain dopamine neurons. In *Psychopharmacology: The Fourth Generation of Progress* (F. E. Bloom and D. J. Kupfer, eds.). Raven Press, New York, pp. 277–244.

Schwartz, J.-C., D., Levesque, M.-P. Martres, and P. Sokoloff (1993). Dopamine D_3 receptor: basic and clinical aspects. *Clin. Neuropharmacol.* 16, 295–314.

Seeman, P. (1995). Dopamine receptors: clinical correlates. In *Psychopharmacology: The Fourth Generation of Progress* (F. E. Bloom and D. J. Kupfer, eds.). Raven Press, New York, pp. 295–302.

Seeman, P. and H. H. M. Van Tol (1994). Dopamine receptor pharmacology. *Trends Pharmacol. Sci.* 15, 264–270.

Sharma, A., M. L. Kramer, P. F., Wick, D. Liu, S. Chari, S. Shim, W. Tan, D. Ouellette, M. Nagata, C. J. DuRand, M. Kotb, and R. C. Deth (1999). D_4 dopamine receptor–mediated phospholipid methylation and its implications for mental illnesses such as schizophrenia. *Mol. Psychiatry* 4, 235–246.

Sibley, D. R. and F. J. Monsma, Jr. (1992). Molecular biology of dopamine receptors. *Trends Pharmacol. Sci.* 13, 61–69.

Sunahara, R. K., H. C. Guan, B. F. O'Dowd, P. Seeman, L. G. Laurier, G. Ng, S. R. George, J. Torchia, H. H. M. Van Tol, and H. B. Niznik (1991). Cloning of the gene for a human dopamine D_5 receptor with higher affinity for dopamine than D_1. *Nature* 350, 614–619.

Tang, L., R. D. Todd, A. Heller, and K. L. O'Malley (1994). Pharmacological and functional characterization of D_2, D_3 and D_4 dopamine receptors in fibroblast and dopaminergic cell lines. *J. Pharmacol. Exp. Ther.* 268, 495–502.

Tarazi, F. I. and R. J. Baldessarini (1999). Dopamine D_4 receptors: significance for moloecular psychiatry at the millennium. *Mol. Psychiatry* 4, 529–538.

Van Tol, H. H. M., J. R. Bunsow, H. C. Guan, R. K. Sunahara, P. Seeman, H. B. Niznik, and O. Civelli (1991). Cloning of the gene for a human dopamine D_4 receptor with high affinity for the antipsychotic clozapine. *Nature* 350, 610–614.

Verhoeff, N. P. L. G. (2001). Imaging of dopaminergic transmission in neuropsychiatric disorders. *Curr. Opin. Psychiatry* 3, 227–239.

Wolf, M. E., A. Y. Deutch, and R. H. Roth (1987). Pharmacology of central dopaminergic neurons. In *Handbook of Schizophrenia. Neurochemistry and Neuropharmacology of Schizophrenia*, vol. 2 (F. A. Henn and L. E. DeLisi, eds.). Elsevier, Amsterdam, pp. 101–147.

Wolf, M. E. and R. H. Roth (1990). Autoreceptor regulation of dopamine synthesis. *Ann. N.Y. Acad. Sci.* 604, 323–343.

Xu, M., X.-T. Hu, D. C. Cooper, R. Moratalla, A. M. Graybiel, F. J. White, and S. Tonegawa (1994). Elimination of cocaine-induced hyperactivity and dopamine-mediated neurophysiological effects in dopamine D_1 receptor mutant mice. *Cell* 79, 945–955.

10

Serotonin (5-Hydroxytryptamine), Histamine, and Adenosine

Serotonin

Of all the neurotransmitters discussed in this book, serotonin remains historically the most intimately involved with neuropsychopharmacology. From the mid-nineteenth century, scientists had been aware that a substance found in serum caused powerful contraction of smooth muscle organs, but over 100 years passed before scientists at the Cleveland Clinic succeeded in isolating this substance as a possible cause of high blood pressure.

At the same time, investigators in Italy were characterizing a substance found in high concentrations in chromaffin cells of the intestinal mucosa. This material also constricted smooth muscular elements, particularly those of the gut. The material isolated from the bloodstream was given the name *serotonin*, while that from the intestinal tract was called *enteramine*. Subsequently, both materials were purified, crystallized, and shown to be 5-hydroxytryptamine (5-HT), which could then be prepared synthetically and shown to possess all of the biological features of the natural substance. The indole nature of this molecule bore many resemblances to the psychedelic drug lysergic acid diethylamide (LSD), with which it could be shown to interact on smooth muscle preparations in vitro. 5-HT is also structurally related to other psychotropic agents (Fig. 10–1).

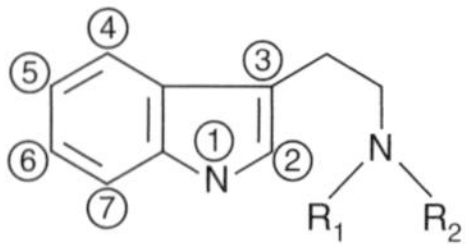

FIGURE 10–1. Structural relationships of the various indolealkyl amines.

Compound	Substitutions
Tryptamine	R_1 and R_2 = H
Serotonin	Tryptamine with 5-hydroxy
Melatonin	5-Methoxy, *N*-acetyl
DMT*	R_1 and R_2 = methyl
DET*	R_1 and R_2 = ethyl
Bufotenine*	5-Hydroxy, DMT
Szara psychotrope*	6-Hydroxy, DET
Psilocin*	4-Hydroxy, DMT
Harmaline*	6-Methoxy, R_1 forms isopropyl link to C_2
5-MT	5-Methoxytryptamine
5,6-DHT	5,6-Dihydroxytryptamine
5,7-DHT	5,7-Dihydroxytryptamine

*Psychotropic or behavioral effects.

When 5-HT was first found within the mammalian central nervous system (CNS), the theory arose that various forms of mental illness could be due to biochemical abnormalities in its synthesis. This line of thought was extended when the tranquilizing substance reserpine was observed to deplete brain 5-HT; throughout the duration of the depletion, profound behavioral depression occurred. As we shall see, many of these ideas are still maintained, although we now have much more evidence with which to evaluate them.

Biosynthesis and Metabolism of Serotonin

Serotonin is found in many cells that are not neurons, such as platelets, mast cells, and the enterochromaffin cells mentioned above. In fact, only about 1%–2% of the serotonin in the whole body is found in the brain. Nevertheless, because 5-HT cannot cross the blood–brain barrier, brain cells must synthesize their own.

For brain cells, the first important step is uptake of the amino acid tryptophan, which is the primary substrate for synthesis. Plasma tryptophan arises primarily from the diet, and unlike the catecholamine precursor tyrosine, elimination of dietary tryptophan can profoundly lower the levels of brain serotonin. In addition, an active uptake process is known to facilitate the entry of tryptophan into the brain, and this carrier process is open to competition from large neutral amino acids, including the aromatic amino acids (tyrosine and phenylalanine), the branched-chain amino acids (leucine, isoleucine, and valine), and others (e.g., methionine and histidine). The competitive nature of the large neutral amino acid carrier means that brain levels of tryptophan will be determined not only by the plasma concentration of tryptophan but also by the plasma concentration of competing neutral amino acids. Thus, dietary protein and carbohydrate content can specifically influence brain tryptophan and serotonin levels by effects on plasma amino acid patterns. Because plasma tryptophan has a daily rhythmic variation in its concentration, it seems likely that this concentration variation could also profoundly influence the rate and synthesis of brain serotonin.

The first step in the synthetic pathway is hydroxylation of tryptophan at the 5 position (Fig. 10–2) to form 5-hydroxytryptophan (5-HTP). The enzyme responsible for this reaction, tryptophan hydroxylase, occurs in low concentrations in most tissues, including the brain; and it was very difficult

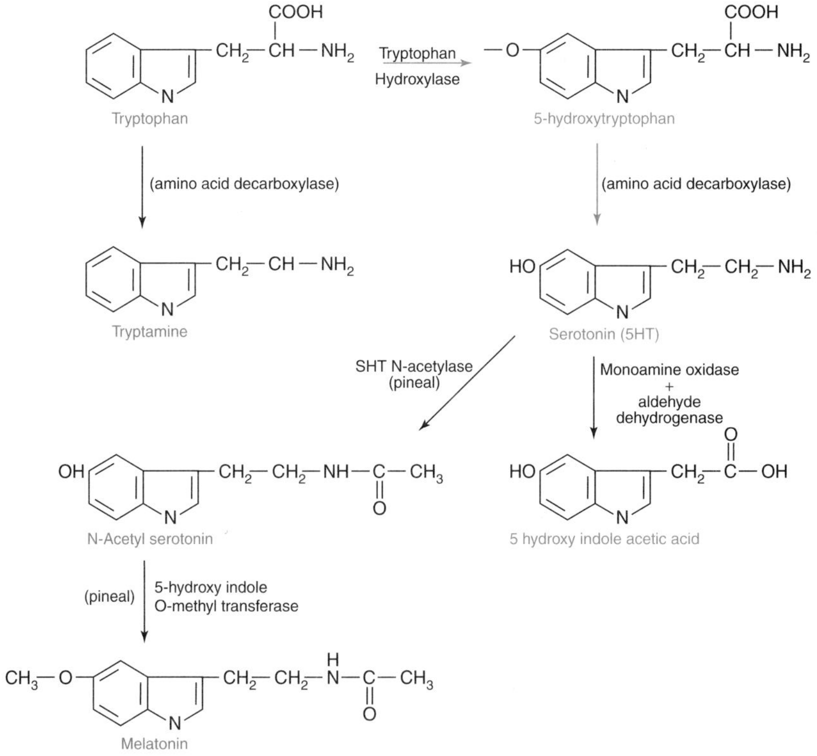

FIGURE 10–2. The metabolic pathways available for the synthesis and metabolism of serotonin.

to isolate for study. After purifying the enzyme from mast cell tumors and determining the characteristic cofactors, however, it became possible to characterize this enzyme in the brain. (Students should investigate the ingenious methods used for the initial assays of this extremely minute enzyme activity.) As isolated from brain, the enzyme appears to have an absolute requirement for molecular oxygen, for reduced pteridine cofactor, and for a sulfhydryl-stabilizing substance, such as mercaptoethanol, to preserve activity in vitro. With this fortified assay system, there is sufficient activity in the brain to synthesize 1 mg of 5-HTP per gram of brain stem in 1 hour. The pH optimum is approximately 7.2, and the K_m for tryptophan is 3×10^{-4} M. Additional research into the nature of the endogenous cofactor tetrahydrobiopterin yielded a K_m for tryptophan of 5×10^{-5} M, which is still above normal tryptophan concentrations. Thus, the normal plasma tryptophan content and the resultant uptake into brain leave the enzyme normally unsaturated with available substrate. Tryptophan hydroxylase appears to be a soluble cytoplasmic enzyme, but the procedures used to extract it from the tissues may greatly alter the natural particle-binding capacity. Investigators examining the relative distribution of particulate and soluble tryptophan hydroxylase have reported that the particulate enzyme may be associated with 5-HT-containing synapses, while the soluble form is more likely to be associated with the perikaryal cytoplasm. The particulate form of the enzyme shows the lower K_m and bears an absolute requirement for tetrahydrobiopterin.

Purified tryptophan hydroxylase has a molecular weight of 52,000–60,000. It is a multimer of identical subunits that can be activated by phosphorylation, Ca^{2+} phospholipids, and partial proteolysis.

Cloning and sequencing of cDNAs for tryptophan hydroxylase have been accomplished. Comparison of the rabbit tryptophan hydroxylase sequence with the sequences of phenylalanine hydroxylase and tyrosine hydroxylase demonstrates that these three pterin-dependent aromatic amino acid hydroxylases are highly homologous, reflecting a common evolutionary origin from a single primordial genetic locus. The pattern of sequence homology supports the hypothesis that the C-terminal two-thirds of the molecules constitute the enzymatic activity cores and the N-terminal one-third constitutes domains for substrate specificity.

The tryptophan hydroxylase step in the synthesis of 5-HT can be specifically blocked by *p*-chlorophenylalanine, which competes directly with the tryptophan and binds irreversibly to the enzyme. Therefore, recovery from tryptophan hydroxylase inhibition with *p*-chlorophenylalanine appears to require the synthesis of new enzyme molecules. In the rat, a single intraperitoneal injection of 300 mg/kg of this inhibitor lowers the brain serotonin content to less than 20% within 3 days, and complete recovery does not occur for almost 2 weeks.

Considerable attention has been directed to the overall regulation of this first enzymatic step of serotonin synthesis, especially in animals and humans treated with psychoactive drugs alleged to affect the serotonin systems as a primary mode of action. These studies have made an important general point that seems to apply to the brain's response to drug exposure in many other cell systems as well as to serotonin: because transmitter synthesis, storage, release, and response are dynamic processes, the acute imbalances produced initially by drug treatments are soon counteracted by the built-in feedback nature of synthesis regulation. Thus, if a drug reduces tryptophan hydroxylase activity, the nerve cells may respond by increasing their synthesis of the enzyme and transporting increased amounts to the nerve terminals.

Mandell and colleagues have provided evidence, for example, that short-term treatment with lithium will initially increase tryptophan uptake, resulting in increased amounts being converted to 5-HT. After 14–21 days of chronic treatment, however, repetition of the measurements shows that while tryptophan uptake is still increased, the activity of the enzyme is decreased so that normal amounts of 5-HT are being made. In this new equilibrium state, the neurochemical actions of drugs like amphetamine and cocaine on 5-HT synthesis rates are greatly reduced, as are their behavioral actions. In this way, the 5-HT system, by shifting the relationship between uptake and synthesis during Li exposure, can be viewed as more stable. This factor may be more fully appreciated when one considers that in the treatment of manic-depressive psychosis a minimum of 7–10 days is usually required before the therapeutic action of Li begins, a period during which the reequilibration of the 5-HT-synthesizing process might undergo restabilization.

Decarboxylation

Once synthesized from tryptophan, 5-HTP is almost immediately decarboxylated to yield serotonin. The enzyme responsible for this conversion is identical with the enzyme that decarboxylates dihydroxyphenylalanine (DOPA, i.e., aromatic amino acid decarboxylase, AADC [EC 4.1.128] or DOPA-decarboxylase). Since this decarboxylation reaction occurs so rapidly and since its K_m (5×10^{-6} M) requires less substrate than the preceding steps, tryptophan hydroxylase is the rate-limiting step in the synthesis of serotonin. Because of this kinetic relationship, drug-induced inhibition of serotonin by interference with the decarboxylation step is not feasible.

It is possible to increase serotonin formation by administering 5-HTP and bypassing the rate-limiting tryptophan hydroxylase step. AADC is widespread in distribution; it is found in the peripheral and central nervous systems associated with catecholamine- and serotonin-containing neurons and in the adrenal and pineal glands. It is also found in the kidney, liver, and various other tissues in which little or no monoamine transmitter is normally produced. Thus, unlike tryptophan administration, which can result in a selective increase in serotonin in serotonin-containing neurons, 5-HTP administration will result in the nonspecific formation of serotonin at all sites containing AADC, including the catecholamine-containing neurons.

Catabolism

The only effective route of continued metabolism for serotonin is deamination by monoamine oxidase. The product of this reaction, 5-hydroxyindoleacetaldehyde, can be further oxidized to 5-hydroxyindoleacetic acid (5-HIAA) or reduced to 5-hydroxytryptophol, depending on the ratio of the oxidized to the reduced form of nicotinamide adenine dinucleotide (NAD^+/NADH) in the tissue. Enzymes have been described in the liver and brain that could catabolize 5-HT without deamination through formation of a 5-sulfate ester. This could then be transported out of the brain, possibly by the acid transport system handling 5-HIAA. The brain contains an enzyme that catalyzes the *N*-methylation of 5-HT using *S*-adenosylmethionine as the methyl donor.

Control of Serotonin Synthesis and Catabolism

Although there is a relatively brief sequence of synthetic and degradative steps involved in serotonin turnover, there is still much to be learned regarding the physiological mechanisms for controlling this pathway. Since tryptophan hydroxylase depends on molecular oxygen, the rate of 5-HT formation could also be influenced by the tissue level of oxygen. In fact, rats permitted to breathe 100% oxygen greatly increase their synthesis of 5-HT. Also, 5-hydroxytryptophan does not inhibit the activity of tryptophan hydroxylase.

If the situation for serotonin were similar to that previously described for the catecholamines, we might also expect that the concentration of 5-HT itself could influence the levels of activity at the hydroxylation step. However, when the catabolism of 5-HT is blocked by monoamine oxidase inhibitors, the brain 5-HT concentration accumulates linearly to levels three times greater than controls, thus suggesting that end-product inhibition by serotonin is, at best, trivial. Similarly, if the efflux of 5-HIAA from the brain is blocked by the drug probenecid (which appears to block all forms of acid transport), 5-HIAA levels also continue to rise linearly for prolonged periods of time, again suggesting that the initial synthesis step is not affected by the levels of any of the subsequent metabolites. Two possibilities, therefore, remain open: the initial synthesis rate may be limited only by the availability of required cofactors or substrate such as oxygen, pteridine, and tryptophan from the bloodstream or the initial synthesis rate may be limited by the other more subtle control features, more closely related to brain activity. Indeed, the evidence suggests that impulse flow may, as in the catecholamine systems, initiate changes in the physical properties of the rate-limiting enzyme tryptophan hydroxylase. Several mechanisms have been postulated for the physiological regulation of tryptophan hydroxylase induced by alterations in neuronal activity within serotonergic neurons. The majority of evidence currently supports the involvement of calcium-dependent phosphorylation in this impulse-coupled regulatory process.

Serotonin Uptake and the Serotonin Transporter

As with the catecholamine-containing neurons, reuptake serves as a major mechanism for the termination of the action of synaptic serotonin. Serotonin nerve terminals possess high-affinity serotonin uptake sites that play an important role in terminating transmitter action and in maintaining transmitter homeostasis. This reuptake of released serotonin is accomplished by a plasma membrane carrier that is capable of transporting serotonin in either direction, depending on the concentration gradient. Although the involvement of transporters in norepinephrine (NE) and serotonin clearance has been appreciated for several decades (see Chapter 9), progress in understanding transporter structure and regulation has been slow, mainly because of difficulties associated with transporter protein purification. However, this has changed with the successful expression and homology-based cloning of the monoamine transporters (NE, dopamine, and serotonin) and the realization that they are members of a large gene family comprised of carriers for other transmitters including γ-aminobutyric acid (GABA) and glycine (see Chapter 6). Expression of these transporters in nonneuronal cells has established useful model systems for analyzing the structural basis of transporter specificity for transmitters and antagonists. The availability of transporter protein has also enhanced the feasibility of obtaining transporter-specific antibodies and nucleic acid probes. Antibodies raised against the cloned serotonin transporter (SERT) as well as DNA and RNA probes derived from it have enabled the localization and expression of SERT in the CNS. The SERT protein is ubiquitous in the CNS, which is consistent with its transport to the nerve terminals of the extensive projections of the serotonin neurons throughout the brain and spinal cord. This is in contrast to SERT mRNA expression, which occurs almost exclusively in the serotonergic cell bodies in the raphe nuclei; especially high levels are found in the median and dorsal raphe. SERT mRNA expression is absent from other brain stem nuclei, including the substantia nigra and locus ceruleus. The availability of transporter-specific antibodies and nucleic acid probes has made it feasible to investigate the endogenous mechanism that acutely regulates monoamine transporters in vivo and to determine whether chronic alterations in transporter genes underlie neuropsychiatric disorders. To date, however, the cloning of SERT has not had as big an impact on the field as the cloning of serotonin receptor subtypes. So far, there is little evidence for SERT heterogeneity. However, this transporter might be subject to regulation. In fact, SERT uptake capacity (V_{max}) is regulated by kinase-linked pathways, particularly those involving protein kinase C (PKC), resulting in transporter phosphorylation and sequestration. Ligand occupancy of the transporter significantly impacts both SERT phosphorylation and sequestration. Ligands such as serotonin and amphetamine, which permeate the transporter, prevent PKC-dependent SERT phosphorylation. In contrast, nontransported antagonists like cocaine and antidepressants are permissive for SERT phosphorylation but effectively block serotonin's effects. Hints of altered SERT gene regulation following hormonal stimulation also suggest that significant information might be acquired from systematic analysis of genomic regulatory elements that control transporter expression. Clearly, a goal for the future is to determine whether hereditary genetic variations in such systems contribute to psychiatric disorders.

Pineal Body

The pineal organ is a tiny gland (1 mg or less in the rodent) contained within connective tissue extensions of the dorsal surface of the thalamus. While physically connected to the brain, the pineal is cytologically isolated for all intents since, as one of the circumventricular organs, it is on the "peripheral" side of the blood-permeability barriers (see Chapter 2) and its innervation arises from the superior cervical sympathetic ganglion. The pineal is of interest for two reasons. First, it contains all of the enzymes required for the synthesis of serotonin plus two enzymes for further processing of serotonin, which are not so pronounced in other organs. The pineal contains more than 50 times as much 5-HT (per gram) as the whole brain. Second, the metabolic activity of the pineal 5-HT enzymes can be controlled by numerous external factors, including the neural activity of the sympathetic nervous system operating through release of norepinephrine. As such, the pineal appears to offer a potential model for the study of brain 5-HT. The pineal also appears to contain many neuropeptides, however, and thus its secretory role remains as clouded as ever.

Actually, the 5-HT content of the pineal was discovered after the isolation of a pineal factor, melatonin, known to induce pigment-lightening effects on skin cells. When melatonin was crystallized and its chemical structure determined to be 5-methoxy-*N*-acetyltryptamine, an indolealkylamine, a reasonable extension was to analyze the pineal for 5-HT itself. The production of melatonin from 5-HT requires two additional enzymatic steps. The first is the *N*-acetylation reaction to form *N*-acetylserotonin. This intermediate is the preferred substrate for the final step, the 5-hydroxyindole-*O*-methyltransferase reaction, requiring *S*-adenosylmethionine as the methyl donor.

The melatonin content, and its influence in supressing the female gonads, is reduced by environmental light and enhanced by darkness. The established cyclic daily rhythm of both 5-HT and melatonin in the pineal is driven by environmental lighting patterns through sympathetic innervation. In animals made experimentally blind, the pineal enzymes and melatonin content continue to cycle but with a rhythm uncoupled from lighting cycles. The adrenergic receptors of the pineal are of the β type, and, as is characteristic of such receptors, their effect on the pineal is mediated by the postjunctional formation of 3′,5′-cyclic adenosine monophosphate (cAMP). Elevated levels of cAMP occur within minutes of the dark phase and lead to an almost immediate activation of 5-HT-*N*-acetyltransferase. The same receptor action also appears to be responsible on a longer time scale for tonic enzyme synthesis. Thus, the proposed use of the pineal as a model applicable to brain 5-HT loses its luster since this regulatory step does not seem to be of functional importance in the CNS. Furthermore, the adrenergic sympathetic nerves also can accumulate 5-HT (which leaks out of pinealocytes) just as they accumulate and bind NE. It remains to be shown whether this is a functional mistake (i.e., secretions of an endogenous false transmitter) or simply a case of mistaken biological identity. It seems likely, however, that it is the NE whose release is required to pass on the intended communications from the sympathetic nervous system since only NE activates the pinealocyte cyclase to start the enzyme regulation cascade.

Localizing Brain Serotonin to Nerve Cells

Serotonin-containing neurons are restricted to clusters of cells lying in or near the midline or raphe regions of the pons and upper brain stem (Fig. 10–3). In addition to the nine 5-HT nuclei (B1–B9) originally described by

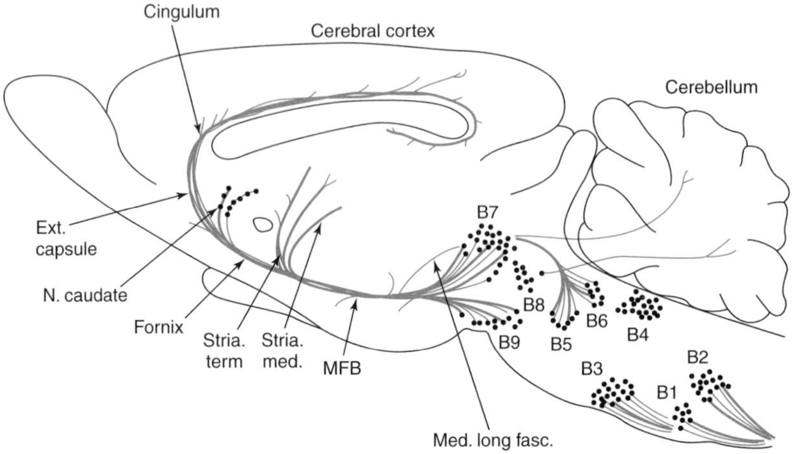

FIGURE 10–3. Schematic diagram illustrating the distribution of the main serotonin-containing pathways in the rat central nervous system. (Modified from Breese, 1975.)

Dahlström and Fuxe, the immunocytochemical localization of 5-HT has detected reactive cells in the area postrema and the caudal locus ceruleus, as well as in and around the interpeduncular nucleus. The more caudal groups, studied by electrolytic or chemically induced lesions, project largely to the medulla and spinal cord. The more rostral 5-HT cell groups (raphe dorsalis, raphe medianus, and centralis superior, or B7–B9 [Fig. 10–4]) are thought to provide the extensive 5-HT innervation of the telencephalon and diencephalon. The intermediate groups may project into both ascending and descending groups, but since lesions here also interrupt fibers of passage, discrete mapping has required analysis of the orthograde and retrograde methods. Immunocytochemical studies have also revealed a far more extensive innervation of the cerebral cortex, which, unlike the noradrenergic cortical fibers, is quite patternless in general.

In part, these studies could be viewed as disappointing in that most raphe neurons appear to innervate overlapping terminal fields and thus are more NE-like than dopamine-like in their lack of obvious topography. Exceptions to this generalization are that the B8 group (raphe medianus) appears to furnish a very large component of the 5-HT innervation of the limbic system, while B7 (or dorsal raphe) projects with greater density to the neostriatum, cerebral and cerebellar cortices, and thalamus (Fig. 10–5).

In the past, attempts to localize 5-HT-containing terminals relied primarily on the uptake of reactive 5-HT analogues or radiolabeled 5-HT. The use of

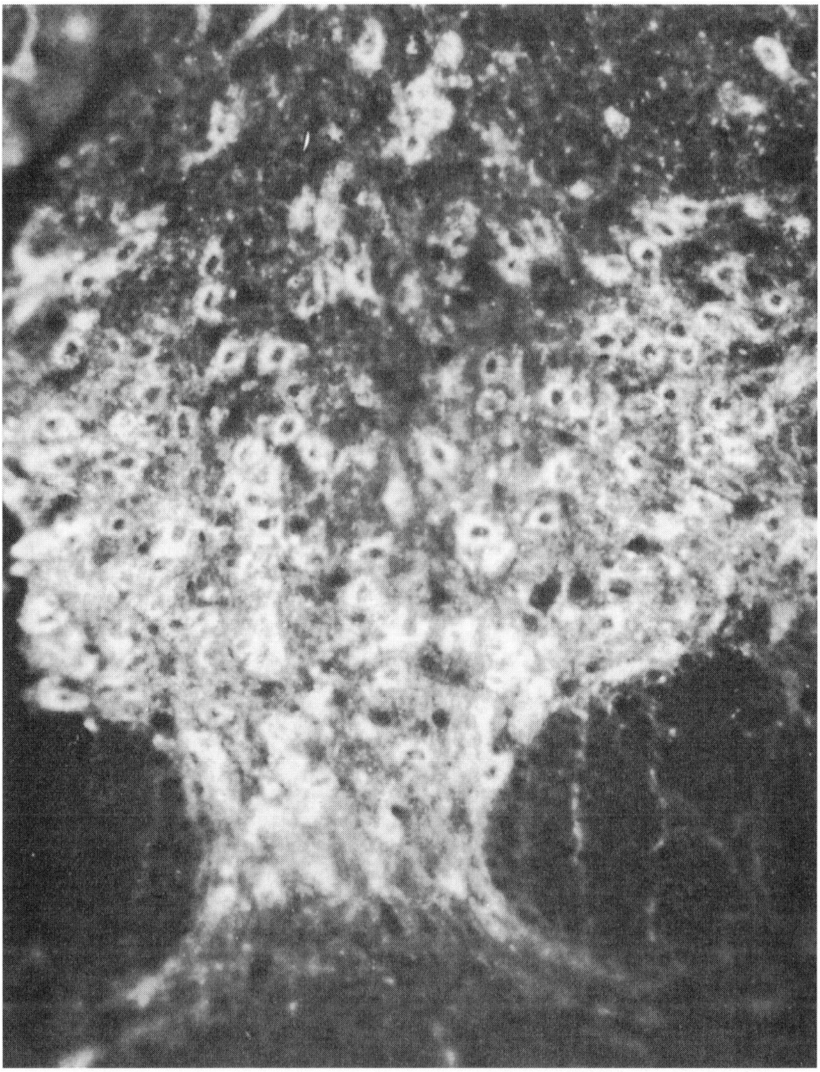

FIGURE 10–4. Fluorescence micrograph of raphe cell bodies in the mid-brain of the rat. This rat was pretreated with L-tryptophan (100 mg/kg) 1 hour prior to death (Courtesy of G. K. Aghajanian, 1995).

labeled 5-HT, electron-dense analogues, or 5-HT-selective toxins (e.g., the dihydroxytryptamines) depended for specificity on the selectivity of the uptake process. This situation has been rectified by immunocytochemistry of endogenous 5-HT (Figs. 10–5, 10–6), employing antibodies directed against serotonin.

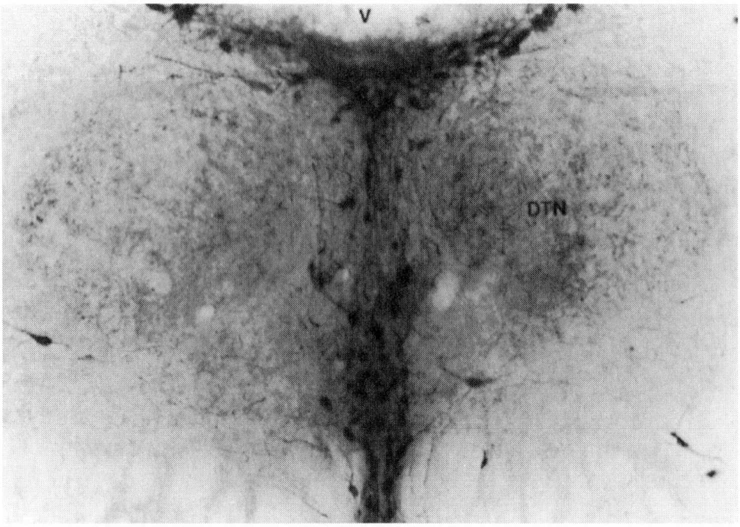

FIGURE 10–5. Photomicrograph of the serotonergic neurons of the caudal portion of the dorsal raphe. Section was stained using an antibody directed against serotonin. The serotonergic innervation of the dorsal tegmental nucleus (DTN) can also be seen. V marks the third ventricle. (Courtesy of A. Y. Deutch.)

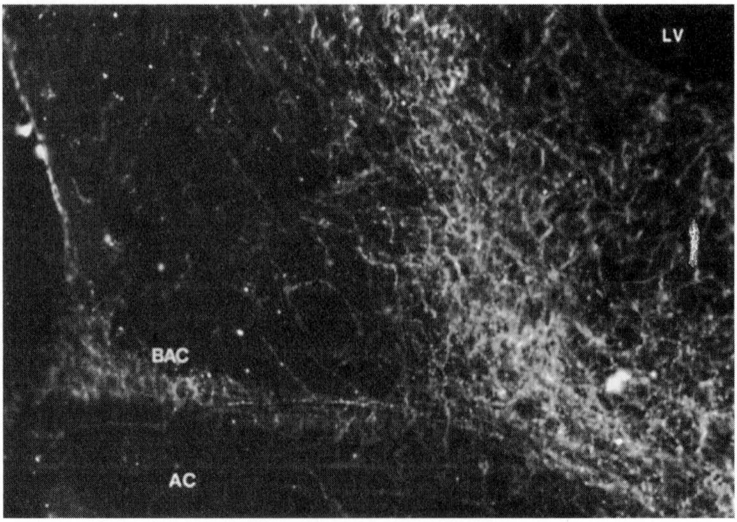

FIGURE 10–6. Darkfield photomicrograph illustrating the serotonergic innervation of the bed nucleus of the stria terminalis, as revealed by immunohistochemical staining with antibody directed against serotonin. The innervation of the bed nucleus of the anterior commissure (BAC) can also be seen above the anterior commissure (AC). LV marks the lateral ventricle. (Courtesy of A. Y. Deutch.)

With this more sensitive technique, it has become clear that the cerebral cortex in many mammals is innervated by two morphologically distinct types of 5-HT axon terminal. Fine axons with small varicosities originate from the dorsal raphe nuclei, and beaded axons with large spherical varicosities arise from the median raphe nuclei. These two types of 5-HT-containing axon have different regional and laminar distributions and appear to be differentially sensitive to the neurotoxic effects of certain amphetamine derivatives, including 3,4-methylenedioxymethamphetamine (MDMA), referred to more commonly as "ecstasy." The fine axons are much more sensitive to the neurotoxic effects than the beaded axons, and the loss of fine axons lasts for months and may be permanent. Beaded axons appear to be resistant and remain unaffected following neurotoxic treatment with MDMA. This finding may be relevant to human studies, which have indicated that individuals using MDMA as a recreational drug may be exposed to dosages approximating those shown to exhibit serotonin neurotoxicity in nonhuman primates. A 26% decrease in the serotonin metabolite 5-HIAA was observed in the cerebrospinal fluid of MDMA users. This indirect evidence of a decrease in serotonin turnover in the brain perhaps reflects destruction or compromised function of this fine serotonin-containing axon system. Further studies of MDMA users seem warranted and could provide important information on the effects of selective loss of this fine axon system in humans. At present, the functional roles played by the fine and beaded axon systems and whether the functions are distinct or similar remain unclear. In serial section analysis of 5-HT terminals in the primate visual cortex, the fine and fat boutons appeared to coexist in the same axon, arguing against distinct 5-HT innervation of this brain region.

Cellular Effects of 5-HT

From the biochemical and morphological data discussed above, we can be relatively certain that the 5-HT of the brain occurs not only within the nerve cells but also within specific tracts or projections of nerve cells. We must now inquire into the effects of serotonin when applied at the cellular level (see Chapter 2). In those brain areas where microelectrophoretically administered 5-HT has been tested on cells that exhibit spontaneous electrical activity, the majority of cells decrease their discharge rate. The effects observed typically last much longer than the duration of the microelectrophoretic current. In other regions, however, 5-HT also causes pronounced activation of discharge rate.

Electrophysiological analyses of 5-HT have focused on neurons of the facial motor nucleus (cranial nerve VII), where innervation by 5-HT fibers has been well documented. With intracellular recordings from these cells, 5-HT is found to produce a slow, depolarizing action accompanied by a modest increase in membrane resistance. As might then be anticipated on biophysical grounds, the combination of depolarization and increased membrane resistance facilitates the response of these neurons to other excitatory inputs. In a rigorous sense, such effects are not exactly in keeping with the emerging characteristics of modulatory actions (see Chapter 2) since here iontophoretic 5-HT changes both membrane potential and resistance on its own. It will be of interest to determine if activation of a 5-HT pathway to these cells at levels that do not in themselves directly change membrane properties will nevertheless modify responses of the target cells to other inputs, analogous to the effects described for noradrenergic connections.

Several specific 5-HT-containing tracts, investigated electrophysiologically, indicate that 5-HT produces mainly, if not exclusively, inhibitory effects. While the pathways do conduct slowly, as would be expected for such fine-caliber axons, the synaptic mediation process appears to be relatively prompt.

Characterization of 5-HT Receptors

The existence of multiple receptors for serotonin in the CNS has been suggested by physiological studies. Radioligand binding studies demonstrating that H^3-5-HT and H^3-spiperone label separate populations of high-affinity binding sites for 5-HT, termed 5-HT_1 and 5-HT_2, respectively, have also provided evidence for multiple receptors in brain tissue. A high correlation between the potencies of 5-HT antagonists in displacing the binding of H^3-spiperone and inhibiting serotonin-induced behavioral hyperactivity, coupled with a lack of such a correlation of H^3-5-HT binding sites, has led to the proposal that inhibition and excitation induced by serotonin are mediated at 5-HT_1 and 5-HT_2 receptors, respectively. Additional binding studies have further subdivided the 5-HT_1 class of recognition sites into 5-HT_{1A} and 5-HT_{1B} subtypes, and compounds selective for these receptor subtypes have been identified. Autoreceptors for 5-HT appear to fall into the 5-HT_{1A} category. Drugs such as 8-hydroxydipropylaminotetralin (8-OH DPAT) and the long-chain substituted piperazines appear to have selective 5-HT_{1A} binding activity, and these 5-HT autoreceptor agonists are effective at inhibiting midbrain raphe neurons. Ketanserin blocks 5-HT receptors that belong to the 5-HT_2 category without having any significant effect on 5-HT_1 receptors. Because of this discrimination, it has been described as a selective 5-HT_2 blocker. However, it is important to note that ketanserin is nonspecific in the sense that it has appreciable affinity for both α_1-adrenergic receptors and histamine H_1 receptors as well. It does have the advantage that its action on 5-HT receptors is as a pure antagonist.

5-HT Receptors

It is more than 40 years since Gaddum and Picarelli (1957), basing their work on the pharmacological properties of serotonin agonists and antagonists, reported evidence for two separate serotonin receptors in peripheral smooth muscle preparations studied in vitro. One they named the *D* receptor because it was blocked by dibenzyline, and the other they called the *M* receptor because the indirect contractile response to 5-HT mediated by the release of acetylcholine from cholinergic nerves in the myenteric plexus could be antagonized by morphine. The advent of receptor-binding studies in the 1970s revealed the existence of multiple binding sites, but while a correlation existed between the pharmacological profile of the 5-HT_2 binding site and the D receptor, no such correlation existed between the 5-HT_1 receptor subtypes identified by receptor-binding techniques and the M receptor. Since the M receptor originally described in the nerves of the guinea pig ileum is pharmacologically distinct from the 5-HT_1 and 5-HT_2 receptors and their numerous subtypes, it was renamed the 5-HT_3 receptor. With the introduction of several extremely potent and highly selective 5-HT_3 antagonists, attention has shifted to this receptor and its possible functions in both the periphery and the CNS, as well as the therapeutic potential for these newly developed 5-HT_3 compounds.

At least eight subtypes of serotonin receptor in brain tissue have been defined and characterized, based on radioligand binding studies. As noted in Chapter 5, it is premature to characterize as a receptor a binding site defined in this manner. Nevertheless, the majority of the original speculations about serotonergic receptor subtypes generated from radioligand binding experiments appear to have been substantiated by many other types of experiment, including, most recently, the cloning of three of these subtypes. Table 10–1 lists the characteristics of these 5-HT receptor subtypes.

Molecular Biology

Since the mid-1980s, a vast amount of new information has become available concerning the various 5-HT receptor subtypes and their functional and structural characteristics. This derives from two main research approaches, operational pharmacology employing selective agonists and antagonists and molecular biology. With the advent of the latter technique, the field of 5-HT receptors has experienced exceptionally rapid growth over the past 5

TABLE 10–1. Characteristics of 5-HT Receptor Subtypes

	5-HT$_1$ Receptors		
	5-HT$_{1A}$	*5-HT$_{1B}$*	*5-HT$_{1D}$*
High-density regions	Raphe nuclei Hippocampus	SN GP	GP SN BG
Selective agonists	R(1)8-OH DPAT 5-CAT	5-CAT	Sumatriptan 5-CAT
Selective antagonists	Spiperone *S*(2)-Pindolol WAY-100135	Isamoltane	Isamoltane GR-127935
Membrane effects	Hyperpolar via opening K channels	?	?
Other functional correlates	Basilar artery Thermoregulation Hypotension Sexual behavior 5-HT Syndrome (Somatodendritic autoreceptor)	Autoreceptor (Release Mod.)	Autoreceptor (Release Mod.)

	5-HT$_2$ Receptors		
	5-HT$_{2A}$	*5-HT$_{2B}$*	*5-HT$_{2D}$*
High-density regions	Layer IV Cortex Hippocampus FMN	Stomach Fundus Cortex	Choroid plexus Medulla Hippocampus
Selective agonists	DOI DOB α-Me5-HT	DOI	DOB α-Me-5-HT
Selective antagonists	MDL-100,907 Ketanserin Ritanserin Spiperone	Ketanserin	Ketanserin Spiperone Metergoline

(*continued*)

Table 10–1. *(Continued)*

	5-HT₂ Receptors		
	5-HT$_{2A}$	*5-HT$_{2B}$*	*5-HT$_{2D}$*
Membrane effects	Depolarization	Depolarization	Depolarization via opening of Ca^{2+} channel
Other functional correlates	?	Vas. cont. Platelet aggregation Paw edema Head twitches	

Other 5-HT Receptor Subtypes			
5-HT$_3$	*5-HT$_4$*	*5-HT$_6$*	*5-HT$_7$*
PN EC Area postrema	Hippocampus Colliculi Ileum	Striatum Accumbens Cortex	Thalamus Hypothalmus Amygdala Heart
α-Me-5-HT	SC-53116	5-CAT LSD	LSD
Zacopride Odanserton Granisotron	SDZ-205557	Lisuride Clozapine	Clozapine Amitryptline
Depolarization			
Transmitter release	Smooth muscle relaxation		
von Bezold-Jarisch reflex	Ionotropic		

α-Me-5-HT, α-methyl-5-hydroxytryptamine; BG, basal ganglia; DOB, 2,5-dimethoxy-4-bromoamphetamine; DOI, 2,5-dimethoxy-4-odoamphetamine; EC, entorhinal cortex; 5-CAT, 5-carboxamidotryptamine; FMN, facial motor neurons; GP, globus pallidus; PN, peripheral neurons; SN, substantia nigra; 5-HT, 5-hydroxytryptamine; LSD, lysergic acid diethylamide; Rel. Mod., release modulation; Vas. cont., vascular contraction.

years and the existence of multiple 5-HT receptors has been unequivocally confirmed. Because medicinal chemistry has lagged behind molecular biology in 5-HT neuropharmacology, there are very few highly selective receptor agonists and antagonists for the individual 5-HT receptor subtypes.

Serotonin receptors are highly heterogeneous, and cloning not only has led to the discovery and recognition of previously unknown receptors but has greatly facilitated their classification. As of the new millenia, there are at least 15 molecularly identified 5-HT receptors, some with splice varients and others with isoforms created by mRNA editing. The majority of the 5-HT receptors belong to the large family of receptors interacting with G proteins, except for the 5-HT_3 receptors, which are ligand-gated ion channel receptors (see Table 10–2). The 5-HT receptors belonging to the G protein receptor superfamily are characterized by the presence of seven transmembrane domains and the ability to alter G protein–dependent processes. The amino acid sequence of the membrane-spanning domains shows the least amount of variability compared with other cloned biogenic amine receptors.

This group of 5-HT receptors can be divided into distinct families based on their coupling to second messengers and their amino acid sequence homology. The 5-HT_1 family contains receptors that are negatively coupled to adenylyl cyclase. The 5-HT_2 family contains three receptors that have striking amino acid homology and the same coupling with second messenger, that is, activation of phospholipase C. The 5-HT receptors that are positively coupled to adenylyl cyclase are a heterogeneous group, including the 5-HT_4, 5-HT_6, and 5-HT_7 subtypes. The 5-HT_5 group contains two types, $5HT_{5A}$ and 5-HT_{5B}, and represents a new family of 5-HT receptors that do not resemble receptors of the 5-HT_1 and 5-HT_2 families in terms of amino acid sequence, pharmacological profile, and transduction system. They are probably coupled to a different effector system. In contrast to the G protein–coupled 5-HT receptors that modulate cell activities via second-messenger systems, 5-HT_3 receptors directly activate a 5-HT-gated cation channel that rapidly and transiently depolarizes a variety of neurons. Like other ligand-gated ion channels, the 5-HT_3 receptor consists of four transmembrane segments and a large extracellular N-terminal region incorporating a cysteine–cysteine loop and potential *N*-glycosylation sites. Other members of this molecular receptor family include GABA, glutamate, glycine, and the nicotinic cholinergic receptors.

A third major molecular recognition site for 5-HT is the transporter proteins. These transporter proteins consist of 12 membrane-spanning proteins and represent a large gene family encoding Na^+ and Cl^--dependent transport proteins. The first 5-HT transporter was identified in rat brain, but 5-HT transporters have been characterized in other tissues, including

TABLE 10–2. Mammalian 5-HT Receptor Subtypes and Their Signal-Transduction Pathways

G Protein–Coupled Receptors	*Receptor Subtype*	*G Protein*	*Effector Pathway*
5-HT_1 family	5-HT_{1A}	G_i	Inhibition of adenylate cyclase
		G_i	Opening of K^+ channel
		G_o	Closing of Ca^{2+} channel
	5-HT_{1B}	G_i	Inhibition of adenylate cyclase
	5-HT_{1D}	G_i	Inhibition of adenylate cyclase
	5-HT_{1E}	G_i	Inhibition of adenylate cyclase
5-HT_2 family	5-HT_{2A}	G_q	Phosphoinositide hydrolysis
	5-HT_{2B}		Phosphoinositide hydrolysis
	5-HT_{2C}		Phosphoinositide hydrolysis
Others	5-HT_4	G_s	Activation of adenylate cyclase
	5-HT_{5A}, 5-HT_{5B}	?	Unknown coupling mechanism
	5-HT_6	G_s	Activation of adenylate cyclase
	5-HT_7	G_s	Activation of adenylate cyclase
Ligand-gated ion channels	5-HT_3	None	Ligand-gated ion channel

platelets, placenta, lung, and basophilic leukemia cells. A number of selective 5-HT uptake blockers have been developed, such as fluoxetine, sertraline, citalopram, and paroxetine (see Fig. 10–7), and several have exhibited clinical utility in the treatment of depression and obsessive-compulsive disorders.

The careful categorization of old and new subtypes of 5-HT receptors should be an important foundation for defining their function. The development of selective agents targeting these receptors, coupled with the molecular approaches highlighted in Chapter 3, such as antisense oligonu-

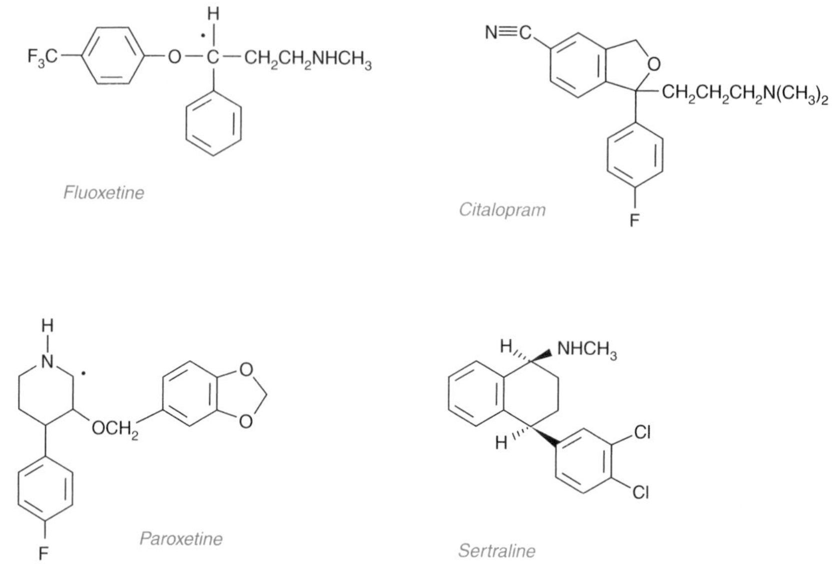

FIGURE 10–7. Selective inhibitors of the serotonin transporter.

cleotides to decrease steady-state levels of target proteins and transgenic animals in which specific receptors have been "knocked out," will undoubtedly enhance our understanding of the function of central serotonergic systems. The existence of a large number of receptors with distinct signaling properties and expression patterns might enable a single substance like 5-HT to generate simultaneously a large array of effects in many discrete brain structures.

Behavioral Aspects of Serotonin Function

5-HT neurons in the brain stem raphe nuclei exhibit spontaneous monotonic activity, discharging in a clock-like manner with an intrinsic frequency of 1 to 5 spikes/second. In the rodent, these monotonic properties are manifested early in development (3 to 4 days before birth). The 5-HT neurons appear to possess a negative feedback mechanism that limits their neuronal activity. As the physiological activity of the 5-HT neuron increases, the local release of 5-HT from dendrites or axonal collaterals acts on somatodendritic 5-HT autoreceptors to inhibit neuronal activity. This autoregulatory mechanism seems to function only under physiological conditions and to be inoperative during periods of low-level activity, but it becomes functional as neuronal activity increases. Dysfunction of this autoregulatory mechanism has been implicated in some forms of human neuropathology, so autoreceptors have become a potentially important site for drug-targeted therapeutic intervention. The combined use of 5-HT_{1B} autoreceptor antagonists and selective 5-HT uptake blockers holds some promise.

In view of the extraordinarily widespread projections and highly regulated pacemaker pattern of activity that is characteristic of serotonin neurons, a broad homeostatic function has been suggested for serotonergic systems. By exerting simultaneous modulatory effects on neuronal excitability in diverse regions of the brain and spinal cord, the serotonergic system is in a strategic position to coordinate complex sensory and motor patterns during varied behavioral states. Single-unit electrophysiological recordings from serotonergic neurons in unanesthetized animals have shown that serotonergic activity is highest during periods of waking arousal, reduced in quiet waking, reduced further in slow-wave sleep, and absent during rapid-eye-movement sleep. An increase in tonic activity of serotonergic neurons during waking arousal would enhance motor neuron excitability via descending projections to the ventral horn of the spinal cord. Suppression of sensory input, conversely, would screen out distracting sensory cues. Cessation of serotonergic neuronal activity during rapid-eye-movement sleep would tend to impede motor function in this paradoxical state in which internal arousal is associated with diminished motor output. Altered function of serotonergic systems has been reported in several psychopathological conditions, including affective illness, hyper-aggressive states, and schizophrenia. There is mounting evidence for impaired serotonergic function in major depressive illness and suicidal behavior. In this connection, several effective antidepressant drugs appear to act by enhancing serotonergic transmission.

The pathophysiology of major affective illness is poorly understood, but several lines of clinical and preclinical evidence indicate that enhancement of 5-HT-mediated neurotransmission might underlie the therapeutic response to different types of antidepressant treatment (see Chapter 13). Table 10–3 shows the effects of long-term administration of different types of antidepressant on the 5-HT system assessed with electrophysiological techniques. Treatment with all of these drugs appears to cause a net increase in 5-HT neurotransmission. Several clinical observations have also provided strong evidence of a pivotal role for 5-HT neurotransmission in depression. For example, a large number of selective 5-HT reuptake inhibitors examined clinically have been found to be effective in major depression. These drugs belong to different chemical families but appear to share a single common property, the ability to inhibit the 5-HT reuptake carrier. In addition, clinical studies show that inhibition of 5-HT synthesis in drug-remitted depressed patients, using either the tryptophan hydroxylase inhibitor *p*-

TABLE 10–3. Effects of Long-Term Administration of Antidepressant Drugs and Electroconvulsive Therapy on the 5-Hydroxytryptamine (5-HT) System Assessed Using Electrophysiological Techniques

Antidepressant Treatment	*Responsiveness of Somatodendritic $5\text{-}HT_{1A}$ Autoreceptors*[a]	*Function of Terminal 5-HT Autoreceptors*	*Function of Terminal α_2-Adrenoceptors*	*Responsiveness of Postsynaptic 5-HT Receptors*[a]	*Net 5-HT Neurotransmission*[b]
Selective 5-HT reuptake inhibitors	Decrease	Decrease	n.c.	n.c.	Increase
Monoamine oxidase inhibitors	Decrease	n.c.	Decrease	n.c. or decrease	Increase
$5\text{-}HT_{1A}$ receptor agonists	Decrease	n.c.	n.d.	n.c.	Increase
Tricyclic antidepressants	n.c.	n.c.	n.d.	Increase	Increase
Electroconvulsive shocks	n.c.	n.c.	n.c.	Increase	Increase

n.c., No change; n.d., not determined.

[a]Assessed by microiontophoresis or systemic injection of 5-HT receptor agonists.

[b]Determined from the firing activity of the presynaptic neurons and the effect of stimulating the ascending pathway on postsynaptic neurons.

SOURCE: Modified from Blier and deMontigny (1994).

chlorophenylalanine or the tryptophan depletion paradigm, produces a rapid relapse of depression. In the latter paradigm, the symptomatology reactivated by the tryptophan depletion was nearly identical to that present before the response to the antidepressant treatment, suggesting a causal relationship.

Signal-Transduction Pathways

Only recently has it been possible to examine the mechanisms of signal transduction in central 5-HT receptor systems, and evidence is emerging that major transducing systems are linked to different 5-HT receptors in the mammalian brain.

In general, two major 5-HT receptor–linked signal-transduction pathways exist: direct regulation of ion channels and a multistep enzyme-mediated pathway. Both pathways require a guanine nucleotide triphosphate (GTP)–binding protein (G protein) to link the receptor to the effector molecule. The 5-HT_1 family of receptors is negatively coupled to adenylyl cyclase via the G_1 family of G proteins. The 5-HT_{1A} receptor is the best characterized of this family. This receptor, in addition to coupling with adenylyl cyclase, is linked directly to voltage-sensitive K^+ channels via G_i-like proteins. This direct coupling with both adenylyl cyclase and the K^+ channel is a recognized characteristic of G_i-linked receptors. Direct coupling to L-type Ca^{2+} channels has also been described as an additional transduction pathway for the G_i-linked family of receptors. The other members of the 5-HT_1 family of receptors have also been shown to be negatively coupled to adenylyl cyclase (Table 10–2). The 5-HT_2 family of receptors, in contrast to the 5-HT_1 family, is coupled to phospholipase C. The G protein involved has not been identified but is assumed to be a member of the G_q family. Phospholipase C activation induces diverse changes in the cell, leading to regulation of numerous cellular processes. All members of this family appear to be coupled primarily to phospholipase C, leading to phosphoinositide hydrolysis. Although stimulation of adenylyl cyclase was the first signal-transduction pathway to be linked to 5-HT, the specific receptors mediating activation of adenylyl cyclase, the 5-HT_4, 5-HT_6, and 5-HT_7 receptors, were identified only recently. The 5-HT_4 receptor is found in rodent brain (hippocampus) and peripheral tissue, including the guinea pig ileum and human atrium. The 5-HT_7 receptor is also found in the brain and heart (see Table 10–1). Another novel receptor, 5-HT_6, is the most recent serotonin receptor to be identified by molecular cloning. The 5-HT_6 receptor has also been shown conclusively to couple positively to adenylyl cyclase and to have a high affinity for tricyclic antidepressant drugs. The interesting distribution of this receptor in the brain coupled with its high affinity for atypical antipsychotic and tricyclic antidepressant drugs has led to significant efforts to learn more about its function and possible role in psychiatric disorders. While our knowledge is far from complete, emerging data suggest that 5-HT_6 receptors regulate cholinergic neurotransmision in the brain rather than the anticipitated modulation of dopaminergic function.

The 5-HT_3 receptor differs from other 5-HT receptors by forming an ion channel that regulates ion flux in a G protein–independent manner. This receptor is a member of a large family of ligand-gated ion channels and thus shares more similarities with the nicotinic cholinergic receptor, which is the prototype of this superfamily, than with the 5-HT_1 and 5-HT_2 receptor families. The 5-HT_3 receptors were first found on peripheral sensory, autonomic, and enteric neurons, where they mediate excitation. The direct demonstration that 5-HT_3 receptors in the guinea pig submucosal plexus are ligand-gated ion channels implies a role for 5-HT as a "fast" synaptic transmitter and fits with their function in the periphery. Molecular studies indicated that the cloned 5-HT_3 receptor protein forms a homomeric subunit, which regulates the gating of cations and thus presumably mediates the rapid and transient depolarization that occurs following 5-HT_3 receptor activation. Immunocytochemical studies have revealed that 5-HT_3 receptor–immunoreactive neurons are broadly distributed throughout the rat brain and spinal cord and suggest that this receptor can subserve significant participation in CNS neurotransmission. However, the neuronal circuits in which this receptor might participate and the functions subserved by it remain to be established.

In general, the pharmacological and electrophysiological characteristics of the cloned receptor are largely consistent with the properties of the native receptors. Despite this consistency, however, a number of biochemical, pharmacological, and electrophysiological studies have suggested that CNS 5-HT_3 receptors exhibit heterogeneity with respect to subtypes and intracellular signal-transduction mechanisms. Thanks to the pharmacological similarities between the 5-HT M receptors and the 5-HT_3 receptors, we have numerous pharmacological agents that exert specific effects on 5-HT_3 receptors, and many studies using these agents have addressed the possible functions of 5-HT_3 receptors. Results obtained from these in vivo studies have also suggested the presence of multiple 5-HT_3-like receptors, spurring on the search for other subunit proteins that might explain this pharmacological heterogeneity. The possibility should be entertained that 5-HT_3 receptors are analogous to the glutamate receptors, which were first characterized as inotropic receptors and later shown to exist also as G protein–linked metabotropic receptors. Clearly, additional studies are required to determine whether the 5-HT_3 receptor heterogeneity can be explained by subtypes of

5-HT_3 receptor from both the inotropic and metabotropic superfamilies. The discovery of multiple subtypes of 5-HT_3 may help to clarify the electrophysiological observation that 5-HT_3-like receptors in the prefrontal cortex elicit a slow depression of cell firing rather than the fast activation expected for a ligand-gated 5-HT_3 receptor.

Recent data suggest that 5-HT_3 receptors are coupled to an ion channel, probably a calcium channel. Thus, they share more similarities with the nicotinic cholinergic receptor than with the 5-HT_1 or 5-HT_2 receptors. The direct demonstration that 5-HT_3 receptors in guinea pig submucosal plexus are ligand-gated ion channels implies a role for 5-HT (and perhaps for other biogenic amines) as a "fast" synaptic transmitter.

Adaptive Regulation

5-HT_{2A} and 5-HT_{2C} receptors adapt to chronic activation by reducing response sensitivity or receptor density as expected, but chronic inactivation does not elicit the opposite adaptive response. Central 5-HT_2 and 5-HT_{2C} receptors seem to be relatively resistent to upregulation. For example, 5-HT_{2A} receptors are not upregulated after denervation of 5-HT neurons or chronic administration of 5-HT antagonists. Instead, chronic administration of 5-HT antagonists results in a paradoxical downregulation of both 5-HT_{2A} and 5-HT_{2C} receptors.

Physiology

Neurophysiological and behavioral studies have benefited from the development and characterization of selective agents for 5-HT receptors. Electrophysiological studies have clearly demonstrated that 5-HT_{1A} receptors mediate inhibition of the raphe nuclei. The firing of serotonergic neurons is tightly regulated by intrinsic ionic mechanisms (e.g., calcium-activated potassium conductance), which accounts for the well-known tonic pacemaker pattern of activity of these cells. The intrinsic pacemaker is modulated by at least two neurotransmitters: *(1)* NE acting through adrenergic receptors accelerates the pacemaker and *(2)* 5-HT acting through somatodendritic 5-HT_{1A} autoreceptors slows the pacemaker. In the hippocampus, which is another anatomical structure containing a high density of 5-HT_{1A} sites, 5-HT_{1A} agonists hyperpolarize CA1 pyramidal cells by opening potassium channels via a pertussis toxin–sensitive G protein. Electrophysiological studies carried out in *Xenopus* oocytes, which express the 5-HT_{1C} receptor, revealed that application of 5-HT causes a detectable inward current that is blocked by mianserin. Activation of the 5-HT_{1C} receptor apparently liberates inositol phosphates, raising intracellular Ca^{2+} levels and leading to the opening of Ca^{2+}-dependent chloride channels. In mammalian systems, two specific neurophysiological effects have been attributed to activation of the 5-HT_2 receptor. 5-HT facilitates the excitatory effects of glutamate in the facial motor nucleus, an action that is antagonized by 5-HT_2 antagonists. Similar data were obtained from intracellular studies that showed that activation of 5-HT receptors caused a slow depolarization of facial motor neurons and increased input resistance, leading to increased excitability of the cell, probably through a decrease in resting membrane conductance to potassium. These data suggest that 5-HT_2 receptors mediate the 5-HT-induced excitation of facial motor neurons.

It has also been shown that 5-HT causes a slow depolarization of cortical neurons that is associated with decreased conductance. The effect can be desensitized by repeated applications of 5-HT and blocked by the selective 5-HT_2 antagonist ritanserin. Thus, the effects of 5-HT on cortical pyramidal neurons share many similarities to the depolarizing effects observed in the facial motor nucleus and appear to be mediated by 5-HT_2 receptors.

5-HT_3 receptors in peripheral nervous tissue mediate excitatory responses to 5-HT and are involved in modulating transmitter release. These receptors are also found in the CNS, where they are present in high density in the entorhinal cortex and area postrema. Release of endogenous dopamine by stimulation of 5-HT_3 receptors in the striatum occurs in a calcium-dependent fashion. Even though the physiological function of 5-HT_3 receptors in the CNS is unclear, the observation that selective 5-HT_3 antagonists possess central activity in rats and primates in anxiolytic and antidopaminergic-like behavioral models has generated considerable excitement and speculation about the potential therapeutic use of 5-HT_3 antagonists and agonists. From animal experiments, it has been speculated that these compounds may be useful in treating schizophrenia, pain, anxiety, drug dependence, and cytotoxic drug-induced emesis; but so far only the analgesic and antiemetic effects have been demonstrated in humans.

Electrophysiology of 5-HT Receptors

Knowledge of the molecular biology of 5-HT receptors has revolutionized electrophysiological approaches to investigating the 5-HT systems in brain since studies can now be directed toward neurons that express specific 5-HT receptor subtypes based on in situ hybridization maps of receptor mRNA expression. This has enabled the diverse electrophysiological actions of 5-HT in the CNS to be categorized according to receptor subtypes and their respective effects or mechanisms of action. Following are a few generalizations from these studies: *(1)* inhibitory effects of 5-HT are mediated by 5-HT_1 receptors linked to the opening of K^+ channels or to the closing of Ca^{2+} channels, both via pertussis toxin–sensitive G proteins; *(2)* facilitative effects of 5-HT that are mediated by 5-HT_2 receptors and involve the closing of K^+ channels can be modulated by the phosphatidylinositol second-messenger system and PKC acting as a negative feedback loop; *(3)* other facilitative effects of 5-HT appear to be mediated by 5-HT_4 and 5-HT_7 receptors by a reduction in certain voltage-dependent K^+ currents mediated through the protein kinase A phosphorylation pathway and thus involving positive coupling of the 5-HT response to adenylyl cyclase; *(4)* fast excitations are mediated by 5-HT_3 receptors through a ligand-gated cationic ion channel that does not require coupling with a G protein or a second messenger. Thus, it is clear that the electrophysiological actions of 5-HT encompass the two major neurotransmitter gene superfamilies, the G protein–coupled receptors and the ligand-gated cationic channels. The end effects are determined by the receptor subtype and its anatomical location (Fig. 10–8).

Relevance to Clinical Disorders and Drug Actions

The diversity of receptors and transduction pathways that underlie the actions of 5-HT, together with the differential expression of 5-HT receptor subtypes in different neuronal and effector cell populations, helps to explain how it is possible for a single transmitter to be linked to such a large array of behaviors, clinical conditions, and drug actions. Alterations in 5-HT function have been implicated in a host of clinical states, including affective disorder, obsessive-compulsive disorder, schizophrenia, anxiety states, phobic disorders, eating disorders, migraine, and sleep disorders (see Table 10–4). There is also a wide range of psychotropic drugs that affect 5-HT neurotransmission, including antidepressants (selective 5-HT uptake blockers, e.g., fluoxetine [see Fig. 10–7]), hallucinogens (e.g., LSD, psilocin [see Fig. 10–1]), anxiolytics (e.g., buspirone), atypical antipsychotics (e.g., clozapine), antiemetics (e.g., ondansetron), appetite-suppressants (e.g., fenfluramine), and antimigraine drugs (e.g., sumatriptan).

Behavioral Effects of 5-HT Systems

One of the first drugs found empirically to be effective as a central tranquilizing agent, reserpine, was employed mainly for its action in the treatment of hypertension. In the early 1950s, when both brain serotonin and the central effects of reserpine were first described, great excitement arose when the dramatic behavioral effects of reserpine were correlated with loss of 5-HT content. However, it was soon found that NE and dopamine content could also be depleted by reserpine and that all of these depletions lasted longer

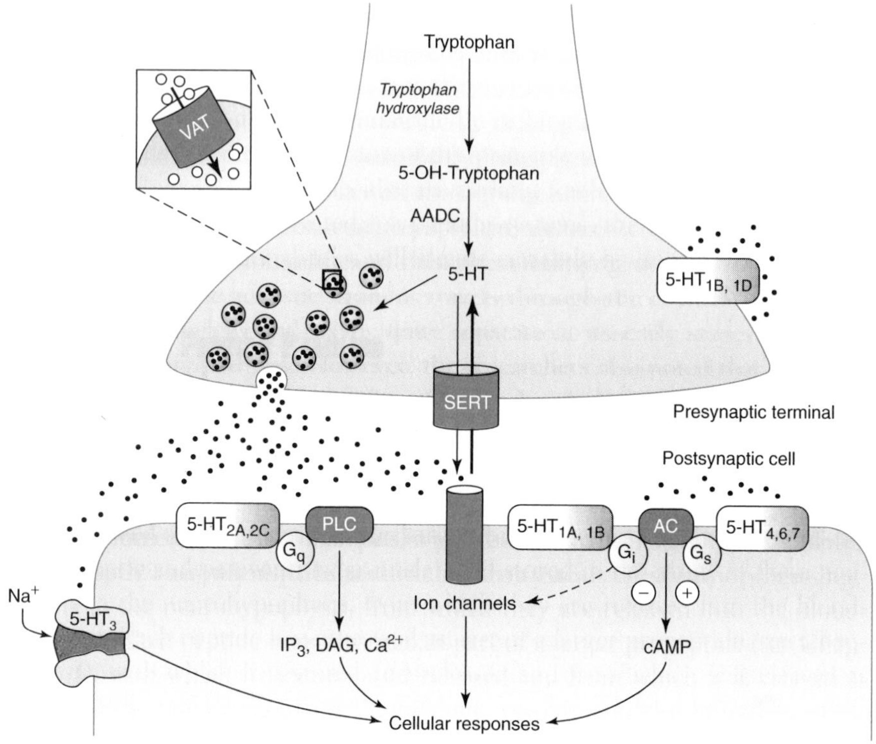

FIGURE 10–8. Model of a serotonin (5-HT) synapse. Tryptophan is taken up into the neuron by an active transport mechanism and converted to 5-OH tryptophan (5-HTP) by the rate-limiting enzyme tryptophan hydroxylase. Aromatic amino acid decarboxylase (AADC) converts 5-HTP to 5-HT. 5-HT is taken up and stored by the VMAT. Once released, 5-HT can interact with as many as 15 different receptors. 5-HT autoreceptors (5-HT_{1B} rodent, and 5-HT_{1D} human), which modulate the stimulus-induced release of 5-HT, are located on the presynaptic terminals. 5-HT_{1A} autoreceptors are located on the 5-HT cell bodies and dendrites where they modulate impulse flow. A number of G protein–coupled 5-HT receptors are located postsynaptically. The 5-HT_3 receptor is also localized postsynaptically but in contrast to the other 5-HT receptors is a ligand-gated ion channel similar to the cholinergic nicotinic receptor. The action of 5-HT is terminated by reuptake into the neuron by SERT. AC, adenylyl cyclase; DAG, diacylglycerol; IP3, inositol triphosphate; PLC, phospholipase C; SERT, plasma membrane serotonin transporter; VMAT, vesicular monoamine transporter. (Modified from Nesther et al., *Molecular Neuropharmacology*.)

TABLE 10–4. Clinical Conditions Influenced by Altered Serotonin Function

Affective disorders	Neuroendocrine regulation
Aging and neurodegenerative disorders	Obsessive-compulsive disorder
Anxiety disorders	Pain sensitivity
Carcinoid syndrome	Premenstrual syndrome
Circadian rhythm regulation	Posttraumatic stress syndrome
Developmental disorders	Schizophrenia
Eating disorders	Sexual disorders
Emesis	Sleep disorders
Migraine	Stress disorders
Myoclonus	Substance abuse

than sedative actions. Brain levels of all three amines remained down for many days, while the acute behavioral effects ended in 48–72 hours. Hence, it became difficult to determine whether it was the loss of brain catecholamine or serotonin that accounted for the behavioral depression after reserpine.

More extensive experiments into the nature of this problem were possible when the drugs that specifically block the synthesis of catecholamines or serotonin were discovered. After depleting brain serotonin content with *p*-chlorophenylalanine, which effectively removed 90% of brain serotonin, investigators observed that no behavioral symptoms reminiscent of the reserpine syndrome appeared. Moreover, when *p*-chlorophenylalanine-treated rats, which were already devoid of measurable serotonin, were treated with reserpine, typical reserpine-induced sedation arose. These results again favor the view that loss of catecholamines could be responsible for the reserpine-induced syndrome. Furthermore, when α-methyl-*p*-tyrosine was given and synthesis of NE was blocked for prolonged periods of time, the animals were behaviorally sedated and their condition resembled the depression seen after reserpine. Students will find it profitable to review in detail the original papers describing the results just summarized and the other theories currently in vogue.

Hallucinogenic Drugs

One of the more alluring aspects of the study of brain serotonin is the possibility that it is this system of neurons through which the hallucinogenic drugs cause their effects. In the early 1950s, the concept arose that LSD might produce its behavioral effects in the brain by interfering with the action of serotonin there as it did in smooth muscle preparations, such as the rat uterus. However, this theory of LSD action was not supported by the finding that another serotonin-blocking agent, 2-bromo-LSD, produced minimal behavioral effects in the CNS. It was subsequently shown that very low concentrations of LSD itself, rather than blocking serotonin action, could potentiate it. However, none of these data could be considered particularly pertinent since all of the research was done on the peripheral nervous system and all of the philosophy was applied to the CNS.

Shortly thereafter, Freedman and Giarman initiated a profitable series of experiments investigating the basic biochemical changes in the rodent brain following injection of LSD. Although their initial studies required them to use a bioassay for changes in serotonin, they detected a small (on the order of 100 ng/g or less) increase in the serotonin concentration of the rat brain shortly after injection of very small doses of LSD. Subsequent studies have shown that a decrease in 5-HIAA accompanies the small rise in 5-HT. Although the biochemical effects are similar to those that would be seen from small doses of monoamine oxidase inhibitors, no direct monoamine oxidase inhibitory effect of LSD has been described. This effect was generally interpreted as indicating a temporary decrease in the rate at which serotonin was metabolized, and it could also be seen with higher doses of less effective psychoactive drugs. In related studies by Costa and co-workers, who estimated the biochemical turnover of brain serotonin, prolonged infusion of somewhat larger doses of LSD clearly promoted a decrease in the turnover rate of brain serotonin.

The next advance in the explanation of the LSD response was made when Aghajanian and Sheard reported that electrical stimulation of the raphe nuclei selectively increased the metabolism of 5-HT to 5-HIAA. This finding suggested that the electrical activity of 5-HT cells could be directly reflected in the metabolic turnover of the amine. Subsequently, the same authors recorded single raphe neurons during parenteral administration of LSD and observed that they slowed down, with a time course similar to the effect of LSD. Thus, following LSD administration, both decreased electrical activity of these cells and decreased transmitter turnover occur.

Of the several explanations originally proposed for this effect, the data now support the view that LSD depresses directly 5-HT-containing neurons at receptors, which may be the sites where raphe neurons feed back 5-HT messages to each other through recurrent axon collaterals or dendrodendritic interactions; at these receptive sites on raphe neurons, LSD is a 5-HT agonist. In other tests on raphe neurons, iontophoretic LSD antagonizes the activity of raphe neurons that show excited responses to either NE or 5-HT. However, at sites in 5-HT terminal fields, when LSD is evaluated as an agonist or antagonist of 5-HT, its effects are considerably weaker than on the raphe neurons. These observations have led Aghajanian to suggest that LSD begins its sequence of physiological and neurochemical changes by acting on 5-HT neurons. In this view, the psychedelic actions of LSD must entail a primary or perhaps total reliance upon decreased efferent activity of the raphe neurons.

As Freedman has indicated, however, several points need further clarification before the description of LSD-induced changes in cell firing and cell chemistry can be incorporated in an explanation of the pharmacological and behavioral effects of LSD. If the effects of LSD could be simply equated with silencing of the raphe and subsequent disinhibition of raphe synaptic target cells, then many aspects of LSD-induced behavioral changes should be detectable in raphe-lesioned animals; but they are not. Typical LSD effects are also seen when raphe-lesioned animals are administered LSD. When normal animals are pretreated with *p*-chlorophenylalanine (in dose schedules that antagonize raphe-induced synaptic inhibition) and given LSD, the effect is accentuated. In contrast to the predictions of the 5-HT silencing effect, the LSD response is further accentuated by concomitant treatment with 5-HTP. The physiological manipulation that most closely simulates the behavioral actions of LSD is stimulation of the raphe, leading to decreased habituation to repetitive sensory stimuli. These data are difficult to reconcile with the view that LSD-induced behavior results from inactivation of tonic 5-HT-mediated postsynaptic actions. An important aspect to be evaluated critically in the continued search for the cellular changes that produce the behavioral effects is the issue of tolerance: LSD and other indole psychotomimetic drugs show tolerance and cross-tolerance in humans, but these properties have not been seen in animal studies at the cellular level.

It is extremely difficult to track down all of the individual cellular actions of a potent drug like LSD and to fit these effects together in a jigsaw puzzle-like effort to solve the question of how LSD produces hallucinations. Similar jigsaw puzzles lie just below the surface of every simple attempt to attribute the effects of a drug or the execution of a complex behavioral task like eating, sleeping, mating, and learning (no rank ordering of author's priorities intended) to a single family of neurotransmitters like 5-HT. While it is clearly possible to formulate hypothetical schemes by which divergent inhibitory systems like the 5-HT raphe cells can become an integral part of such diverse behavioral operations as pain suppression, sleeping, thermal regulation, and corticosteroid receptivity, a very wide chasm of unacquired data separates the concept from the documentary evidence needed to support it.

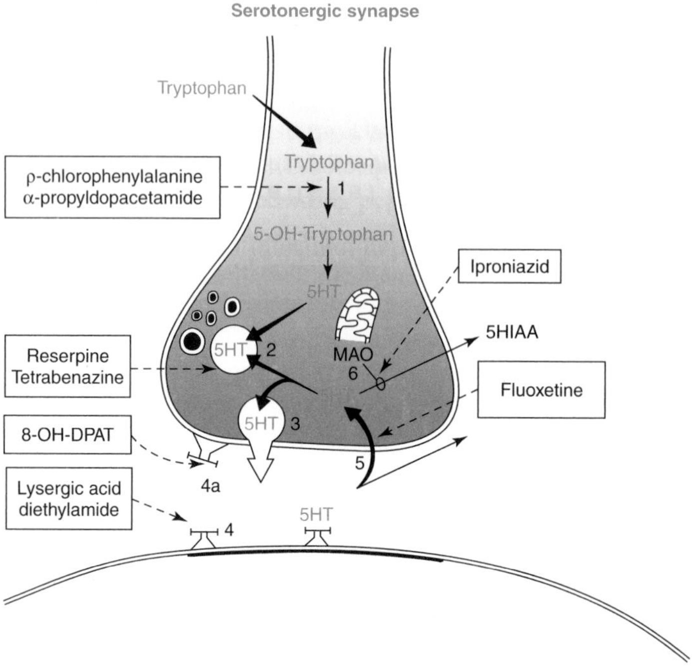

Figure 10–9. Schematic model of a central serotonergic neuron indicating possible sites of drug action.

Site 1: *Enzymatic synthesis:* Tryptophan is taken up into the serotonin-containing neuron and converted to 5-OH-tryptophan by the enzyme tryptophan hydroxylase. This enzyme can be effectively inhibited by p-chlorophenylalanine and α-propyldopacetamide. The next synthetic step involves the decarboxylation of 5-OH-tryptophan to form serotonin (5-hydroxytryptamine [5-HT]).

Site 2: *Storage:* Reserpine and tetrabenazine interfere with the uptake–storage mechanism of the amine granules, causing a marked depletion of serotonin.

Site 3: *Release:* At present, there is no drug available which selectively blocks the release of serotonin. However, lysergic acid diethylamide (LSD), because of its ability to block or inhibit the firing of serotonin neurons, causes a reduction in the release of serotonin from the nerve terminals.

Site 4: *Receptor interaction:* LSD acts as a partial agonist at serotonergic synapses in the central nervous system (CNS). A number of compounds have also been suggested to act as receptor-blocking agents at serotonergic synapses, but direct proof of these claims at the present time is lacking.

Site 4a: *Autoreceptor interaction:* Autoreceptors on the nerve terminal appear to play a role in the modulation of serotonin release. 8-Hydroxydipro-plaminotetraline (8-OH-DPAT) acts as an autoreceptor agonist at serotonergic synapses in the CNS.

Site 5: *Reuptake:* Considerable evidence suggests that serotonin may have its action terminated by being taken up into presynaptic terminals. The tricyclic drugs with a tertiary nitrogen, such as imipramine and amitriptyline, appear to be potent inhibitors of this uptake mechanism.

The gap is even wider than it seems since we do not at present have the slightest idea of the kinds of method that can be used to convert correlational data (raphe firing associated with sleeping-stage onset or offset) into proof of cause and effect. While previous editions of this guide to cellular neuropharmacology have provided a superficial overview of the behavioral implication of 5-HT neuronal actions, we now relinquish that effort until the cellular bases become more readily perceptible.

Actions of Other Psychotropic Drugs

Subsequent studies by Aghajanian and colleagues have indicated that other drugs that can alter 5-HT metabolism can also disturb the discharge pattern of the raphe neurons. Thus, monoamine oxidase inhibitors and tryptophan, which would be expected to elevate brain 5-HT levels, also slow raphe discharge, while clinically effective tricyclic drugs (antidepressants such as imipramine, chlorimipramine, and amitriptyline) also slow the raphe neurons and elevate synaptic 5-HT levels locally by inhibition of 5-HT reuptake. At present, it remains unclear whether these metabolic changes in 5-HT after psychoactive drugs are administered reflect changes in the cells that receive 5-HT synapses or in the raphe neurons themselves. A significant question is whether these circuits mediate the therapeutic responses of the drugs or are the primary pathological site of the diseases the drugs treat. One etiological view of schizophrenia, for example, might be based on the production of an abnormal indole, such as the hallucinogenic dimethyltryptamine, a compound that can be formed in human plasma from tryptamine. Some evidence indicates that schizophrenic patients have abnormally low monoamine oxidase activity in their platelets, which could permit the formation of abnormal amounts of plasma tryptamine and consequently abnormal amounts of dimethyltryptamine. Figure 10–9 summarizes possible sites of drug interaction in a hypothetical serotonin synapse in the CNS.

In this chapter, we have encountered one of the more striking examples of an intensively studied brain biogenic amine, and there is every reason to believe that it is an important synaptic transmitter. Still to be determined are the precise synaptic connections at which this substance transmits information and the functional role these connections play in the overall operation

---

Site 6: *Monoamine oxidase (MAO):* Serotonin present in a free state within the presynaptic terminal can be degraded by the enzyme MAO, which appears to be located in the outer membrane of mitochondria. Iproniazid and clorgyline are effective inhibitors of MAO. 5-HIAA, 5-hydroxyindoleacetic acid.

of the brain with respect to both effective and other multicellular interneuronal operations. The central pharmacology of serotonin, however, remains poorly resolved. Specific inhibition of uptake and synthesis are possible, but truly effective and selective postsynaptic antagonists remain to be developed.

HISTAMINE

Since the beginning of this century, when Henry Dale demonstrated that histamine is an endogenous tissue constituent and a potent stimulator of a variety of cells, this substance has been thought to act as a transmitter or neuromodulator. However, direct evidence in support of this idea has accumulated relatively slowly.

The challenge posed by histamine to neuropharmacologists has led to a vigorous chase across meadows of enticing hypotheses surrounded by bogs of confusion and dubious methodology. At last, more than 70 years after its isolation from the pituitary by J. J. Abel, the role of histamine in the brain seems to be approaching a resolution that this amine occurs in two types of cells: mast cells and magnocellular neurons in the posterior hypothalamus.

Most of our previous understanding of the synthesis and degradation of histamine in the brain was based on attempts to simulate in brain tissue data obtained from more or less homogeneous samples of peritoneal mast cells (Fig. 10–10). However, since mast cells were not supposed to be found in healthy brains and since histamine does not cross the blood–brain barrier, it had long been assumed that histamine could also be formed by neurons and, hence, be considered as a neurotransmitter. In fact, attempts to develop drugs that were histamine antagonists, in the mistaken belief that battlefield shock was compounded by histamine release, were a key to the subsequent development of antipsychotic drugs. Furthermore, as every hay-fever sufferer knows, present-day antihistamine drugs clearly produce substantial CNS actions such as drowsiness and hunger. Finally, in retrospect, the increased content of histamine found in transected degenerating sensory nerve trunks was likely an artifact of mast cell accumulation rather than a peculiar form of neurochemistry. Viewing the other data on CNS histamine through that same retrospectroscope, as they have been illuminated by the innovative experiments of Schwartz and associates, we now have a rather compelling case that histamine qualifies as a putative central neurotransmitter in addition to its role in mast cells.

The major obstacles in elucidating the functions of histamine in the brain have been the absence of a sensitive and specific method to demonstrate putative histaminergic fibers in situ, the lack of suitable methods to measure histamine and its catabolites, and problems in the characterization of hista-

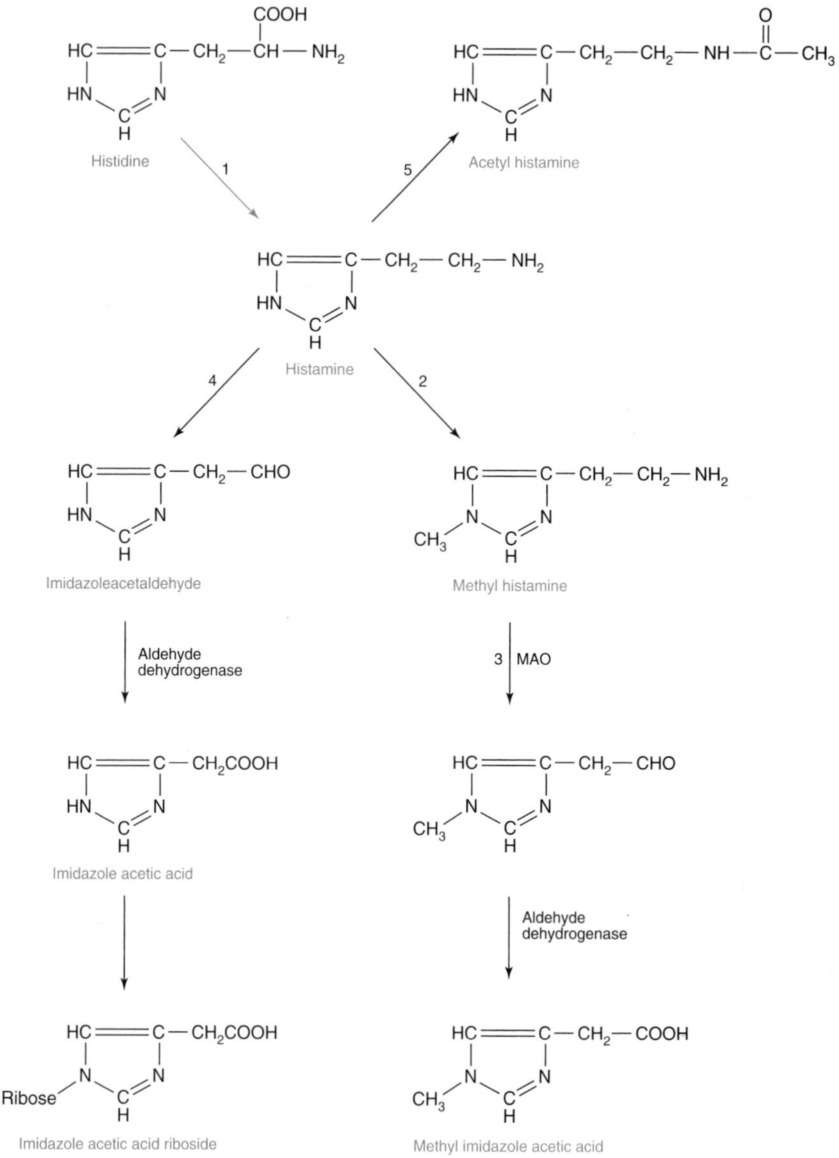

FIGURE 10–10. Metabolism of histamine. *(1)* Histidine decarboxylase. *(2)* Histamine methyltransferase. This is the major pathway for inactivation in most mammalian species. *(3)* Monoamine oxidase. *(4)* Diamine oxidase (histaminase). *(5)* Minor pathway of histamine catabolism.

mine receptors in the nervous system. Some progress has been made in overcoming these impediments, and evidence has accumulated to support the hypothesis that histamine functions as a neurotransmitter or neuromodulator in the brain.

Histamine Synthesis and Catabolism

Because histamine penetrates the blood–brain barrier so poorly, brain histamine arises from histamine synthesis in situ from histidine. Active transport of histidine by brain slices has been demonstrated, and, because histidine loading has been shown to elevate brain histamine, histidine transport could be a controlling factor in brain histamine synthesis.

Two enzymes are capable of decarboxylating histidine in vitro: L-AADC (i.e., DOPA-decarboxylase) and the specific histidine decarboxylase. The pH optimum, affinity for histidine, effects of selective inhibitors, and regional distribution of histamine-synthesizing activity indicate that the specific histidine decarboxylase is responsible for histamine biosynthesis in the brain (Fig. 10–10). The instability of histidine decarboxylase and its low activity in adult brain have precluded purification of the enzyme from this source, but fetal liver histidine decarboxylase has been purified to near homogeneity. Studies of the pH optimum, cofactor (pyridoxal phosphate), requirements, inhibitor sensitivity, and antigenic properties have demonstrated that the brain and fetal enzymes have similar properties. Antibodies to histidine decarboxylase have been used to map the distribution of this enzyme in the brain by immunohistochemical techniques.

Estimates of the K_m of histidine for brain histidine decarboxylase are much higher than the concentration of histidine in plasma or brain, suggesting that it is not saturated with substrate in vivo and consistent with the observations that administration of L-histidine increases brain histamine levels.

There are surprisingly few selective inhibitors of histidine decarboxylase. Most of the effective inhibitors act at the cofactor site, and although they reduce histamine formation, they also inhibit other pyridoxal-requiring enzymes.

A selective inhibitor of histidine decarboxylase, α-fluoromethylhistidine, has been identified. This agent acts selectively and irreversibly by formation of a covalent linkage with the serine residue at the active site of the enzyme. It is more selective and potent than other decarboxylase inhibitors and does not inhibit DOPA and glutamate decarboxylase or other histamine-metabolizing enzymes, such as histamine *N*-methyltransferase and *N*-acetylhistamine deacetylase. This agent may prove useful for manipulating in vivo actions of histamine.

Histamine is metabolized by two distinct enzymatic systems in mammals. It is oxidized by diamine oxidase to imidazoleacetaldehyde and then to imidazoleacetic acid and methylated by histamine methyltransferase to produce methylhistamine. Mammalian brain lacks the ability to oxidize histamine and nearly quantitatively methylates it. Methylhistamine undergoes oxidative deamination by either diamine oxidase or monoamine oxidase. The affinity is higher for diamine oxidase, but in brains that lack diamine oxidase, methylhistamine is oxidized primarily by monoamine oxidase type B.

The major route of catabolism of orally injested histamine is via *N*-acetylation of bacteria in the gastrointestinal tract to form *N*-acetylhistamine.

Histamine-Containing Cells

Although there are fluorogenic condensation reactions that can detect histamine in mast cells by cytochemistry, the method has never been able to demonstrate histamine-containing nerve fibers or cell bodies. Because the histamine content of mast cells is quite substantial, however, it has been possible to use the cytochemical method to detect mast cells in brain and peripheral nerve. From such studies, Schwartz et al. has estimated that the histamine content of brain regions and nerve trunks that show about 50 ng histamine/g (i.e., every place except the diencephalon) could be explained on the basis of mast cells. Moreover, the once inexplicable rapid decline of brain histamine in early postnatal development can also be attributed to the relative decline in the number of mast cells in the brain during development.

Mast cell histamine shows some interesting differences from what we shall presume is neuronal histamine. In mast cells, histamine levels are high, turnover is relatively slow, and the activity of histidine decarboxylase is relatively low; moreover, mast cell histamine can be depleted by mast cell–degranulating drugs (48/80 and polymyxin B). In brain, only about 50% of the histamine content can be released with these depletors and, for that which remains, the turnover is quite rapid. The activity of histidine decarboxylase is also much greater. Even better separation of the two types of cellular histamine dynamics comes from studies of brain and mast cell homogenates. In the adult brain, a significant portion of histamine is found in the crude mitochondrial fraction that is enriched in nerve endings and released from these particles upon hypo-osmotic shock. In cortical brain regions and in mast cells, most histamine is found in large granules, which sediment with the crude nuclear fraction; this histamine, unlike that in the hypothalamic nerve endings, has the slow half-life characteristic of mast cells.

Despite the encouraging result that histamine may be present in fractions of brain homogenates containing, among other broken cellular elements, nerve endings, it is difficult to parlay that information into a direct documentation of intraneuronal storage. More promising steps in that direction were taken in studies where specific brain lesions were made. Lesions of the medial forebrain bundle region produced a loss of forebrain NE and 5-HT along a time course parallel to nerve fiber degeneration in experiments that were done when our understanding of CNS monoaminergic systems was not as far along as it is for histamine today. When such lesions were placed unilaterally and specific histidine decarboxylase activity followed, a 70% decline in forebrain histidine decarboxylase activity was found at the end of 1 week. The decline was not due to the concomitant loss of monoaminergic fibers since lesions made by hypothalamic injections of 6-hydroxydopamine or 5,7-dihydroxytryptamine did not result in histidine decarboxylase loss. The most reasonable explanation would be that the lesion interrupted a histamine-containing diencephalic tract; the histamine and decarboxylating activity on the ipsilateral cortex was reduced only about 40%, suggesting that the pathway may make diffuse contributions to nondiencephalic regions. However, subsequent studies of completely isolated cerebral cortical "islands" showed complete loss of histamine content. As mentioned earlier, antibodies to histidine decarboxylase have also been utilized to map the distribution of this enzyme in rodent brain by immunohistochemical techniques. Steinbusch and coworkers have developed an immunohistochemical method for the visualization of histamine in the CNS using an antibody raised against histamine coupled to a carrier protein. Results obtained with this technique are in general agreement with studies on the immunohistochemical localization of histidine decarboxylase. The highest density of histamine-positive fibers is found in brain regions known to contain high histamine levels, such as the median eminence and the premammillary regions of the hypothalamus.

Lesion and immunohistochemical studies indicate the presence of a major histamine-containing pathway emanating from neurons in the posterior basal hypothalamus and reticular formation and ascending through the medial forebrain bundle to project ipsilaterally to broad areas of the telencephalon (Fig. 10–11). A descending pathway originating from the hypothalamus and projecting to the brain stem and spinal cord has also been suggested.

The majority of histamine-containing perikarya are comprised of a continuous group of magnocellular neurons numbering about 2000 in the rat and confined primarily to the tuberal region of the posterior hypothalamus, collectively named the *tuberomammillary nucleus*. A similar organization has also been described in humans, although histamine-containing neurons are more abundant (>64,000) and occupy a greater portion of the hypothalamus. Like other monoaminergic neurons, the histaminergic system consists of long, highly divergent, mostly unmyelinated, varicose fibers projecting in a diffuse manner to many telencephalic, mesencephalic, and cerebellar structures (Fig. 10–11). Individual neurons appear to project to widely divergent areas but make few typical synaptic contacts. Histamine neurons are unique in that they are characterized by the presence of an unusually large number of biological markers normally associated with other neurotransmitter systems, including glutaminic acid decarboxylase, adenosine deaminase, substance P, galanin, met-enkephalin, and brain natriuretic peptide. Tuberomammillary neurons also contain monoamine oxidase B, the major enzyme responsible for deamination of *N*-methylhistamine, a major metabolite of histamine in the CNS. Unraveling the possible roles of such a large number of putative cotransmitter peptides in a single population of magnocellular neurons in the posterior hypothalamus (the tuberomammillary nucleus) poses an exciting challenge.

Histamine Receptors

Three classes of histamine receptors have been identified in vertebrates and named in the order of their discovery: H_1, H_2, and H_3. Both H_1 and H_2 receptor DNA have been cloned. They belong to the superfamily of receptors with seven transmembrane domains and are coupled to guanylnucleotide-sensitive G proteins. The autoradiographic distribution of the three histamine receptors has been mapped in primate brain. H_1 receptors are particularly abundant in the neocortex, while H_2 and H_3 receptors, although present in the cerebral cortex, are enriched in caudate and putamen. H_3 receptors are predominant in basal ganglia, with the highest density found in the globus pallidus. Histamine receptors are also found in the hippocampus

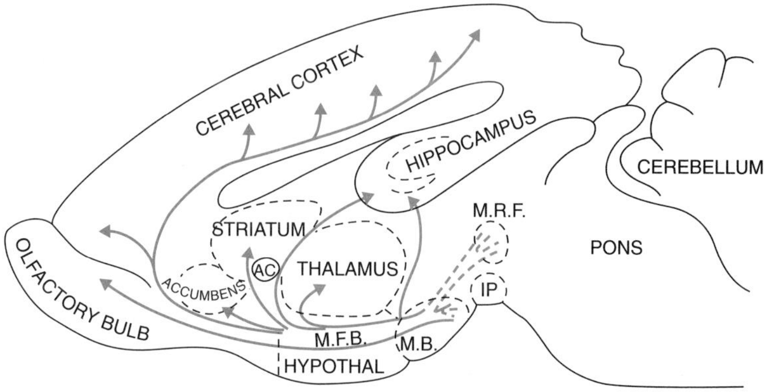

Figure 10–11. Schematic illustration of the distribution of histamine-containing neurons in brain. M.R.F., mesencephalic reticular formation; M.B., mammillary bodies; M.F.B., medial forebrain bundle. (Modified from Schwartz and Arrang, 2002.)

and cortical areas. Although the presence of H_3 receptors in brain has been appreciated since the mid-1980s, the molecular identity of this receptor has remained elusive. Thus, this receptor has been characterized in the CNS largely on the basis of radioligand binding and pharmacological studies with subtype-specific agents. In general, it is conjectured that the histamine H_3 receptor functions as an autoreceptor coupled to G_i proteins, analogous to dopamine D_2, adrenergic α_2, and serotonergic $5HT_{1A}$ autoreceptors. H_3 receptors also appear to function as heteroreceptors regulating the release of other neurotransmitters from their nerve terminals.

The cloning and functional expression of the human H_3 receptor were finally achieved in 1999. The molecular structure of the H_3 receptor demonstrates that it, like the other histamine receptors, belongs to the superfamily of G protein–coupled receptors. With the availability of the cDNA for the H_3 receptor, it will now be possible to develop chemical and biological reagents to address a number of questions concerning the possible role of H_3 receptors in the CNS using molecular biology.

A new histamine receptor, H_4, has been reported and characterized, adding a new chapter to the histamine story. This gene encodes a 390–amino acid G_i-coupled receptor with 40% homology to the H_3 receptor and a similar intron–exon structure. The localization of this receptor is quite restricted, and it does not appear to occur in the CNS. Expression appears localized to spleen, thymus, intestine, and immunologically active cells such as neutrophils, eosinophils, and T cells, suggesting a role in the regulation of immune function. If confirmed, this could offer a novel therapeutic potential for histamine receptor ligands in allergic and inflammatory diseases.

H_1 receptors are coupled to inositol phospholipid hydrolysis and involved in a variety of responses, including contraction of smooth muscle, increased capillary permeability mediated by contraction of terminal venules, hormone release, and brain glycogenolysis induced by a rise in the intracellular concentration of free calcium ions. H_2 receptors are positively coupled to adenylyl cyclase via a G_s protein and mediate most of their effects by changes in the intracellular levels of cAMP. The functional responses mediated by H_2 receptors include smooth muscle relaxation, gastric acid secretion, positive chronotropic and inotropic actions on cardiac muscle, and inhibitory effects on the immune system.

A third class of histamine receptors, H_3, was initially identified as an autoreceptor controlling histamine synthesis and release from histamine-containing neurons in rat cerebral cortex. Subsequently, the H_3 receptor was shown to inhibit the presynaptic release of other monoamines and peptides from brain and peripheral tissue. Functional studies have provided evidence for presynaptic release modulating H_3 receptors on noradrenergic, serotonergic, dopaminergic, cholinergic, and peptidergic neurons. H_3 receptors are believed to provide a major regulatory mechanism involved in the control of histaminergic activity under physiological conditions. In vivo microdialysis studies have demonstrated that administration of selective H_3 receptor agonists reduces histamine release and turnover. Administration of H_3 receptor antagonists, in contrast, enhances histamine turnover and release, suggesting that H_3 autoreceptors are normally under tonic stimulation by endogenous histamine.

H_1 Receptor Agonists and Antagonists

Because of its high affinity and good receptor selectivity, mepyramine remains the agent of choice as a selective H_1 antagonist. A number of H_1 antagonists exist as optical isomers, and in many instances the (+) stereoisomer exhibits higher potency than the (−) stereoisomer. (+)-Chlorpheniramine is such an example. One of the most potent H_1 antagonists is the geometric isomer transtriprolidine, which has a K_d of 0.1 nM. In humans, a major side effect associated with administration of H_1 antagonists is a high degree of sedation, which has been attributed to the ability of the antagonists to occupy H_1 receptors in the brain. Most of the classic H_1 antihistamines (e.g., diphenhydramine, mepyramine) readily cross the blood–brain barrier and elicit a significant degree of sedation. A number of new compounds, however, penetrate poorly into the CNS and thus are relatively nonsedating. Terfenadine (Fig. 10–12) is an example of such an agent. Highly selective agonists do not exist for H_1 receptors. 2-Methylhistamine and 2-thiazoylethylamine show some selectivity, but their relative potency at H_1 receptors differs by no more than an order of magnitude from that exhibited at H_2 receptors.

H_2 Receptor Antagonists

Following the successful use of H_2 receptor antagonists like cimetidine in the treatment of peptic ulcers, a number of these agents have become available. The structures of several H_2 antagonists are illustrated in Figure 10–13. Although a number of more potent agents have been developed, only ranitidine and tiotidine lack antagonist activity on H_3 receptors. These agents have limited access to the CNS. The only selective H_2 receptor antagonist that effectively penetrates into the CNS is zolantidine. This compound has been used extensively in experimental animals but has not been marketed for clinical use. Dimaprit is a fairly selective H_2 agonist that discriminates well between H_2 and H_1 or H_3 receptors. Impromidine is one of the most potent H_2 receptor agonists available. Like dimaprit, it has minimal agonist effects on H_1 receptors, but it has more potent H_3 receptor antagonist properties.

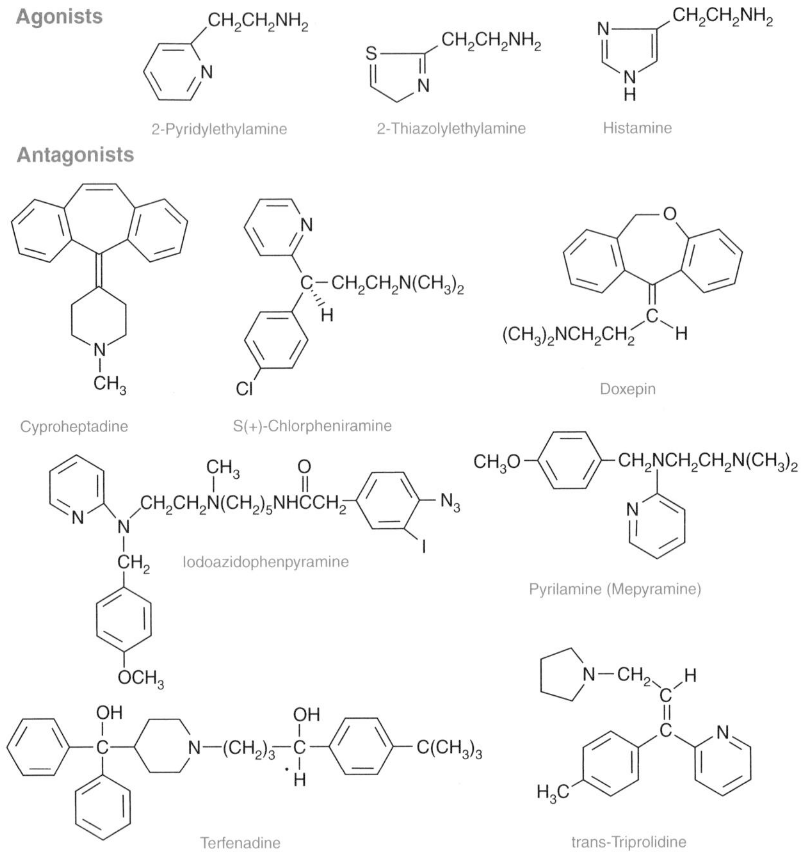

FIGURE 10–12. H_1 agonists and antagonists.

Amthamine is also a more potent agonist at H_2 receptors than diamaprit and does not affect H_1 receptors. However, this agent exerts weak agonist activity at the H_3 receptor and thus does not exhibit optimal subtype selectivity.

H_3 Receptor Antagonists and Agonists

H_3 receptor antagonists and agonists are illustrated in Figure 10–14. Thioperamide is a potent H_3 receptor antagonist that exhibits good receptor specificity. This drug, unlike other H_3 antagonists, crosses the blood–brain barrier and has been useful in evaluating the behavioral role played by H_3 receptors in the CNS. A new antagonist, clobenpropit, is the most potent H_3 receptor antagonist available for investigating these receptors. *Nα*-Methylhistamine and *Nα-Nβ*-dimethylhistamine are both potent H_3 receptor agonists, but neither shows any marked discrimination between the three histamine receptors. However, substitution of methyl groups in the α or β position of the side chain of the histamine molecule produces agents that exhibit a high degree of selectivity for the H_3 receptors. For example, $R\alpha$-methylhistamine and $R\alpha,S\beta$-dimethylhistamine are 15–20 times more potent than histamine as H_3 agonists but possess only about 1% of the activity of histamine on H_1 and H_2 receptors. Imetit is a potent and selective H_3 receptor full agonist that is effective both in vivo and in vitro and 60 times more potent than histamine. Table 10–5 summarizes the pharmacology of histamine receptors and their effector pathways.

In summary, there is compelling evidence that histamine, in addition to its presence in mast cells, exists in neurons, where it probably functions as a neurotransmitter in the CNS. The brain has a similar nonuniform distribution of histamine, a specific histidine decarboxylase, and methylhistamine (the major metabolite of brain histamine). The cerebellum has the lowest levels, whereas the hypothalamus is most enriched. Hypothalamic synaptosomes contain histamine and histidine decarboxylase, suggesting a localization in

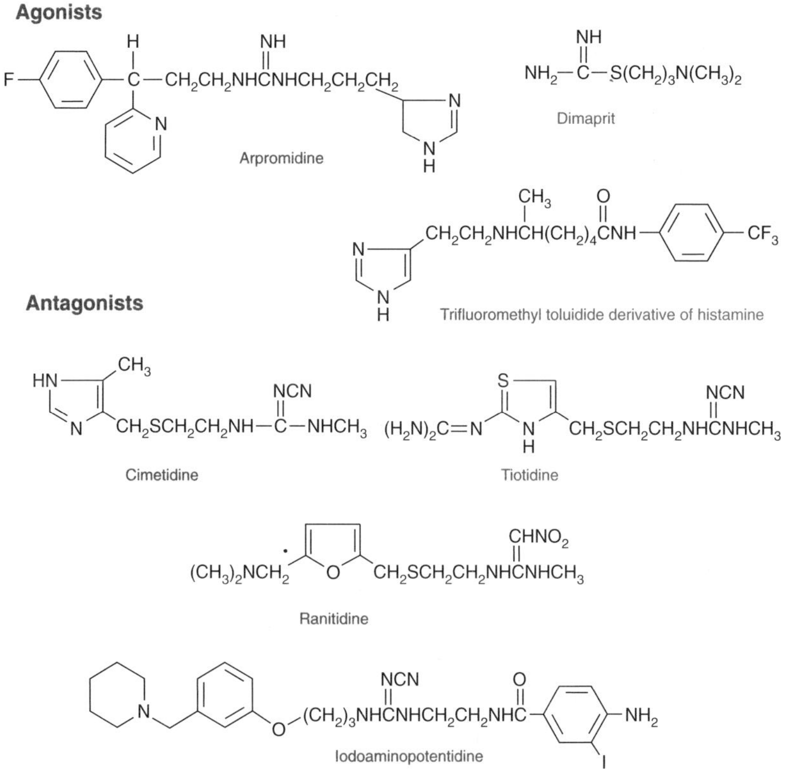

FIGURE 10–13. H_2 agonists and antagonists.

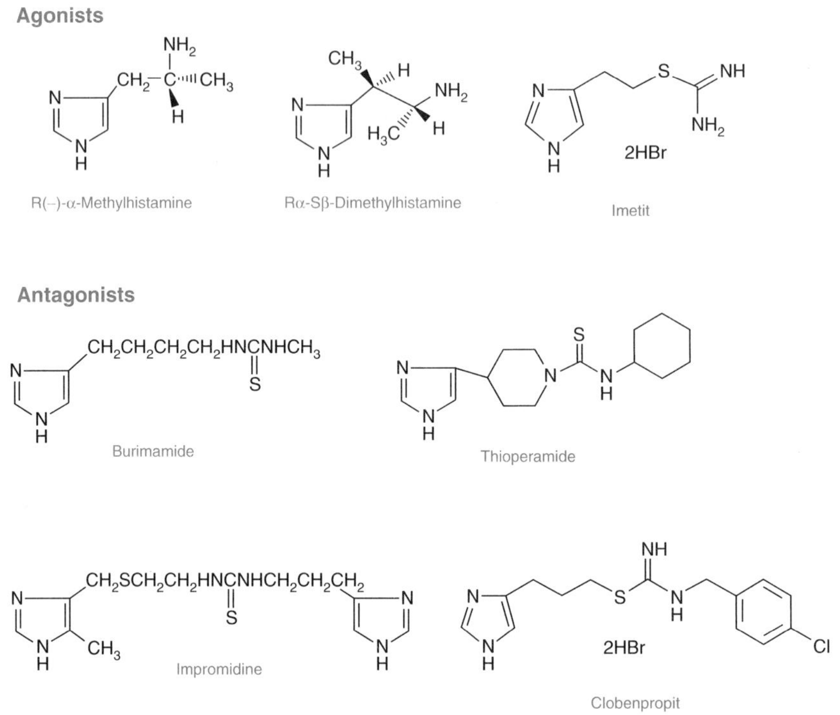

FIGURE 10–14. H_3 agonists and antagonists.

nerve endings. Depolarization of hypothalamic slices causes a calcium-dependent release of histamine, and tetrodotoxin sensitive histamine release from neurons has been demonstrated in vivo. The turnover of brain histamine is quite rapid, as with other biogenic amine transmitters.

The actions of histamine in the CNS appear to be mediated by three classes of receptors. These three types, H_1, H_2, and H_3, can be distinguished in the CNS by their pharmacology, their localization, and the intracellular responses they mediate. Neuropharmacological studies of brain histaminergic systems have been hampered by the lack of appropriate pharmacological agents. In the last few years, parenterally active drugs, H_2 antagonists, H_3 agonists, and antagonists have become available; and their effects on behavioral, neuroendocrine, and vegetative functions are being studied in animals.

Lesions of the midbrain or caudal hypothalamus cause a progressive reduction in forebrain histamine and histidine decarboxylase, consistent with the existence of an ascending histamine-containing system, which has now been directly visualized by immunohistochemical techniques. Continued application of these immunohistochemical techniques for visualization of his-

TABLE 10–5. Pharmacology of Histamine Receptors

Receptor Subtype	H_1	H_2	H_3
Selective agonists	2-(*m*-Bromophenyl) histamine	Amthamine Impromidine[a] (also an H_3 antagonist)	(*R*)-α-Methylhistamine[c] (*R*)-α,(*S*)β-Dimethylhistamine[c] Imetit[c]
Selective antagonists	Mepyramine[a] Triprolidine[a]	Ranitidine[b] Tiotidine[b]	Thioperamide[c] Clobenpropit[c]
Effector pathways	↑ IP_3/DAG	↑ cAMP	(? $G_{i/o}$)

IP_3/DAG, inositol triphosphate/diacylglycerol; cAMP, cyclic adenosine monophosphate.

[a]See Fig. 10–11.

[b]See Fig. 10–12.

[c]See Fig. 10–13.

tidine decarboxylase-like and histamine-like activity in the brain should provide more detailed maps of histamine-containing neuron systems in the brain in the near future. However, we are still far from being able to speak with conviction of specific histaminergic circuits and their relation to behavior; thus, the functions of histamine in brain are highly speculative.

The high levels of histamine and the presence of histamine receptors in the hypothalamus, together with the known ability of histamine to alter food and water intake thermoregulation, autonomic activity, and hormone release, indicate a possible role for histamine in these vegetative functions. Also, because of the presence of ascending histamine-containing fibers originating from the reticular formation, coupled with the sedative nature of H_1 antagonists and the activation effects of histamine on some brain cells, a possible role for histamine in arousal has been suggested.

Despite numerous speculations based largely on observations of the response to locally applied histamine, only a few physiological roles for histamine are well documented. These include arousal, control of appetite, regulation of pituitary hormone secretion, and control of vestibular reactivity. The therapeutic role for histaminergic drugs is fairly limited, and the primary pharmacological actions of these agents are often associated with side effects ascribed to other therapeutic agents. For example, several antipsychotic (e.g., clozapine and thioridazine) and antidepressant (e.g., doxepin and amitriptyline) drugs with sedative and weight-gain side effects display potent H_1 antagonist properties. Histamine antagonists have been used therapeutically for only a limited number of conditions. H_1 antagonists like meclizine are the most commonly used anti-motion-sickness drugs. However, it is still uncertain whether H_1 blockade is primarily responsible for the effectiveness of these agents. Some H_1 antagonists, notably meclizine and dimenhydrinate, are of benefit in vestibular disturbances such as Meniere's disease. H_1 receptor antagonists are also the ingredients of a number of proprietary remedies sold over the counter as sleeping pills for the treatment of insomnia. These preparations are generally ineffective at the recommended dosages, although some sensitive individuals may derive benefit. H_1 antagonists have a well-established and valued place in the symptomatic treatment of hypersensitivity reactions (allergic rhinitis and conjunctivitis). However, their effect is purely palliative, being confined to suppression of the symptoms attributed to the antigen–antibody-induced release of histamine. Despite popular belief, H_1 antagonists are without value in the treatment of the common cold.

We would be remiss not to mention the therapeutic usefulness of the H_2 antagonists. The main therapeutic use of these agents is in the treatment of gastric and duodenal ulcers. H_2 antagonists are very effective at lowering the basal and nocturnal secretion of gastric acid and that which is stimulated by meals. They both reduce the pain and the consumption of antacids and hasten the healing of ulcers. The incidence of side effects with these H_2 antagonists (cimetidine and ranitidine) is low and can probably be attributed, in part, to the limited function of H_2 receptors in organs other than the stomach and to their poor penetration of the blood–brain barrier.

Adenosine

Considerable interest has developed over the past two decades in the possible role of adenosine and adenine nucleotides, such as adenosine triphosphate (ATP), as transmitters or neuromodulators in the CNS. While it was known more than 70 years ago that adenosine and ATP elicited potent pharmacological effects on the nervous system (Drury and Szent-Gyorgy, 1929), evidence that endogenous purines regulated the activity of the nervous system was not available until recently. The seminal work of Rall, Sattin, McIlwain, and others demonstrated that adenosine stimulated the formation of cAMP in brain slices and that methylxanthines such as a theophylline and caffeine were competitive pharmacological antagonists of this response. These findings were instrumental in developing much of the interest in adenosine in the mid-1980s. Adenosine and adenosine analogues are potent activators of cellular function, altering neuronal activity and affecting behavior as well. Highly specific, high-affinity binding sites for adenosine have also been identified in the CNS, and it is thought that most, if not all, of the physiological actions of adenosine are mediated via the interaction of extracellular adenosine with these G protein–coupled receptor sites. Nevertheless, despite the fact that adenosine is a potent neuroactive substance, it meets few of the criteria considered essential to identify an endogenous substance as a candidate for a neurotransmitter. Adenosine is neither stored nor released as a classical neurotransmitter. In contrast to traditional neurotransmitters, it has poorly defined pathways leading to its synthesis and degradation and appears to act more as a neuromodulator to fine-tune synaptic function. Adenosine is released by most cells, including neurons and glial cells, and modulates the activity of the nervous system by acting presynaptically (facilitating or inhibiting transmitter release), postsynaptically, and/or nonsynaptically. To date, four adenosine receptor subtypes have been cloned, A_1, A_{2A}, A_{2B}, and A_3, each of which exhibits a unique tissue distribution, ligand affinity, and signal-transduction mechanism. Adenosine mediates it physiological effects through activation of high-affinity receptors (A_1 and A_{2A}; see Chapter 4). Activation of lower-affinity receptors (A_{2B} and A_3) is believed to be involved in mediating pathological actions since these subtypes become activated only when adenosine levels increase into the micromolar range during periods of hypoxia, ischemia, or inflammation. Thus, the pathophysiological role of these adenosine receptor subtypes may be very different from that of A_1 and A_2 in that they may act as endogenous regulators largely under conditions of more severe challenge.

In addition to exerting an effect on nerve cells directly, adenosine receptor activation influences the action of other neurotransmitters and neuromodulators indirectly. A convincing case has been made for adenosine receptors having a critical function in regulating the activation of multiple receptors that affect neurotransmitter release and synaptic function. In particular, they regulate *N*-methyl-D-aspartate and metabotropic glutamate receptors, nicotinic receptors, and peptidergic receptors such as calcitonin gene–related peptide and vasoactive intestinal polypeptide receptors. Thus, adenosine has been speculated to behave as a modulator to fine-tune the actions of other transmitters and neuromodulators.

Selected References

Serotonin

Aghajanian, G. K. and E. Sanders-Bush (2002). Serotonin. In *Neuropsychopharmacology The Fifth Generation of Progress* (K. L. Davis, D. Charney, J. T. Cole and C. Nemeroff, eds) Lippincott Williams and Wilkins, Philadelphia, PA, pp. 15–34.

Barker, E. L. and R. D. Blakely (1995). Norepinephrine and serotonin transporters: molecular targets of antidepressant drugs. In *Psychopharmacology, Fourth Generation of Progress* (F. E. Bloom and D. J. Kupfer, eds.). Raven Press, New York, pp. 321–333.

Barnes, N. M. and T. Sharp (1999). A review of central 5-HT receptors and their function. *Neuropharmacology* 38, 1083–1152.

Blakely, R. D., S. Ramamoorthy, S. Schroeter, Y. Quian, S. Apparsondarum, A. Galliand and L. DeFelice (1998). Regulated phosphorylation and trafficking of antidepressant-sensitive serotonin transporter proteins. *Biol. Psychiatry* 44, 169–178.

Blier, P. and C. deMontigny (1994). Current advances and trends in the treatment of depression. *Trends Pharmacol. Sci.* 15, 220–226.

Branchek, T. A. and T. P. Blackburn (2000). 5-HT_6 receptors as emerging targets for drug discovery. *Annu. Rev. Pharmacol. Toxicol.* 40, 319–334.

Fitzpatrick, P. F. (2000). The aromatic amino acid hydroxylases. *Adv. Enzymol.* 74, 235–294.

Glennon, R. A. and M. Dukat (1995). Serotonin receptor subtypes. In *Psychopharmacology, Fourth Generation of Progress* (F. E. Bloom and D. J. Kupfer, eds.), Raven Press, New York, pp. 415–430.

Heninger, G. R. (1997). Serotonin, sex and psychiatric illness. *Proc. Natl. Acad. Sci. U.S.A.* 94, 4823–4824.

Hoyer, D., D. E. Clarke, J. R. Fozard, P. R. Hartig, G. R. Martin, E. J. Mylecharane, P. R. Saxena, and P.P.A. Humphrey (1994). International Union of Pharmacology classification of receptors for 5-hydroxytryptamine (serotonin). *Pharmacol. Rev.* 46, 157–203.

Lovenberg, T. W., B. M. Baron, L. deLecea, J. D. Miler, R. A. Porsser, M. A. Rea, P. E. Foye, M. Racke, A. L. Slone, B. W. Siegel, P. E. Danielson, J. G. Sutcliffe, and M. G. Erlander (1993). A novel adenylyl cyclase–activating serotonin receptor (5-HT_7) implicated in the regulation of mammalian circadian rhythms. *Neuron* 11, 449–458.

Mamounas, L. A., C. A. Mullen, E. O'Hearn, and M. E. Molliver (1992). Dual serotonergic projection to forebrain in the rat: morphologically distinct 5HT axon terminals exhibit differential vulnerability to neurotoxic amphetamine derivatives. *J. Comp. Neurol.* 314, 558–586.

Maricq, A. V., A. S. Peterson, A. J. Brake, R. M. Myers, and D. Julius (1991). Primary structure and functional expression of the $5HT_3$ receptor, a serotonin-gated ion channel. *Science* 254, 432–437.

Morales, M., E. Battenberg, and F. E. Bloom (1998). Distribution of neurons expressing immunoreactivity for the $5HT_3$ receptor subtype in the rat brain and spinal cord. *The Journal of Comparative Neurology* 385, 385–401.

Murphy, D. L., C. Wichems, Q. Li, and A. Heils (1999). Molecular manipulations as tools for enhancing our understanding of 5-HT neurotransmission. *Trends Pharmacol. Sci.* 20, 246–252.

Owens, M. J. and C. B. Nemeroff (1994). The role of serotonin in the pathophysiology of depression: focus on the serotonin transporter. *Clin. Chem.* 40, 288–295.

Peroutka, S. J. (1994). Molecular biology of serotonin (5-HT) receptors. *Synapse* 18, 241–260.

Ramamoorthy, S. and R. Blakely (1999). Phosphorylation and sequestration of serotonin transporters differentially modulated by psychostimulants. *Science*, 285, 763–766.

Ricaurte, G. A., J. Yuan, and U. D. McCann (2000). (±)3,4-Methylenediozymethamphetamine ("ecstasy")-induced serotonin neurotoxicity: studies in animals. *Neuropsychobiology* 42, 5–10.

Sanders-Bush, E. and H. Canton (1995). Serotonin receptors: signal transduction pathways. In *Psychopharmacology, Fourth Generation of Progress* (F. E. Bloom and D. J. Kupfer, eds.). Raven Press, New York, pp. 431–442.

Saudou, F. and R. Hen (1994). 5-Hydroxytryptamine receptor subtypes in vertebrates and invertebrates. *Neurochem. Int.* 25, 503–532.

Steinbusch, H. W. M. and A. H. Mulder (1984). Serotonin-immunoreactive neurons and their projections in the CNS. In *Handbook of Chemical Neuroanatomy. Classical Transmitters and Transmitter Receptors in the CNS*, vol. 3, part II (A. Björklund, T. Hökfelt, and M. Kuhar, eds.), Elsevier, New York, pp. 68–118.

Histamine

Hill, S. J. (1990). Distribution, properties, and functional characteristics of three classes of histamine receptor. *Pharmacol. Rev.* 42, 45–83.

Honrubia, M. A., M. T. Vilaro, J. M. Palacios, and G. Mengod (2000). Distribution of the histamine H_2 receptor in monkey brain and its mRNA localization in monkey and human brain. *Synapse*, 38, 343–354.

Lovenberg, T. W., B. L. Roland, S. J. Wilson, X. Jiang, J. Pyati, A. Huvar, M. R. Jackson and M. G. Erlander (1999). Cloning and functional expression of the human histamine H_3 receptor. *Molecular Pharmacology*, 55, 1101–1107.

Martinez-Mir, M. I., H. Pollard, J. Moreau, J. M. Arrang, M. Ruat, E. Traiffort, J. C. Schwartz, and J. M. Palacios (1990). Three histamines receptors (H1, H2 and H3) visualized in the brain of human and non-human primates. *Brain Research* 526, 322–327.

Onodera, K., A. Yamatodani, T. Watanabe, and H. Wadas (1994). Neuropharmacology of the histaminergic neuron system in the brain and its relationship with behavioral disorders. *Prog. Neurobiol.* 42, 685–702.

Panula, P. and M. S. Airaksinen (1991). The histaminergic neuronal system as revealed with antisera against histamine. In *Histaminergic Neurons: Morphology and Function* (T. Watanabe and H. Wada, eds.). Boca Raton, FL, CRC Press, pp. 127–144.

Schwartz, J. C. and J. M. Arrang (2002). Histamine. In *Neuropsychopharmacology The Fifth Generation of Progress* (K. L. Davis, D. Charney, J. T. Cole and C. Nemeroff, eds.) Lippincott Williams & Wilkiins, Philadelphia, PA, pp. 179–190.

Schwartz, J. C. and H. L. Haas, eds. (1992). *The Histamine Receptor*. New York, Wiley-Liss.

Staines, W. A. and J. I. Nagy (1991). Neurotransmitter coexistence in the tuberomammillary nucleus. In *Histaminergic Neurons: Morphology and Function* (T. Watanabe and H. Wada, eds.). Boca Raton, FL, CRC Press, pp. 163–176.

Steinbusch, H. W. M. and A. H. Mulder (1984). Immunohistochemical localization of histamine in neurons and mast cells in brain. In *Handbook of Chemical Neuroanatomy. Classical Transmitters and Transmitter Receptors in the CNS*, vol. 3, part II. (A. Björklund, T. Hökfelt, and M. J. Kuhar, eds.). Elsevier, New York, pp. 126–140.

Tohyama, M., R. Tamiya, N. Inagaki, and H. Takagi (1991). Morphology of histaminergic neurons with histidine decarboxylase as a marker. In *Histaminergic Neurons: Morphology and Function* (T. Watanabe and H. Wada, eds.). Boca Raton, FL, CRC Press, pp. 107–126.

Vizuete, M. L., E. Traiffort, M. L. Bouthenet, M. Ruat, E. Souil, J. Tardivel-Lacombe, and J.-C. Schwartz (1997). Detailed mapping of the histamine H_2 receptor and its gene transcripts in guinea-pig brain. *Neuroscience* 80(2), 321–343.

Wouterlood, F. G. and H.W.M. Steinbusch (1991). Afferent and efferent fiber connections of histaminergic neurons in the rat brain: comparison with dopaminergic, noradrenergic and serotonergic systems. In *Histaminergic Neurons: Morphology and Function* (T. Watanabe and H. Wada, eds.). Boca Raton, FL, CRC Press, pp. 145–162.

Adenosine

Bischofberger, N., K. A. Jacobson and D. K. J. E. von Lubitz (1997). Adenosine A_1 receptor agonists as clinically viable agents for treatment of ischemic brain disorders. *Ann. N.Y. Acad. Sci.* 825, 23–29.

Krugel, U., H. Kittner, and P. Illes (2000). Mechanisms of adenosine 5′-triphosphate–induced dopamine release in the rat nucleus accumbens in vivo. *Synapse* 39, 222–232.

Nyce, J. W. (1999). Insight into adenosine receptor function using antisense and gene knockout approaches. *Trends Pharmacol. Sci.* 20, 79–83.

Ongini, E., M. Adami, C. Ferri, and R. Bertorelli (1997). Adenosine A_{2A} receptors and neuroprotection. *Ann. N.Y. Acad. Sci.* 825, 30–48.

Williams, M. and M. F. Jarvis (2000). Purinergic and pyrimidinergic receptors as potential drug targets. *Biochem. Pharmacol.* 59, 1173–1185.

Zimmermann, H. (1994). Signalling via ATP in the nervous system. *Trends Neurosci.* 10, 420–426.

11

Neuroactive Peptides

After more than 30 years of ever incisive examination, neuropeptides remain dear to the hearts of neuropharmacologists. For those hoping eventually to understand the role of neuropeptides, the quest to devise new medications to enhance or antagonize their actions remains intense. Rapid advances in molecular biology and the regulation of gene expression have accelerated the determination of peptide amino acid sequences. In addition, the characterization of their receptors by cloning and transfection studies (see Chapter 3) and the successful synthesis of both peptidic and nonpeptidic antagonists for these receptors fuel this attraction. New peptides continue to be discovered, often through attempts to characterize the as yet undiscovered natural ligands for cloned but orphaned receptors. The new student entering this field might assume that the peptide story must be nearly all told. Alas, that conclusion would be erroneous. As in most of the fields discussed in this book, better data beget better and more complex questions.

Almost all peptides made by neurons will affect specific target cells of the central nervous system (CNS) and peripheral nervous system: neuronal, glial, smooth muscle, glandular, and vascular. The census of peptides already exceeds several dozen, with no obvious end in sight. However, the actual molar concentration of any given peptide in the brain is maximally two to three orders of magnitude lower than that of the monoamines, acetylcholine (ACh), and amino acids; and their receptors respond at correspondingly lower concentrations. The specific signaling advantages of peptides are not yet obvious. Furthermore, with very few exceptions, no peptides have been linked selectively to a given functional system in the brain or correlated in any exclusive manner with any pathological states. Thus, the challenge of devising new drugs for neurological or psychiatric disorders based on these peptides is considerably more difficult than just designing potent congeners that can tweak their receptors on or off. It requires that we have some reason to do that. Nevertheless, there are strong implications that the biological properties evoked by peptide-mediated signals will influence future pharmacological development.

Some Basic Questions

What Is Different about Peptides?

Peptide-secreting neurons differ in their biology from neurons using amino acids and monoamines in ways other than the molecular structure of their transmitter. Amino acid or monoamine transmitters are formed from diet-derived amino acids by one or two intracellular enzymatic steps. The end product of these enzymatic actions is the active transmitter molecule, which is then stored in the nerve terminal until release. After release, the transmitter (or choline in the case of ACh) can be reaccumulated back into the nerve terminal by the energy-dependent, active reuptake property, thus conserving the requirement for de novo synthesis. Peptide-secreting cells employ a somewhat more formidable approach: synthesis directed by mRNA can take place only on ribosomes and, thus, only in the perikaryon or dendrites of a neuron. Furthermore, all known neuropeptides arise from larger prohormone forms (inactive precursors). The biologically active, secreted form is cleaved from these prohormones by the actions of selective proteolytic enzymes.

Thus, the process for peptides starts with ribosomal synthesis of the prohormone, which is then packaged into vesicles in the smooth endoplasmic reticulum and transported from the perikaryon to the nerve terminals for eventual release. These steps in the molecular processing of secretory products are described in detail in Chapter 3.

Insofar as the present data reveal, peptide-releasing neurons share certain basic properties with all other chemically characterized neurons. Transmitter release is Ca^{2+}-dependent. Postsynaptically, all fully characterized receptors for neuropeptides are G protein–coupled, as described in Chapter 4. The effects of neuropeptides thus are mediated indirectly by regulating ion channel properties through second messengers such as Ca^{2+}, cyclic nucleotides, or inositol triphosphates (see Chapter 5). Furthermore, like the monoamines, peptides identical to those extracted from the nervous system are often made by nonneural secretory cells in endocrine glands and the cells lining hollow viscera like the intestinal tract. Some observers of the peptide scene emphasize characteristics of neuropeptide action that are extrapolations from the actions of endocrine peptide hormones. Thus, they point to targets that are distant from release sites, release into portal or other circulatory systems, are of higher potency, and have a longer time course of effects. Although endocrine hormones (peptides or amines) can produce effects in the nanomolar range or below, we do not yet know the time course or thresholds for the same neuropeptides to act on their target cells in the brain or ganglia. In amphibian autonomic ganglia, at least one peptide does act in such a local endocrine, or paracrine, fashion, producing synaptic potentials that last for hundreds of milliseconds. These effects are observed with an endogenous peptide that is very similar to hypothalamic luteinizing hormone–releasing hormone and produced on a nonsynaptic neuronal target by diffusing at least 10–100 μm from preganglionic synaptic release sites, without apparently acting at all on the immediately adjacent postsynaptic neuron. Whether these particular paracrine properties are ways around the simple amphibian autonomic anatomy or a prototype for all central and peripheral peptide actions remains debatable.

How Are New Peptides Found?

Before considering the interesting relationships among the many known peptides, we can anticipate future events by examining the recurrent pattern of events that seems to underlie every wave of peptide discovery and characterization. From this historical overview, the student can see that new peptides do not just happen into existence. The primary necessity for the discovery of a new peptide has classically been the recognition of a biological function controlled by an unknown factor. This biological action is then employed as the bioassay for the subsequent steps of the discovery process. This approach was used extensively for the characterization of gastrointestinal peptide hormones in the 1940s and 1950s. However, in their pursuit of the hypothalamic hormones that regulate secretion from the anterior pituitary, the process was perfected and greatly increased in power by Roger Guillemin and Andrew Schally and their teams during the 1960s and 1970s. Typically, a bioassay system was used to detect enrichment of an active factor when extracts of a tissue thought to make the factor were put through a series of chemical manipulations (i.e., differential extractions, molecular sieves, etc.). For some factors, the purification of nanograms of active peptide required hundreds of thousands of brains. The peptidic nature of the purified material was inferred by the loss of biological activity after treatment with peptidases (carboxypeptidases or endopeptidases). The cleaved peptide fragments were then sequenced, and the entire factor was eventually reconstructed.

With modern technology, many of those historical steps can be rapidly set aside.

When a sequence has been achieved and matched against the amino acid composition of the pure factor, it is necessary to synthesize the peptide and determine if the synthetic replicate matches the actions and potencies of the purified natural factor. Only when the replicate's properties match those of the factor can the natural peptide be declared "identified." At this point, the factor is given a name that reflects its actions in the original bioassay.

However, this is far from the end of the discovery process. When the chemical structure of a peptide has been identified and synthetically replicated, two new opportunities arise. The synthetic material can be tested for physiological and pharmacological properties. At the same time, the synthetic material can be used to generate antibodies. The antibodies are then employed for quantitative measurements (i.e., radioimmunoassay) and to determine the qualitative distribution of the peptide by immunocytochemistry. Those data then frequently reveal that the peptide has a much broader distribution and, therefore, ostensibly, a much more general series of actions than was originally presumed. Autoradiographic mapping of potential response sites (first by ligand binding for peptides and more recently by in situ hybridization for the mRNAs of their receptors) is generally the last step in the molecular and cellular characterization. Curiously, these receptor maps reveal a flagrant disregard for the cellular maps of peptides themselves. This mismatch strengthens in some minds the concept of paracrine diffusion for peptide actions at a distance. Empowered by the assays and knowledge of the chemical neuroanatomy of the peptide-containing circuitry, the peptide or an analog of it is subsequently proposed to be the cause or cure of a major mental illness. In many of the more recent reports, the peptide isolated and found to be active is, in fact, not the only active form contained in the tissue. In many cases, better extraction methods confirmed by radioimmunoassay reveal larger molecules with full or even greater potency. Although the general rule is that active peptides are synthesized from larger precursor hormones, when prohormone processing gives rise to active peptides larger than the form first found, the larger forms cannot really be viewed as "precursors" since they have potent activity in their own right.

As more and more peptides have been identified within the nervous system and from other sources, their internal amino acid sequence similarities allow us to consider them as members of one or another family grouping of peptides. Family groupings are useful because they indicate that certain sequences of amino acids have had considerable evolutionary conservation, presumably because they provide unambiguous signals between secreting and responding cells.

Two aspects of peptide family groupings merit appreciation. *(1)* Some peptide families employ separate propeptide genes or mRNAs yet share one or more short peptide sequences that are expressed by different cell groups in the same species. *(2)* In other peptide families, the peptides expressed by homologous cell groups share generally identical peptide sequences, with minor to moderate variations. The first category, exemplified by the opioid peptide and the tachykinin peptide families, implies that the natural ligand's receptors may diverge into subtypes that can discriminate the fine differences in peptide sequences. Although some individual peptides have already earned their own receptor subtypes (vasopressin, somatostatin), in other cases, such as the tachykinins and the opioid peptides, the multiple receptor subtypes predicted the later definition of multiple natural ligands. In cases in which the propeptide form gives rise to multiple different agonists, it is possible, but by no means yet broadly supported, that more than one peptide messenger is released from the possible array. Some dreamy-eyed pundits have theorized that different branches of a neuron might have the option of changing the mix of cleavage products available for release, perhaps under steroid-driven conditions.

The latest additions to the list of neuroactive peptides have been discovered by several strategies. In some cases, the discovery hinged on specific common general features, such as a C-terminally amidated sequence (seen in the secretin/vasoactive intestinal peptide [VIP] and pancreatic polypeptide families among others). In other cases, discovery was based on the detection of potential cleavage fragments when the propeptide sequence was deduced by molecular cloning of the mRNA or gene (e.g., the γ-melanocyte-stimulating hormone [MSH] found within proopiomelanocortin mRNA or calcitonin gene–related peptide). Most recently, cloning of the abundant G protein–coupled receptors from the brain has identified new peptides that can activate the once-orphaned receptors (e.g., orphanin FQ). By refinements of differential display hybridization and searching for region-specific brain peptides by mRNA detection, still other peptides have been identified (cortistatin and hypocretin) Although these historical discoveries are individually entertaining stories, what the peptides do and how to make useful drugs based on these actions remain elusive.

Transmitters or Not?

In Chapter 2, we considered the kinds of evidence needed to demonstrate that a substance found in nerve cells is the transmitter that those neurons secrete at their synapses (in brief: *[1]* synthetic capacity, *[2]* presence, *[3]* activity-dependent release, and *[4]* mimicry of the effects of the presynaptic neuron). One other criterion is especially pertinent to the objectives of this book. Drugs that can simulate or antagonize the effects of presynaptic neuronal activation must have an identical influence on the effects of the substance applied exogenously to the target cell.

In the case of peptides, the opportunities for supplying these data would at first glance seem better than they are with the many categories of small molecules discussed in previous chapters. Once the peptide has been isolated and its structure determined, immunocytochemical localization of the peptide usually follows promptly, as does the demonstration that, with the right sensitivity assay, the peptide can be released from brain slices by a Ca^{2+}-dependent, depolarization process. The development of a suitable synthetic ligand or receptor mRNA in situ probe helps define receptor distribution, possible actions on presumptive synaptic targets, their mechanisms of regulation, and the discrimination of receptor subtypes. In several peptide families, synthetic agonists and antagonists have documented behavior-altering actions of the synthetic peptide, establishing, by inference, a potential role of the endogenous form in behavior and possibly in disease states. Recent advances here include short-term inactivation of peptide effects by intracerebral injection of antisense oligonucleotides and longer-term regulation of peptide synthesis in transgenic animals by overexpression or gene knockouts of specific peptides or their receptors. These molecular tools have accelerated the early partial fulfillment of transmitter criteria for peptides generally. However, despite intense efforts, there are very few peptide-containing circuits within the CNS for which the complete array of transmitter criteria have been satisfied. Perhaps for the present, it would serve our knowledge best to refer to the role played by neuropeptides as being a modulatory one (see Chapter 5).

They Are Not Alone

A once-predominant principle in neurotransmitter lore held that a given neuron operated by one and only one transmitter. According to this concept (attributed to Henry Dale, one of the pioneer transmitter discoverers), one aspect of a neuron's phenotype, as classic as its size, shape, location, circuitry, and synaptic function, was its neurochemical designation as "GABAergic," "cholinergic," "adrenergic," or whatever. All card-carrying pharmacologists knew that autonomic neurons came in only two color-coded categories: adrenergic and cholinergic. Then, a curious finding began to repeat itself: neuroactive peptides started showing up in autonomic neurons, where there was no need for additional transmitters. Initially viewed by many as an oddity, the idea of coexisting peptides in central as well as peripheral neurons that are already occupied by an amino acid or monoamine or even another neuropeptide has now gained wide acceptance (see Table 11–1). Actually, what Dale formulated as a principle was that a given neuron would release the same neurotransmitter from all of its many synapses. There being at the time very few other viable chemicals to consider as transmitters, this was scarcely heretical. However, Dale's principle was subsequently restated by Eccles, to hold that a given neuron used only one transmitter, which is almost, but not quite, the same thing and established the straw man that the neuropeptides blew away.

The notion of coexistence of other transmitters with peptides leads one to ask whether there is any such thing as a truly "peptidergic" neuron or, rather, only neurons in which a peptide or two is there to expand the armamentarium of messenger molecules available. Observations on cultured cells transfected to express peptide and amine receptors suggest that, in some cases, G protein–coupled heteromeric forms could allow for dual convergent effects.

Such a revolutionary concept then forces us to ask a simple question. If the peptide is not the sole signaling molecule but, rather, a minor, fractional percent of the messenger-signaling capacity of the nerve terminal, how might

TABLE 11–1. Examples of Neuroactive Peptides Coexisting with Other Transmitters

Transmitter	*Peptide*	*Location*
GABA	Somatostatin	Cortical and hippocampal neurons
	Cholecystokinin	Cortical neurons
Acetylcholine	VIP	Parasympathetic and cortical neurons
	Substance P	Pontine neurons
Norepinephrine	Somatostatin	Sympathetic neurons
	Enkephalin	Sympathetic neurons
	NPY	Medullary and pontine neurons
	Neurotensin	Locus ceruleus
Dopamine	CCK	Ventrotegmental neurons
	Neurotensin	Ventrotegmental neurons
Epinephrine	NPY	Reticular neurons
	Neurotensin	Reticular neurons
Serotonin	Substance P	Medullary raphe
	TRH	Medullary raphe
	Enkephalin	Medullary raphe
Vasopressin	CCK, dynorphin	Magnocellular hypothalamic neurons
Oxytocin	Enkephalin	Magnocellular hypothalamic neurons

GABA, γ-aminobutyric acid; VIP, vasoactive intestinal polypeptide; NPY, neuropeptide Y; CCK, cholecystokinin; TRH, thyrotropin-releasing hormone.

peptides affect signals transmitted by the coexisting amino acid or monoamine? Except for some recent work with opioid peptide circuits in the hippocampus (see Opioid Peptides below), "We don't know" remains the easy answer. However, in the autonomic nervous system, where Tomas Hökfelt and his peptide liberation team have found at least one peptide in every sympathetic, parasympathetic, or enteric neuron they have studied, some solid peptide–monoamine interactions are emerging. In the salivary gland, low-frequency stimulation of the parasympathetic nerves releases ACh, but VIP is also released when the cholinergic nerves are stimulated at high frequency. VIP augments parasympathetic control of salivation by increasing glandular blood flow. Neuropeptide Y (NPY), found in many sympathetic neurons, is also released at higher frequencies of activation and sensitizes smooth muscle target cells to their adrenergic signals. Other peptide actions, although not necessarily between coexisting amino acid or monoamine transmitters, are worth noting. For example, VIP greatly accentuates the cAMP response of cortical neurons to low doses of norepinephrine, and galanin can change the release of ACh in the ventral hippocampus.

One way to recast these players in the proper perspective might be to consider that a peptide usually embellishes what the primary transmitter for a neuronal connection seeks to accomplish. Such an effect may be to strengthen or prolong the primary transmitter's actions, especially when the nerve is called on to fire at higher than normal frequencies. The peptide may provide part of the intraneuronal signal to alter the rate of production of the primary transmitter. There may still be places in which the peptide itself can fulfill all of the effects of a primary transmitter. Of course, accepting coexistences leads to new logistical problems for peptide-using neurons. How do such neurons coordinate synthesis of the peptide with their small-molecule transmitters? (Recall that the latter can be made within the synapse and are often reaccumulated after release but that peptides can be made only with ribosomes and not in nerve terminals.) How do such neurons decide when the secreted transmission should include the peptide?

Other important concepts for pharmacology also emerge: for example, the effects of peptide-directed drugs will depend not only on the location of the receptors and their actions but also on the context of its coexisting signals. Thus, the pharmacology of NPY or neurotensin may be most appropriately understood as a part of the total picture of central noradrenergic pharmacology, given the degree to which these two peptides may participate in that monoaminergic neuron's transmission. Given the relatively modest number of response mechanisms on which peptide messengers may operate through G protein–coupled receptors, a great enrichment of signaling possibilities becomes attainable through the interplay between frequency-dependent and diffusion-dependent release and response sites.

In terms of the development of drugs active on the nervous system, research on peptides has given us a humbling view of the riches that may await mining in the inner depths of the brain. The rest of this chapter reviews some emerging trends in neuropeptide research, followed by pharmacologically relevant data for the grand peptide families, a few individual noteworthy peptides, and a newly recognized large family of neuroactive and "glia-active" peptides, the cytokines, which may figure prominently in pharmacological attacks on neurodegenerative disorders.

Emerging Trends in Neuropeptide Research

Observers of the active peptide parade may find it useful to monitor certain recurring themes in several of the brain and autonomic peptide systems. They illustrate potential principles that could apply to peptide-signaling systems generally.

1. *Peptides are dynamically regulated.* With current technology, the dynamic synthesis and catabolism of neuropeptides has been estimated by measuring simultaneously mRNA, the precursor propeptide, and the signaling form. These studies have helped clarify the effects on peptide synthesis of drug treatments or other experimental perturbations that alter the levels of the coexisting amines in a given class of neurons. Furthermore, examination of the genes for some neuropeptides has revealed regulatory elements that make gene transcription sensitive to gonadal hormones (oxytocin, vasopressin, and proenkephalin) or to intracellular second messengers (VIP, proenkephalin, neurotensin, corticotropin-releasing hormone, and prosomatostatin), providing direct proof of independence from their cotransmitters.
2. *Postrelease cleavage products may provide different signals.* Based on the absence of any specific data, it had been assumed that, despite the enormous logistical prelude to constructing a propeptide and then cleaving and storing the active form, its secretion was the end of the line. In several peptide systems, however, pharmacological evidence suggests that catabolic forms of the secreted peptide (i.e., fragments of the secreted form) may have real signal value, both on the neuron that secreted the full peptide and on the postsynaptic neuron. The data suggest that such shorter forms may participate in fine-tuning the levels of interaction between the peptides and the effects of other coexisting transmitters.
3. *Nonpeptidic agonist and antagonist drugs have arrived.* The third emerging trend is less an example of peptide physiology than an expression of appreciation for the achievements of medicinal chemists in

synthesizing rigid nonpeptidic molecules that can act as agonists and (experimentally more important) as antagonists of the peptide's receptors. While it was clear that nature could do this in the form of the basic morphine molecule by making an agonist for opioid peptide receptors, extension of this principle to other peptides was slow. However, new molecules are coming forth, and as nonpeptidic antagonists are created for peptide systems, the ability to determine their functional roles will almost certainly be improved.

The Grand Peptide Families

Vasopressin and Oxytocin

These two highly similar nonapeptides (see Fig. 11–1) with internal 1,6-disulfide bridges are the original mammalian peptide "neurohormones." They are synthesized in the neuronal perikarya, found in the large neurons of the supraoptic and paraventricular nuclei, and stored in the axons of these neurons in the neurohypophysis, from which they are released into the bloodstream. Each peptide is synthesized as part of a larger propeptide (see Chapter 3), with which it is stored and released and from which it is cleaved as part of the release process. In the kidney, vasopressin (also known as antidiuretic hormone) facilitates distal tubular water reabsorption, while oxytocin stimulates epididymal and uterine muscle contraction.

The oxytocin and vasopressin peptides are expressed within the classic magnocellular neurons of the paraventricular and supraoptic nuclei, whose axons form the neurohypophysis, where the peptides are secreted into the bloodstream for their peripheral targets. The same peptides are also expressed within the parvocellular neurons of these two nuclei for secretion into the pituitary portal circulation, where vasopressin can act synergistically with corticotropin-releasing hormone (see below) to release ACTH. In addition, vasopressin is expressed within a subset of neurons of the suprachiasmatic nu-

CYFQNCPRG* Vasopressin

CYIQNCPLG* Oxytocin

FIGURE 11–1. Molecular sequence of oxytocin and vasopressin, in which amino acid names are symbolized by the standard single-letter symbols and * indicates an amidated C terminus. (Key to amino acid single-letter symbols: A, Ala; R, Arg; N, Asn; D, Asp; C, Cys; Q, Gln; E, Glu; G, Gly; H, His; I, Ile; L, Leu; K, Lys; M, Met; F, Phe; P, Pro; S, Ser; T, Thr; W, Trp; Y, Tyr; V, Val.)

cleus; outside the hypothalamus, vasopressin is also expressed within some neurons of the bed nucleus of the stria terminalis (which project to the lateral septum, medial amygdala, and periaqueductal gray, among others), the medial amygdala (which project to the ventral hippocampus), the septum, and reputedly within neurons of the locus ceruleus. While neurons in these target regions are definitely responsive to the peptides infused locally, documentation of synaptic actions is still lacking. An interesting but mysterious aspect of the biology of these neurons is their capacity to transport the mRNAs for their peptides into their axons. This transport is more easily detected during periods of functional load (postpartum or salt load), when the mRNA is also transported back again to the perikaryon, where the messages might either be translated or in some manner regulate translation.

The behavioral actions of vasopressin are quite impressive, although the mechanisms accounting for them remain unclear. These effects are described in somewhat greater detail in Chapter 13.

Analogs of vasopressin and oxytocin have now been developed with selective agonist or antagonist properties; these molecules, often shorter and more stable than the native peptide, should make useful pharmacological probes. Based on the selective actions of synthetic vasopressin analogs, two subtypes of peripheral receptor have been characterized: V_1 receptors, which mediate responses on arteriolar walls, and V_2 receptors, which mediate the effects on the renal tubules. A V_1 receptor agonist has been reported to maintain long-term potentiation in neurons of the lateral septal nucleus (see Chapter 13). The effects of vasopressin analogs on either memory or learning behaviors or on secretory responses mediated through the median eminence do not adhere to either receptor class and may indicate that more subtypes will be characterized. At least four arginine vasopressin receptors have been cloned. The dynamic regulation of oxytocin and vasopressin expression is sensitive to gonadal steroids, thus providing a gender-specific neuronal expression pattern that is turned on at puberty and that can be further regulated through reproductive function. Novice neuropeptide researchers have an opportunity left behind by the marauding peptide pioneers: what is the function of oxytocin in the male brain?

The Tachykinin Peptides

In 1931, von Euler and Gaddum discovered an unexpected pharmacologically active substance in extracts of brain and intestine, which they later named substance P because the activity was present in the dried acetone powder of the extract. Although studied intermittently through bioassays, substance P remained somewhat shaded in obscurity until almost 40 years later, when

Leeman and co-workers purified a sialogogic peptide from hypothalamic extracts while looking for the still elusive corticotropin-releasing factor. This sialogogic peptide turned out to have an amino acid content and pharmacological profile identical to substance P, and, when finally purified, sequenced, and synthesized, it was identified as substance P, an undecapeptide (Fig. 11–2). Knowledge of the structure and availability of large amounts of the synthetic material permitted the development of radioimmunoassays and immunocytochemical tests, which were then used to map the brain and assay its substance P content. Substance P is present in small neuronal systems in many parts of the CNS, and upon subcellular fractionation, it is found in the vesicular layer. It is especially rich in neurons projecting into the substantia gelatinosa of the spinal cord from the dorsal root ganglia and has even been proposed to be the transmitter for primary afferent sensory fibers. While it is very potent as a depolarizing substance in direct tests of spinal cord excitability (about 200 times stronger on a molar basis than GLU), its long duration of action prompted caution in accepting substance P as the primary sensory transmitter. Other peptides have also been identified in dorsal root ganglion cells, and GLU released from cultures of these cells also produces a prompt, powerful excitatory action on spinal dorsal horn neurons. If GLU and substance P were cotransmitters for some sensory fibers, substance P could be viewed as prolonging and intensifying the transmission into the cord.

Radioimmunoassays and immunocytochemistry also show brain regions other than the spinal cord to be rich in substance P, especially the substantia nigra, caudate putamen, amygdala, hypothalamus, and cerebral cortex. Tests with iontophoretic application in these regions generally indicate excitatory actions, again with a long duration. The dynamic regulation of substance P within the rat striatum has been informative. Substance P–containing neurons of the striatum project to the dopamine neurons of the substantia nigra in reciprocal circuitry. Dopamine antagonist or 6-(OHDA) treatment of the substantia nigra leads to a substantial drop in substance P content;

RPKPQQFFGLM*	Substance P
DV**PK**SD**QF**V**GLM***	Kassinin
H**K**TDS**F**V**GLM***	Neurokinin A
pE **P**SKDA**F** I **GLM***	Eledoisin
DMHDF**F**V**GLM***	Neurokinin B

FIGURE 11–2. Structural homologies between the peptides of the tachykinin family, presented according to the schema of Figure 11–1. pE, pyroglutamate.

analysis of the peptide, propeptide, and mRNA indicates that this drop is due to decreased gene expression and decreased release. After treatment with methamphetamine, substance P synthesis is accelerated. After axotomy, the levels of substance P and several of its coexisting neuropeptides plunge in dorsal root ganglion neurons but rise in axotomized sympathetic ganglioneurons. In the human neurological disease Huntington's chorea, characterized by profound movement disorders and psychological changes, substance P levels in the substantia nigra are considerably reduced while alternative neuropeptides are expressed.

Tachykinin research has been one of the windfall areas benefiting from the application of molecular biological methods. The cloning of the precursor forms of the three main tachykinin peptides in the mammalian CNS and their first receptor subtypes has helped to clarify their relationships to each other and to other neurotransmitters. There are two mammalian tachykinin genes. The neurokinin A (NKA) gene (on human chromosome 7) can produce three forms of mRNA through alternative splicing of its many exons. The least prevalent form encodes only substance P, and the other two forms encode both substance P and NKA (the term the tachykinin nomenclature committee prefers for the peptide previously called *substance K*). In general, the tissue and cellular distributions of substance P and NKA are similar. A second tachykinin gene, located on human chromosome 12, encodes the neurokinin B (NKB) precursor, previously called *neuromedin K*. Receptors for all three peptides have been cloned, and all are G protein–coupled receptors (see Chapter 6). The receptors are called NK_1, NK_2, and NK_3. Substance P is more potent than NKA or NKB at the NK_1 site but much less potent than NKA or NKB at the NK_2 site. At the NK_3 site, NKB is slightly more potent than NKA or Substance P. Inositol phosphate breakdown is a major transducing pathway for at least the NK_1 and NK_2 receptors. After activation, either physiologically or pharmacologically, NK_1 receptors are internalized into the smooth endoplasmic reticulum of receptor-bearing dendrites (see Fig. 11–3). Highly selective nonpeptidic agonists and antagonists have helped to define the potential behavioral actions of substance P. However, drugs acting at the human NK_3 receptor do not work well on rat NK_3, eliminating the most frequently used animal model. Based on the rich expression of NK_1 receptors on 5-HT, NE and DA-containing neurons, an NK_1 antagonist originally developed for its antiemetic properties was clinically tested as an antidepressant and found in initial clinical trials to be equipotent with a serotonin-selective reuptake inhibitor. The field is now waiting for replication of this exciting finding. However, NK_1 antagonists have had no effect on pain, even though knockout of the receptors or the dorsal horn neurons expressing them does reduce chronic inflammatory pain. However, dorsal horn afferent fibers contain multiple other peptides (see Fig. 11–4).

HSDAVFTDNYTRLRKOMAYKKYLNSILN *	VIP
HSDGI**FTD**S**Y**S**R**Y**RKQMA**V**KKYL**AAVL *	PACAP
HA**D**G**VFT**SDFS**RL**LG**Q**LSA**KKYL**E**S**LI *	PHI 27
HA**D**G**VFT**SDFS**RL**LG**Q**LSA**KKYL**E**S**MI *	PHM 27
YADI**FT**NS**Y**RKYLG**Q**LSAR**K**K**L**LQD ...	GHRH 1–24
HSQGT**FT**SD**Y**SKYLDSRRAQDFVQWLMNT	Glucagon
HSDGT**F**YSELS**RLR**DSARLQRL**L**QGLY *	Secretin

FIGURE 11–3. The vasoactive intestinal peptide (VIP)–related peptide family represented by their single-letter amino acid symbols. The sequences of PHI-27, PHM-27, growth hormone–releasing hormone 1–24, glucagon, and secretin that match those of VIP are indicated in bold letters. For an interesting exercise, the reader may wish to construct a complementary table in which matches to glucagon are highlighted.

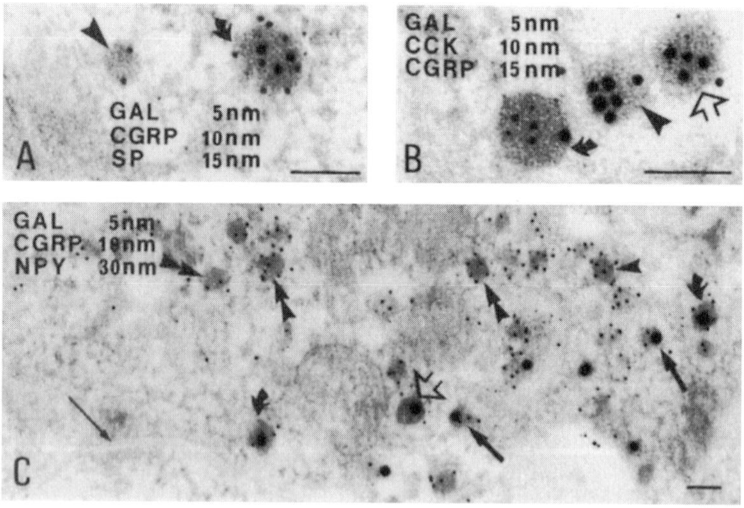

FIGURE 11–4. Electron micrographs of large dense-core vesicles (LDCVs) in calcitonin gene–related peptide (CGRP)–immunoreactive primary afferent terminals in lamina II of the dorsal horn of rat lumbar spinal cord after triple-immunogold staining for galanin (GAL)-, CGRP-, and substance P (SP)-like immunoreactivity (LIs) (A); for GAL-, cholecystokinin (CCK)-, and CGRP-LIs (B); or for GAL-, CGRP-, and neuropeptide Y (NPY)-LIs (C). The size of gold particles for each peptide is indicated in the micrographs. The different examples of colocalization of neuropeptides in LDCVs are shown. (A) Curved arrow indicates 3-labeled LDCVs. Arrowhead, CGRP alone. (B) Curved arrow indicates 3-labeled LDCVs. Arrowhead, CGRP alone; open arrow, CGRP plus growth hormone. (C) Curved arrow indicates 3-labeled LDCVs. Arrowhead, GAL and CGRP; double arrowheads, CGRP alone; open arrow, GAL + NPY; thick arrows, CGRP + NPY; thin arrow points to a synapse. Bars indicate 100 nM. (From X. Zhang and T. Hökfelt.)

VIP-Related Peptides

VIP, a peptide composed of 29 amino acids, was originally isolated from porcine intestine, using the C-terminal amidation strategy of Mutt. It was originally named for its ability to alter enteric blood flow. Establishment of its sequence revealed marked structural similarities to glucagon, secretin, and another peptide of gastric origin that inhibits muscular contraction, gastric inhibitory peptide. These structural similarities constitute the VIP-related peptide family (see Fig. 11–5), named for the member most enriched in brain. The development of synthetic VIP for preparation of immunoassays and cellular localization revealed that VIP exists independently of its cousins in the gut and pancreas and that it was prominent in many regions of the CNS and the autonomic nervous system. In parasympathetic nerves to the cat salivary gland, VIP coexists with ACh and is apparently released with ACh as part of an integrated command to activate secretion and to increase blood flow through the gland. In the CNS, VIP-reactive neurons are among the most numerous of the chemically defined cells of the neocortex, and they exhibit there a very narrow radial orientation, suggesting that they innervate cellular targets located wholly within a single cortical column assembly. In peripheral structures, VIP-reactive nerves innervate the gut, especially sphincter regions, as well as the lung and possibly even the thyroid gland.

Additional members were added to this important family of peptides when Guillemin and colleagues isolated the long-sought growth hormone–releasing hormone (GHRH) peptide and recognized the homologous shared amino acid sequences. Two later members of the family are termed *PHI-27* and *PHM-27*, to reflect their size, 27 amino acids long, and the one-letter initials of their N-terminal (H = histidine) and C-terminal (isoleucine or methionine) peptides (where *P* stands for peptide, not proline). PHI-27 was initially isolated by Mutt and associates from gut extracts while searching for other C-terminally amidated peptides, a property all active members of this

YPS**KPDNPGEDAPAEDLARYY**S**ALRHYINLITRQRY** *	NPY
YPA**KP**EA**PG**QN**A**SPQQ**L**S**RYY**AS**LRHY**L**NL**V**TRQRY** *	PYY
G**P**SQ**P**TY**PG**D**DAP**V**EDLIR**F**Y**DN**L**QQ**Y**L**N**VV**TR**H**RY** *	APP
A**P**LE**P**VY**PG**DN**A**TP**E**QM**A**Q**Y**AAD**LR**R**YIN**M**LTR**P**RY** *	HPP

FIGURE 11–5. The pancreatic polypeptide family represented by their single-letter amino acid symbols. The sequences of peptide YY (PYY), avian pancreatic polypeptide (APP), and human pancreatic polypeptide (HPP) that match those of NPY are indicated in bold letters.

family share. PHM-27 was identified from the deduced sequence of the pro-VIP mRNA in cloning experiments. Growth hormone–releasing hormone has a relatively limited neuronal distribution concentrated in the hypothalamus and median eminence. No cellular maps for PHI or PHM in the brain have yet been reported.

The next addition to the VIP-related peptide family is or pituitary adenylate cyclase–activating peptide (PACAP), which has a 1000-fold greater capacity to activate adenylate cyclase in pituitary cultures relative to VIP. PACAP is a C-terminally amidated, 27-residue peptide with a very high conservation of the VIP sequence (see Fig. 11–5), which is also expressed in the brain; a 38-residue, C-terminally extended form has also been isolated from the brain and may be the predominant form. Two forms of PACAP receptor have been cloned, and all receptors for peptides in this family show strong structural similarity.

To reverse a trend, molecular cloning of the mRNA for the precursor of secretin, another member of this peptide superfamily, indicates that this specific peptide is not actually found in the brain, as had been proposed based on immunocytochemistry with polyclonal antibodies some years ago. Nevertheless, fueled in part by reports that systemic secretin can improve the behavioral problems of autistic children, there has been renewed interest in the biological actions of secretin in the brain. Most recently, a novel member of this family, found to be enriched in hypothalamus by differential expression patterns, was identified twice in rapid order by different strategies. In the first, sequencing of hypothalamus-enriched mRNA revealed a pair of peptides (named *hypocretin-1* and *-2*) with sequence similarity to different domains of secretin than seen with the previously identified members. Then, independently, a group screening for natural ligands for orphan G-protein receptors identified the same peptide and demonstrated a potent appetitive action, for which it was named *orexin A* and *B*. The hypothalamic neurons expressing the gene for hypocretin/orexin are held to be engaged in both appetite and blood pressure regulation. This same peptide gene was associated with individuals exhibiting inheritable narcolepsy.

Pancreatic Polypeptide–Related Peptides

The pancreatic polypeptides were recognized in extracts of pancreatic islets in the mid-1970s, and those from pigeon, pig, and human pancreas were found to be highly homologous 36-residue peptides with amidated C termini (see Fig. 11–6). Antisera against these peptides recognized cells and fibers in the CNS and autonomic nervous system, but these immunocytochemical observations were rightly qualified as "pancreatic polypeptide-like immunoreactivity" because when the same sera were used in their highly dilute form for radioimmunoassay, only scant extractable material was detected. When Mutt and colleagues applied their C-terminal amidated peptide isolation strategy to these extracts, two more members of this family were identified. The first one, found in gut, was given the name PYY (a neuropeptide with tyrosine [single-letter amino acid symbol Y] residues at both the N and C termini). However, because the one found in brain had the same length and the same N- and C-terminal tyrosines, it was named NPY. In rapid order, this peptide's distribution was described in rodent and human brains, showing NPY to be one of the most extensive central peptide systems. Particularly high amounts were found in the hypothalamus, limbic system, and neocortex. In many places, including the autonomic nervous system, NPY

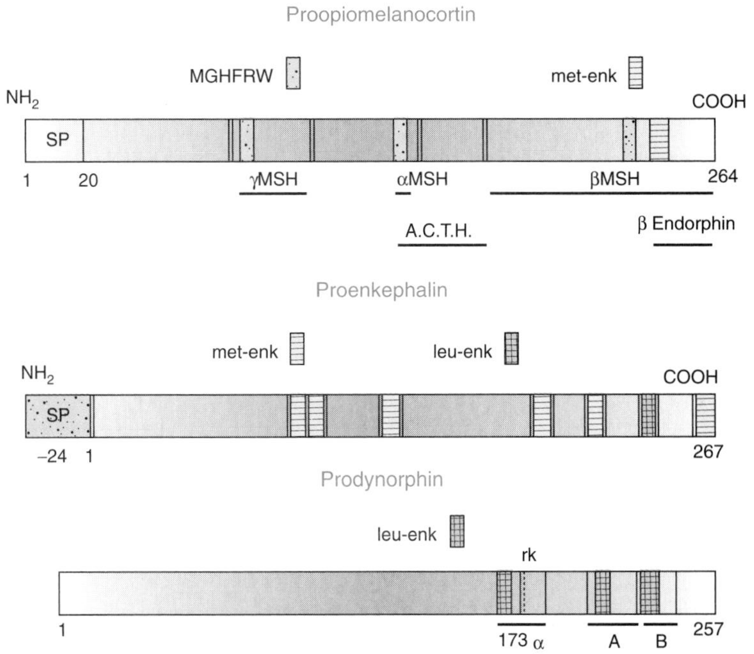

FIGURE 11–6. Structural relationships among the prohormone, precursor forms of the three major branches of the opioid peptides, depicted as a bar diagram, whose length in amino acid residues is indicated by the small number at the corresponding C termini. Locations of the repeating peptide sequences are indicated. Basic amino acid sequences that constitute consensus cleavage sites for processing are indicated as single or double vertical lines within the bar. Proopiomelanocortin and proenkephalin have consensus signal peptides indicated at their N termini. MSH, melanocyte-stimulating hormone; ACTH, corticotropin.

coexists with either NE or epinephrine. NPY increases the sensitivity of sympathetically innervated smooth muscle to NE and is one of the most potent natural vasoconstrictors known. Upon sequencing of the gene in many species, from humans to fish, the structure of the NPY mRNA is one of the most highly conserved genes ever reported, surpassing even the conservation seen in insulin. Within the C terminus of the pre-NPY peptide, a second potential cleavage product, known as the C-terminal peptide of NPY, (or CPON) is almost as highly conserved. Since CPON will be formed to liberate NPY, it is frustrating that there are still no reports of its actions more than a decade after its description.

NPY, especially in the autonomic nervous system, documents the signal value of shortened versions of the secreted peptide. Three types of receptor have been detected pharmacologically. The postsynaptic Y_1 requires the full NPY(1–36) for action and enhances responses to coreleased NE. The presynaptic Y_2 reacts as well to N-terminally truncated forms of NPY and reduces NE release. Central injection of NPY suggests active roles in anxiety reduction and appetite stimulation, partially confirmed by the anxiogenic effects of intraventricularly injected antisense RNA oligonucleotide for the Y_1 receptor but not the one for Y_2. During fasting, hypothalamic levels of NPY rise. Nevertheless, mice in whom the NPY gene is knocked out continue to eat normally, although in genetically obese rodents lacking leptin, knockout of NPY expression significantly reduces the obesity.

Opioid Peptides

An *endorphin* is any endogenous substance (i.e., one naturally formed in the living animal) that exhibits the pharmacological properties of morphine. When this term was first coined in mid-1975, it was a useful abstraction covering "morphine-like factors" from brain extracts and spinal fluids that were active in opiate ligand-displacement assays or opiate-sensitive smooth muscle assays. Within a year, a highly competitive effort resulted in the purification, isolation, sequencing, and synthetic replication of not only one but nearly a half-dozen peptides that deserved the term *endorphin*.

The incredible explosion of work on this class of peptides began with attempts to isolate and characterize the receptor to opiates as part of a molecular biological approach to the question of narcotic addiction. When specific binding assays were developed, substantial evidence was accumulated independently and almost simultaneously by Sol Snyder's group at Johns Hopkins, Eric Simon's group at New York University, and Lars Terrenius' group at Uppsala University that a high-affinity binding site in synaptic membranes showed stereoselective opioid recognition properties. A whole new approach to neurotransmitter identification took shape when John Hughes, Hans Kosterlitz, and their colleagues in Aberdeen demonstrated that extracts of brain contain a substance that can compete in the opiate receptor assays and show opioid activity in in vitro smooth muscle bioassays.

In late 1975, this endogenous opioid activity was attributed to two pentapeptides, named *enkephalins*, which shared a tetrapeptide sequence, YGGF, varying only in the C-terminal position: hence, they were called Met^5-enkephalin and Leu^5-enkephalin. Perhaps even more dramatic than the announcement that the brain contained not one but two opioid peptides was the realization that the entire structure of Met^5-enkephalin is contained within a 91–amino acid pituitary hormone, β-lipotropin (β-LPH), whose isolation and sequence was reported several years earlier by C. H. Li but whose function was unknown.

Several groups then reported the isolation, purification, chemical structures, and synthetic confirmation of three additional endorphin peptides: α-endorphin (β-LPH_{61-76}), γ-endorphin (β-LPH_{61-77}), and β-endorphin (β-LPH_{61-91}, also called C fragment). Numerous claims and counterclaims were parried across the symposium stages for several months as to one person's peptide being another person's artifact or precursor.

Subsequent research has greatly clarified the molecular and genomic relationships between the three major branches of the opioid peptide family (see Fig. 11–7): the proopiomelanocortin (POMC)–derived peptides, the proenkephalin-derived peptides, and the prodynorphin-derived peptides. The superfamily has been further expanded with the heptadecapeptide orphanin FQ, also called nociceptin, identified as an activator of an opioid-like orphan receptor. The natural peptide shows no affinity for opioid receptors. Likewise, opioid peptides and ligands do not bind to the orphanin receptor. Nevertheless, microinjection of orphanin into the periaqueductal gray blocks the anti-nociceptive effects of morphine injected into the same site. While it is known from receptor knockout studies that the m-opiate receptor is required for the effects of exogenous morphine, none of the originally described opioid peptides showed impressive affinity for this receptor. This mystery may have been partially solved by the identification of two novel endogenous tetrapeptides that show very little similarity to the key N-terminal YGGF of all the others. Other structurally related natural peptides lack opioid receptor activity altogether, such as the invertebrate cardioacceleratory peptide FMRF-amide and its mammalian counterparts neuropeptides FF and AF); nevertheless, neuropeptide expression is altered under experimental chronic pain and reduces the response to centrally injected opiates. Furthermore, some opioid-acting peptides have been found exogenously in milk and in plant proteins and have been called *exorphins*.

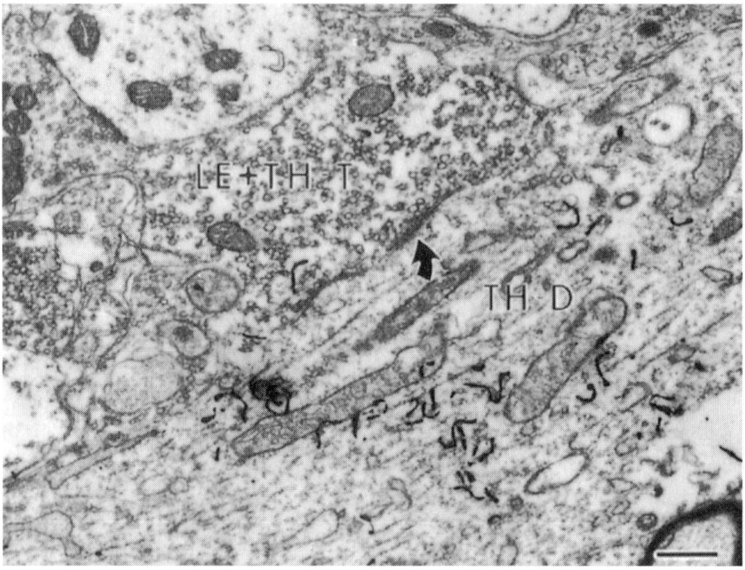

FIGURE 11–7. An ultrastructural basis for functional interaction between central opioid-containing neurons and central catecholamine neurons is provided by this dual label, autoradiographic image obtained by immunocytochemistry. Nerve terminals exhibit leu-enkephalin immunoreactivity, and one also exhibits tyrosine hydroxylase immunoreactivity (small autoradiographic grains, terminal labeled LE + TH-T). The latter makes a symmetric synaptic contact on a dendrite (TH-D) bearing many autoradiographic grains, indicating extensive tyrosine hydroxylase content. Scale bar lower right = 0.5 μm. (Modified from Milner et al., 1989.)

Chemical and Cellular Relationships among the Opioid Peptides

1. The proopiomelanocortin (POMC) peptides are expressed independently in the anterior pituitary, the intermediate lobe of the pituitary, and one main cluster of neurons in the area of the arcuate nucleus of the hypothalamus. The major endorphin agonist produced from POMC is the 31–amino acid C-terminal fragment β-endorphin, the most potent of the natural opioids. N-terminal fragments of β-endorphin are much less potent, and analogs with no N-terminal tyrosine (so-called des-Tyr versions) lose all opioid activity, although in some hands des-Tyr peptides are reputed to be active behaviorally. In the corticotropin-secreting cells of the anterior pituitary, POMC is processed largely to corticotropin and to an inactive form of β-endorphin; in intermediate lobe cells and arcu-

ate neurons, the same precursor is processed to α-MSH and active β-endorphin. A third MSH-containing heptapeptide component, discovered during the cloning and sequencing of the POMC mRNA, suggests that yet another end product may be possible.

2. The enkephalin pentapeptides Met^5-enkephalin and Leu^5-enkephalin are expressed in wholly separate neuronal systems from the POMC neurons and are much more pervasively distributed throughout the CNS and peripheral nervous system, including the adrenal medulla and enteric nervous system. Cloning and sequencing of the mRNA for the proenkephalin, starting with mRNA from adrenal medullary tissue, produced an unexpected dividend in that the precursor exhibits multiple copies of the two peptides in almost exactly the 6:1 ratio of Met^5- to Leu^5-enkephalins that had been described in regional brain and gut assays. This solved the mystery of the two similar peptides.
3. The prodynorphin peptides consist of C-terminally extended forms of Leu^5-enkephalin arising from a different gene and from a different mRNA that encodes for production of four major peptides: dynorphin A, dynorphin B, and two neoendorphins, α and β. These C-terminally extended peptides act as potent opioid agonists without cleavage down to the enkephalin pentapeptide form, and upon mapping they were found to represent a third separated series of rather generally distributed central and peripheral neurons. The amino acid sequence of orphanin bears greater resemblance to dynorphin than to any of the other opioid peptides (see Fig. 11–7).

Each of the separate classes of neurons containing β-endorphin, enkephalin, or dynorphin peptides has distinct morphological features. The β-endorphin-containing neurons are long projection systems that fall within the general endocrine-oriented systems of the medial hypothalamus, diencephalon, and pons. The proenkephalin-derived peptides and the prodynorphin-derived peptides are generally found in neurons with modest to short projections, groups of which are widespread and some of which innervate presumptive dopaminergic neurons, potentially providing a basis for some endogenous reward properties (see Fig. 11–8). In some regions, the enkephalin-derived and dynorphin-derived peptides show intriguing relationships. For example, enkephalin-containing neurons project from the entorhinal cortex to the molecular layer of the dentate gyrus of the hippocampus, while dynorphin-containing neurons project from the dentate gyrus to the CA3 pyramidal cells. In the spinal cord, intrinsic interneurons contain dynorphin peptides, while descending long axons from the pons and medulla contain enkephalin pep-

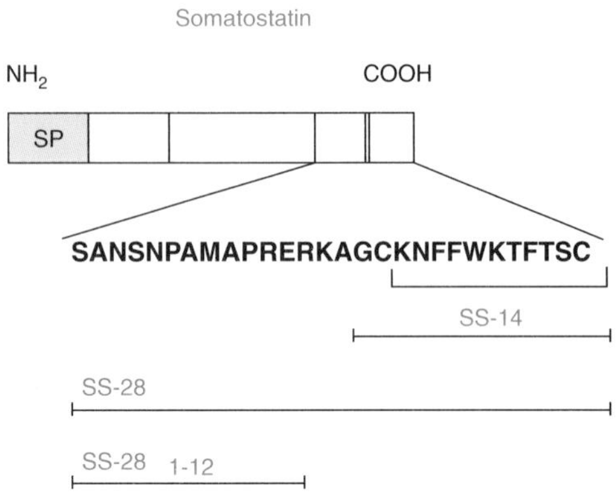

FIGURE 11–8. Structure of the prohormone form of somatostatin, with the location and sequence of the 28–amino acid residue form at the C terminus and the relative sequences and locations of the different forms of somatostatin indicated.

tides. In addition, all of these peptide systems presumably contain other nonpeptidic transmitters that have not yet been rigorously identified as well as some peptides that have.

Physiological and Pharmacological Effects

Traditional (i.e., precloning) pharmacology produced at least three schemes of endorphin–receptor classification based on *(1)* the comparative actions of morphine or of morphine-like drugs with mixed agonist–antagonist actions on the dog spinal cord (the μ, κ, and σ scheme of Martin and Gilbert), *(2)* the relative effects of endogenous and synthetic opioids on various smooth muscle in vitro assays (the μ and δ scheme of Kosterlitz or the μ, δ, and ϵ scheme of Herz), or *(3)* guanosine triphosphate–modifiable ligand binding. Remarkably, given the impetus for their discovery, none of the opioid receptors was affected by chronic exposure to morphine. The best concordance between receptor subtypes and opioid peptide systems remains that of the dynorphin-derived peptides reported by Goldstein and Chavkin, to act on the κ opioid receptors in both gut and brain.

The membrane mechanisms transducing the various classes of opioid receptors are not well understood, but several categories of opioid action have been characterized at the molecular level. These include inhibition of adenylate cyclase, inhibition of the N-type Ca^{2+} conductance that underlies transmitter release, enhancement of the M current K^+ conductance, as well as activation of the inward rectifier and delayed rectifier K^+ conductances. Intracellular recordings of gut neurons and brain neurons in vitro indicate that these cells respond by hyperpolarization to acute administration of opioids mediated by an increased K^+ conductance that secondarily depresses Ca-dependent spike activity. In the dorsal horn, neither membrane potential nor conductance is impressively altered but responsiveness to depolarizing synaptic potentials is reduced.

Iontophoretic tests with enkephalins and β-endorphin suggest that neurons throughout the CNS can be influenced by these peptides at naloxone-sensitive receptors. In general, responses tend to be depressant, except in the hippocampus, where excitatory actions are so profound that hippocampal seizures can be induced. However, the mechanism of this excitation is the exception that proves the rule, as it is based on inhibition of GABA-ergic inhibitory interneurons bearing the opiate receptors and produces excitation by disinhibition. In such tests, all endorphins show qualitatively equivalent effects and similar onset of actions, but β-endorphin may be more potent and longer lasting due to its slower hydrolysis. Peptides with no N-terminal Tyr are much weaker, but the potency of endomorphins with only the N-terminal Tyr in common suggests that there is much more to be learned.

This picture of opioid receptor pharmacology changed irrevocably within a few short months, beginning in the fall of 1992 when first one (the δ receptor) and then all three of the opioid peptide receptor subtypes were cloned. As predicted, all were similar G protein–coupled receptors with some very surprising similarities to other peptide receptors outside the opioid peptide family. Considering all of the potent and receptor subtype-specific agonists and antagonists and their assessment in the hippocampus, it was only a matter of time before the well-defined opioid peptide-containing circuits there were subjected to analysis of their roles in synaptic transmission.

The results of these studies (see review by Wagner and Chavkin, 1994) have paid off in what is arguably the best evidence yet for central synaptic actions of any neuropeptide. High-frequency stimulation of the perforant path will activate dynorphin-containing dentate granule cells. This in turn leads to a naloxone-sensitive suppression of IPSPs recorded from CA3 pyramidal cells. This initial claim of synaptic action was soon attributed to presynaptic opioid suppression of NE release. A second synaptic effect was then detected as a blunted excitatory response to perforant pathway stimulation in the dentate granule cells themselves (the ones containing the dynorphin). However, this latter effect turned out to be a highly novel mechanism in which dendritic dynorphin acted as a retrograde transmitter to reduce presynaptic release of glutamate. Extended analysis of the ability of naloxone to block hippocampal long-term potentiation (LTP, see Chapter 12) provided the best evidence for more conventional synaptic actions. Here, μ antagonists block the LTP induced in CA3 neurons by mossy fiber stimulation, and both μ- and δ-receptor antagonists block the LTP produced by stimulation of the lateral perforant pathway (a proenkephalin-containing pathway). Furthermore, LTP produced in dentate granule cells by high-frequency stimulation in the hilus of the hippocampus is enhanced with κ agonists, and this effect is blocked with antidynorphin antibodies or κ-selective antagonists.

A minor disappointment in the field of opioid peptide research has been its failure to deliver on the promise of providing insight into the nature of drug dependence. Despite many attempts, little significant change in the concentrations of any of the opioid peptides or their receptors can be detected with drug dependence. Studies of morphine-dosed animals show initial increases in β-endorphin mRNA, leading to a C-terminally shortened form that may be either a very weak agonist or, more interestingly, an opioid antagonist. Surprisingly, NMDA antagonists can block opiate tolerance formation. Chronic naltrexone treatments, which up-regulate ligand binding sites for the shorter opioid peptides, also increase their mRNAs by severalfold in the rat striatum and, to lesser degrees, in other opioid-sensitive brain regions; however, the concentration of peptides encoded by the mRNAs is not changed.

Although the behavioral effects of these peptides once attracted much attention, the area has not maintained that momentum. The enkephalins produce only transient analgesia after direct intracerebroventricular injection; in such tests, β-endorphin is 50–100 times more potent than morphine on a molar basis. The fact that blood-borne peptides do not penetrate into the brain well or survive the gauntlet of peptidases to which they are exposed in the process has spurred the search for more effective analogs, although surpassing the morphine analogs has been difficult. The sheer number of whole-animal effects that have been interpreted as endorphinergic physiology on the basis of naloxone effects has grown considerably but remains open to more comprehensive analysis. Among the proposed physiological properties that may be regulated by one or another of the endorphin substances are blood pressure, temperature, feeding, sexual activity, and lymphocyte mitosis, along with pain perception and memory. All of this seems like a lot to ask of one peptide family, but then, who are we to ask what a peptide may, or can, do for us? As we said before, don't blink or you'll miss the next blazing development.

Individual Peptides Worth Tracking

We conclude this chapter with brief comments on a few of the other neuroactive peptides that are under active investigation. Around those we have selected, bodies of research are consolidating and the neuropharmacological implications seem strong.

Somatostatin (Somatotropin Release–Inhibiting Factor)

In 1971, workers began to look for other goodies in their hard-won hypothalamic extracts. Vale, Brazeau, and Guillemin tried their extracts for potency in releasing growth hormone from long-term cultured anterior pituitary cells and found to their amazement a factor that inhibited even basal growth hormone release in minute amounts. They named this factor somatostatin. Isolation and purification studies eventually culminated in the characterization of this molecule as a tetradecapeptide with a disulfide bridge between Cys^3 and Cys^{14} (Fig. 11–8). Radioimmunoassays, immunocytochemistry, and whole-animal tests of the synthetic peptide made it clear that somatostatin was doing more than just inhibiting growth hormone release. Somatostatin was found to be widely distributed in the gastrointestinal tract and pancreatic islets, localized in the latter tissue to the δ cells by immunocytochemistry. Somatostatin in islets can apparently suppress the release of both glucagon and insulin, and in diabetic humans, who have no insulin, suppression of glucagon can be an important element in determining insulin requirements. The mechanisms of this suppression are not known but may be similar to the effect of somatostatin on growth hormone release from pituitary, an action accompanied by suppression of TSH release.

When somatostatin is injected intracerebroventricularly, animals show decreased spontaneous motor activity, reduced sensitivity to barbiturates, loss of slow-wave and REM sleep, and increased appetite. Radioimmunoassays and immunocytochemistry of rat brain regions show that somatostatin is largely concentrated in the mediobasal hypothalamus, with much smaller amounts present in a few other brain regions (see Fig. 11–9). Immunocytochemical studies led to the surprising finding that somatostatin-reactive cells and fibers are also present in dorsal root ganglia; the autonomic plexi of the intestine; and the amygdala, hippocampal formation and neocortex. Avian and amphibian brains contain considerably more somatostatin than mammalian brains. Physiological tests on the isolated frog spinal cord suggest that it may facilitate the transmission of dorsal root reflexes after a long latent period. Iontophoretic tests on rat neurons indicate a relatively common depressant action that is brisk in onset and termination. At the level of cellular electrophysiology, somatostatin hyperpolarizes neurons in a potent manner and can dynamically open the so-called M current channel, which cholinergic muscarinic receptors close in order to excite hippocampal and other neurons. The net result of somatostatin's actions in the intact brain is

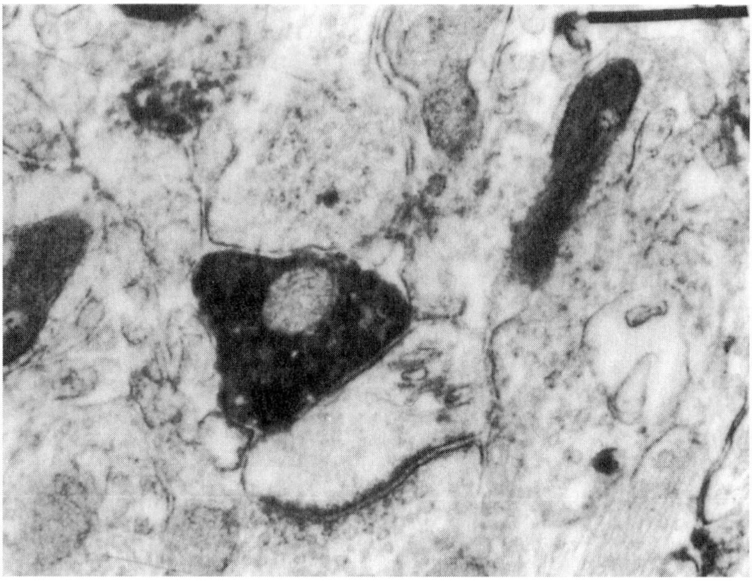

FIGURE 11–9. Ultrastructural immunoperoxidase localization of somatostatin in a nerve terminal making synaptic contact with an unreactive small dendritic spine in the rat neocortex. Calibration bar = 0.5 μ. (Unpublished micrographs from Battenberg and Bloom, The Scripps Research Institute.)

to enhance responsiveness to ACh. The important missing datum is what somatostatin does to the responses to GABA, its frequent but not universal coexistence partner.

An endogenous large somatostatin, extended at the N terminus by an additional 14 amino acids, also exists in both brain and gut and shares equipotency for many somatostatin actions. In rodent and primate neocortex, the N-terminally extended forms reveal a far more extensive neuronal density, which at least overlaps with the intrinsic GABAergic cortical neurons. In early-onset Alzheimer's disease, cortical somatostatin content is depleted, and somatostatin-28 (SS-28)–immunoreactive neuritic processes are involved in the formation of the hallmark Alzheimer's lesion, the plaque. Cysteamine, a drug developed to treat the metabolic disorder known as cystinosis, will selectively deplete somatostatin-14 (SS-14) without altering brain levels of SS-28 or SS-28_{1-12}, suggesting that these peptides may be stored in separable subcellular compartments.

As we have seen with the other peptides, receptor subtypes are the norm, and here cloning studies have raised the ante well above what was expected from the physiology: five different cloned receptors have now been reported.

Two of them align fairly well on distributional maps: $SRIF_1$ was expressed in dentate granule cells, striatum, locus ceruleus, and deeper layers of neocortex and showed homologous desensitization, while $SRIF_2$ was seen more diffusely in cortex, was more abundant in the CA1 field of the hippocampus, and did not desensitize. The third cloned form matches neither, while the fourth form binds SS-28 with greater affinity than SS-14.

Although somatostatin was an unexpected by-product of the race to identify the positive regulators of anterior pituitary release, its pharmacological and diagnostic potentials promise an important future. Somatostatin deficits in cerebrospinal fluid have been linked to affective disorder and to Alzheimer's disease. Long-lasting analogs of somatostatin have already been clinically applied in the control of pituitary hypersecretion of growth hormone. Evidence from cell lines transfected with various G protein–coupled receptor mRNAs suggests that under some conditions the somatostatin 5 receptor may form functional heterodimers with the DA_2 receptor.

Quite surprisingly, in a search for genes expressed in the cortex and hippocampus, a closely related neuropeptide, cortistatin, has been identified and is restricted to sparse GABAergic interneurons. This 14-residue peptide shares 11 amino acids with somatostatin and binds to all five cloned somatostatin receptors when expressed in transfected cells. Cortistatin also depresses neuronal activity, antagonizes some effects of ACh, but can be distinguished from somatostatin. Cortistatin mRNA accumulates during sleep deprivation.

Cholecystokinin

Among the gut hormones with the longest histories are gastrin and cholecystokinin (CCK). Subsequently, when their molecular structures were determined, a striking homology was revealed between the C termini. After some confusing interludes, it became recognized that the gastrin-like material extractable from nervous tissue was caused by antisera cross-reacting with the C-terminal octapeptide of CCK, whose sole tyrosine is sulfated. As it turns out, there is more CCK-8 in the brain than there is in the gut, but the N-terminally extended forms found in plasma, especially after eating, arise from the periphery. CCK is among the most promiscuous of the coexisting peptides, being found with dopamine and neurotensin in the substantia nigra and ventrotegmental area; with VIP, NPY, and GABA in thalamocortical and thalamostriatal connections; and with substance P and 5-HT in medullary neurons. It is also seen in the dorsal root ganglia with several other peptides (see Fig. 11–4). Interestingly, when animals are given morphine, although enkephalins do not change, brain CCK is depleted while cerebrospinal fluid CCK rises. Exogenous CCK has opioid antagonist properties. In iontophoretic tests on spontaneously active hippocampal pyramidal cells, CCK-8 potently activates firing, a potentially synaptic action compatible with the CCK contained within the dentate granule cell mossy fiber synapses to these neurons.

Two (and only two, a rarity in this business) forms of CCK receptor were characterized pharmacologically for their responsivity to the sulfated (CCK_A, known to be rich in the pancreas) or unsulfated (CCK_B, known to be rich in the stomach) forms of gastrin and CCK. Both forms have now been cloned and are also expressed in brain. In fact, the CCK_B receptor is the predominant brain form, while CCK_A is expressed by neurons in the nucleus of the solitary tract, the area postrema, the posterior hypothalamus, and the nucleus accumbens. CCK pharmacology was for a time the leader in de novo development of nonpeptidic antagonists. Antagonists of the CCK_A form have been linked functionally to enhancement of opiate analgesia and to centrally elicited postprandial satiety. Antagonists of the CCK_B form have been linked functionally to anxiety disorder, which is supported by the observation that intravenous CCK tetrapeptide or CCK-8 is anxiogenic in humans and experimental animals. Low doses of systemic or iontophoretic benzodiazepines block CCK-induced cell firing.

Neurotensin

Neurotensin, a tridecapeptide (see Fig. 11–10) and one of the few that are not C-terminally amidated, became quite interesting to neuropsychopharmacologists after its inauspicious origins as a by-product discovery during the search for corticotropin-releasing factor (CRF). In screening for what this structurally independent peptide might do, researchers found that its

SQEPPISLDLTFHLLREVLEMTKADQLAQQAHSNRKLDIA	CRF (OVINE)
SEEPPISLDLTFHLLREVLEMARAEQLAQQAHSNRKMEII	CRF (RAT)
pEGPPISIDLSLELLRKMIEIEKQEKEKQQAANNRLLDTI	SAUVAGINE
SCNTATCVTHRLAGLLSRSGGVVKDNFVPTNVGSEAF*	"CGRP"
DYMGWMDF*	CCK_8

Figure 11–10. Homologies between ovine and rodent corticotropin-releasing factor (CRF) and sauvagine are indicated by bold single-letter amino acid symbols. Below are indicated the amino acid sequences of calcitonin gene–related peptide (CGRP) and the C-terminal octapeptide of cholecystokinin (CCK), with a sulfated tyrosine at position 2. pE, pyroglutamate.

neuropharmacological properties after intracisternal administration included potent analgesia, hypothermia, accentuation of barbiturate and ethanol sleeping time, and increased release of growth hormone and prolactin. Mapping studies showed that the largest amounts of neurotensin are present in the anterior and basal hypothalamus, the nucleus accumbens and septum, the midbrain dopamine neurons (along with CCK), and selected scattered neurons in the spinal cord and brain stem. This led to the speculation that with its profile of sedation, hypothermia, and analgesia, neurotensin might be an "endogenous antipsychotic peptide." However, biochemical neuropharmacologists really began to pay attention when it was found that treatment of rats with conventional antipsychotic drugs increased levels of neurotensin and its mRNA in both striatum and nucleus accumbens, while comparable treatment with the so-called atypical antipsychotic drugs (i.e., those that do not cause the extrapyramidal symptom complex known as tardive dyskinesia, see Chapter 13) only increased neurotensin peptide and mRNA in nucleus accumbens. Furthermore, this change in gene expression would occur whether D_2 dopamine receptors were blocked or not; D_1 agonists produced similar effects. While it has been reported that cerebrospinal fluid levels of neurotensin are reduced in unmedicated schizophrenics and recover somewhat with treatment, the story remains incomplete as to its role in either pathogenesis or treatment. The human and rat neurotensin receptors have been cloned and show similar pharmacology, with inositol triphosphate hydrolysis as the transducing mechanism. Cellular tests of neurotensin action suggest that it may produce excitation, but the mechanisms are not yet defined.

Corticotropin-Releasing Factor

The search for the CRF ended in 1981 when Wylie Vale and colleagues completed the purification and sequence analysis of the peptide taken from extracts of sheep hypothalamus. Thus ended a three-decade-long search for a mystery factor that put the secretion of corticotropin and other anterior pituitary hormones under neuronal control and allowed the brain to replace the pituitary as the master gland of the body. The remaining member of the original four horsemen of the hypophysiotrophic–hypothalamic control system was lassoed a year later when Vale and Roger Guillemin independently reported the isolation, sequencing, and synthetic replication of growth hormone–releasing hormone. Nonparticipants in this highly competitive struggle probably had mixed emotions when these quests ended since along the way the world of neuropeptide discovery was at least triply blessed first by substance P, then somatostatin, and eventually neurotensin, all of which were discovered by teams intent on CRF.

When CRF was in hand, it turned out to be an interesting 41–amino acid peptide (see Fig. 11–10) structurally related to a rather obscure frog skin peptide, sauvagine, and to another peptide, urotensin, obtained from an even more obscure organ, the urohypophysis of certain fish. The cellular origin of mammalian CRF was traced to that portion of the paraventricular nucleus that projects to the median eminence but not into the posterior pituitary. Later, these and other CRF-containing circuits were identified as innervating a very extensive group of neurons in the pons and medulla, cortex, and amygdala, perhaps representing a stress-related circuitry. CRF is an extremely potent ACTH secretogogue, but its effects are significantly augmented by vasopressin (produced by the adjacent magnocellular paraventricular neurons) as well as by NE and angiotensin.

In addition to its premier action in the regulation of corticotropin secretion, synthetic CRF has equally potent effects on neurons in vitro and in vivo, increasing the frequency of action potentials in cells that fire in bursts, such as the hippocampal pyramidal cells. Even modest doses of the peptide can induce seizure-like activity within the limbic system. At still smaller doses given intracerebroventricularly, CRF is a potent activator of spontaneous locomotion and can produce an anxiogenic response (opposite to the effects of antianxiety drugs like benzodiazepines and ethanol) in different behavior tests. Clearly, the complete physiological effects of CRF at the pituitary and at other neuronal sites represent a much more comprehensive basis for the mobilization of bodily systems during stress.

Initial efforts to define CRF receptors focused on ligand binding autoradiography, which more or less replicated the tissue patterns of CRF immunoreactivity, with highest binding in the cortex, amygdala, hippocampus, and pons. CRF is also expressed in an as yet unexploited circuit crying for functional analysis, a projection from the inferior olive to cerebellar cortex containing CRF and glutamate. In the final days of pregnancy, maternal blood is very high in apparently high molecular weight forms of CRF believed to arise in placenta and yet without signs of excessive corticotropin secretion generally. Efforts to solve this puzzle led to the definition of a CRF-binding protein, also well expressed in brain, although with a distribution different from that of CRF fibers.

The Vale group subsequently characterized two CRF receptors and developed potent CRF antagonist peptides, knockout mice for CRF, their receptors, and mice overexpressing CRF. In addition, they characterized a highly related peptide that is probably the endogenous ligand for CRF2, termed *urocortin* (for its effects, which are similar to urotensin in appetite suppression; see Fig. 11–10).

A Reader's Guide to Peptide Poaching

Until more is learned about the functions, sites, and mechanisms of action of any one peptide, we will have trouble formulating even tentative roles for these substances in the neurotransmitter–neurohormonal regulation of central drug actions.

Only the stern hand of our stingy editor prevents us having the space to discuss in any depth other important neuropeptides. The next three would be galanin (a free ice cream cone to the first reader who deduces how it got its name; clue: it has alanine at its C terminus), a neuroendocrine peptide that has expression and pharmacological interest for those focused on depression, pain regulation, and possibly epilepsy; cocaine and amphetamine–regulated transcript (CART), a widely expressed neuropeptide of still uncertain amino acid sequence length, whose 55–102 fragment is potent at suppressing feeding and may be among those brain systems responsive to leptin; and the cytokines and chemokines, an increasingly important family of peptides for neuropharmacologists. Cytokines are a large and diverse family of polypeptide regulators, produced widely throughout the body by cells of many embryological lines. In general, cytokines interact as a network with synergistic, additive, or opposing actions. Within the immune system, macrophages and activated T lymphocytes are the major producers of the macrophage-derived cytokines interleukin (IL)-1A, IL-1B, and IL-6 and tumor necrosis factor α. These are the cytokines that have received the most attention for potential regulatory roles in nervous system inflammation (as with the early dementia of CNS human immunodeficiency virus infection) and in recovery from traumatic injury. Had they been discovered in brain by neuropharmacologists, they might have "growth factor" names instead of cytokine numbers. Chemokines are cytokines whose expression and receptors may naturally control the interstitial migration of lymphocytes and macrophages into all organs, including the brain.

As the story of neuroactive peptides unfolds (the sheer number of peptides available for pursuit makes the going slower owing to the division of the available workforce), interested students should be alert for answers to the following questions: Are there general patterns of circuitry? Are specific peptidases amenable to selective pharmacological intervention, or do neuronal and glial peptidases read only the dipeptides, whose bonds they are about to cleave? Do the peptides act presynaptically, postsynaptically, or at both sites? How are peptidases specifically activated either to release neuroactive peptides from precursors or to terminate the activity of the peptide? Can the peptides modulate the release or response to transmitters of other neurons or those with which they coexist in a given neuron? Can the promise of peptide chemistry deliver useful antagonists to prove the identity of receptors with nerve pathway stimulation? Does the presence of receptors to other centrally active drugs indicate that still more endogenous peptides should be sought? By now it should have occurred to the reader that nothing in the neurosciences is simple.

Selected References

General sources

Asensio, V. C. and I. L. Campbell (1999). Chemokines in the CNS: plurifunctional mediators in diverse states. *Trends Neurosci.* 22. 504–512.

Bloom, F. E. (1990). Peptides: regulators of cell function in brain and beyond. In *FIDIA Research Foundation Neuroscience Award Lectures.* vol. 4. Raven Press, New York, pp. 229–268.

Branchek, T. A., K. E. Smith, C. Gerald, and M. W. Walker (2000). Galanin receptor subtypes. *Trends. Pharmacol. Sci.* 21, 109–116.

Campbell, I. L. (1998). Transgenic mice and cytokine actions in the brain: bridging the gap between structural and functional neuropathology. *Brain Res. Brain Res. Rev.* 26, 327–336.

Civelli, O., R. K. Reinscheid, and H. P. Nothacker (1999). Orphan receptors, novel neuropeptides and reverse pharmaceutical research. *Brain Res.* 848, 63–65.

Elmquist, J. K., C. F. Elias, and C. B. Saper (1999). From lesions to leptin: Hypothalamic control of food intake and body weight. *Neuron* 22, 221–232.

Hökfelt, T., C. Broberger, Z. Q. Xu, V. Sergeyev, R. Ubink, and M. Diez (2000). Neuropeptides—an overview. *Neuropharmacology* 39, 1337–1356.

Hökfelt, T., M.-N. Castel, P. Morino, X. Zhang, and Å. Dagerlind (1994). General overview of neuropeptides. In *Psychopharmacology: The Fourth Generation of Progress* (F. E. Bloom and D. J. Kupfer, eds.). Raven Press, New York, pp. 483–492.

Kask, K., M. Berthold, and T. Bartfai (1997). Galanin receptors: involvement in feeding, pain, depression and Alzheimer's disease. *Life Sci.* 60, 1523–1533.

Poyner, D., H. Cox, M. Bushfield, M. J. Treherne, and M. K. Demetrokopoulos (2000). Neuropeptides in drug research. *Prog. Drug Res.* 54, 121–149.

Oxytocin and vasopressin

Alescio-Lautier, B. and B. Soumireu-Mourat (1998). Role of vasopressin in learning and memory in the hippocampus. *Prog. Brain Res.* 119, 101–119.

Bohus, B. and D. de Wied (1998). The vasopressin deficient Brattleboro rats: a natural knockout model used in the search for CNS effects of vasopressin. *Prog. Brain Res.* 119, 555–573.

de Wied, D. (1983). The importance of vasopressin memory. *Trends Neurosci.* 7, 62–64.

Gash, D. M. and G. J. Thomas (1983). What is the importance of vasopressin in memory processes? *Trends Neurosci.* 6, 197–198.

Jirikowski, G. F., P. P. Sanna, D. Macjiewski-Lenoir, and F. E. Bloom (1992). Reversal of diabetes insipidus in Brattelboro rats: intrahypothalamic injection of vasopressin mRNA. *Science* 255, 996–998.

Koob, G. F., R. Dantzer, F. Rodriguez, F. E. Bloom, and M. Le Moal (1985). Osmotic stress mimics the effects of vasopressin on learned behaviour. *Nature* 315, 750–752.

Neurotensin

Carraway, R. and S. E. Leeman (1975). The amino acid sequence of a hypothalamic peptide, neurotensin. *J. Biol. Chem.* 250, 1907–1912.
Kinkead, B., E. B. Binder, and C. B. Nemeroff (1999). Does neurotensin mediate the effects of antipsychotic drugs? *Biol. Psychiatry* 46, 340–351.
Kobayashi, R. M., M. Brown, and W. Vale (1977). Regional distribution of neurotensin and somatostatin in rat brain. *Brain Res.* 126, 584–590.
Vincent, J. P., J. Mazella, and P. Kitabgi (1999) Neurotensin and neurotensin receptors. *Trends Pharmacol. Sci.* 20, 302–309.

Tachykinin peptides

Maggi, C. A. (2000). The troubled story of tachykinins and neurokinins. *Trends. Pharmacol. Sci.* 21, 173–175.
Maggio, J. E. (1988). Tachykinins. *Annu. Rev. Neurosci.* 11, 13–28.
Nawa, H., H. Kotani, and S. Nakanishi (1984). Tissue specific generation of two preprotachykinin mRNAs from one gene by alternative RNA splicing. *Nature* 312, 729–734.
Nicoll, R. A., C. Schenker, and S. E. Leeman (1980). Substance P as a transmitter candidate. *Annu. Rev. Neurosci.* 3, 227–268.
Rupniak, N. M. J. and M. S. Kramer (1999). Discovery of the antidepressant and anti-emetic efficacy of substance P receptor (NK1) antagonists. *Trends. Pharmacol. Sci.* 20, 485–490.
Vanden Broeck, J., H. Torfs, J. Poels, W. Van Poyer, E. Swinnen, K. Ferket, and A. De Loof (1999). Tachykinin-like peptides and their receptors. A review. *Ann. N.Y. Acad. Sci.* 897, 374–387.

VIP-related peptides

Dicicco-Bloom, E., N. Lu, J. E. Pintar, and J. Zhang (1998). The PACAP ligand/receptor system regulates cerebral cortical neurogenesis. *Ann. N.Y. Acad. Sci.* 865, 274–289.
Guillemin, R., P. Brazeau, P. Bohlen, F. Eseh, N. Ling, and W. B. Wehrenberg (1982). Growth hormone releasing factor from a human pancreatic tumor that caused acromegaly. *Science* 218, 585–587.
Hosoya, M., H. Onda, K. Ogi, Y. Masuda, Y. Miyamoto, T. Ohtaki, H. Okazaki, A. Arimura, and M. Fujino (1993). Molecular cloning and functional expression of rat cDNAs encoding the receptor for pituitary adenylate cyclase activating polypeptide (PACAP). *Biochem. Biophys. Res. Commun.* 194, 133–143.
Kopin, A. S., M. B. Wheeler, and A. B. Leiter (1990). Secretin: structure of the precursor and tissue distribution of the mRNA. *Proc. Natl. Acad. Sci. U.S.A.* 87, 2299–2303.
Lundberg, J. M., B. Hedlund, and T. Bartfai (1982). Vasoactive intestinal polypeptide enhances muscarinic ligand binding in cat submandibular gland. *Nature* 295, 147–149.

Magistretti, P. J., J. R. Cardinaux, J. L. Martin (1998). VIP and PACAP in the CNS: regulators of glial energy metabolism and modulators of glutamatergic signaling. *Ann. N.Y. Acad. Sci.* 816, 213–225.

Matsumoto, Y., M. Tsuda, and M. Fujino (1993). Regional distribution of pituitary adenylate cyclase activating polypeptide (PACAP) in the rat central nervous system as determined by sandwich-enzyme immunoassay. *Brain Res.* 602, 57–63.

Peyron, C., D. K. Tighe, A. N. van den Pol, L. de Lecea, H. C. Heller, J. G. Sutcliffe, and T. S. Kilduff (1998). Neurons containing hypocretin (orexin) project to multiple neuronal systems. *J. Neurosci.* 18, 9996–10015.

Rivier, J., J. Spiess, M. Thorner, and W. Vale (1982). Characterization of a growth hormone releasing factor from a human pancreatic islet tumor. *Nature* 300, 276–278.

Tatemoto, K. and V. Mutt (1981). Isolation and characterization of the intestinal peptide porcine PHI (PHI-27), a new member of the glucagon-secretin family. *Proc. Natl. Acad. Sci. U.S.A.* 78, 6603–6607.

Yung, W. H., P. S. Leung, S. S. Ng, J. Zhang, S. C. Chan, and B. K. Chow (2001). Secretin facilitates GABA transmission in the cerebellum. *J. Neurosci.* 21, 7063–7068.

Pancreatic polypeptide–related peptides

Allen, Y. S., T. E. Adrian, J. M. Allen, K. Tatemoto, T. J. Crow, S. R. Bloom, and J. M. Polak (1983). Neuropeptide Y distribution in the rat brain. *Science* 221, 877–879.

El Majdoubi, M., A. Sahu, S. Ramaswamy, and T. M. Plant (2000). Neuropeptide Y: a hypothalamic brake restraining onset of puberty in primates. *Proc. Natl. Acad. Sci. U.S.A.* 97, 6179–6184.

Gehlert, D. R. (1999). Role of hypothalamic neuropeptide Y in feeding and obesity. *Neuropeptides* 33, 329–338.

Hökfelt, T., C. Broberger, X. Zhang, M. Diez, J. Kopp, Z. Xu, M. Landry, L. Bao, M. Schalling, J. Koistinaho, S. J. DeArmond, S. Prusiner, and J. Gong (1998). Neuropeptide Y: some viewpoints on a multifaceted peptide in the normal and diseased nervous system. *Brain Res. Brain Res. Rev.* 26, 154–166.

Mutt, V. (1983). New approaches to the identification and isolation of hormonal polypeptides. *Trends Neurosci.* 6, 357–360.

Wahlestadt, C. and M. Heilig (1994). Neuropeptide Y and related peptides. In *Neuropsychopharmacology: The Fourth Generation of Progress* (F. E. Bloom and D. J. Kupfer, eds.). Raven Press, New York, pp. 543–551.

Wahlestedt, C., E. M. Pich, G. F. Koob, F. Yee, and M. Heilig (1993). Modulation of anxiety and neuropeptide Y-Y1 receptors by antisense oligodeoxynucleotides. *Science* 259, 528–531.

Opioid peptides

Akil, H., C. Owens, H. Gutstein, L. Taylor, E. Curran, and S. Watson (1998). Endogenous opioids: overview and current issues. *Drug Alcohol Depend.* 51, 127–140.

Darland, T., M. M. Heinricher, and D. K. Grandy (1998). Orphanin FQ/nociceptin: a role in pain and analgesia, but so much more. *Trends. Neurosci.* 21, 215–221.

Goldstein, A., S. Tachibana, L. I. Lowney, and L. Hold (1979). Dynorphin—(1–13), an extraordinarily potent opioid peptide. *Proc. Natl. Acad. Sci. U.S.A.* 76, 6666–6670.

Hughes, J., T. W. Smith, H. W. Kosterlitz, L. A. Fothergill, G. A. Morgan, and H. R. Morris (1975). Identification of two related pentapeptides from the brain with potent opiate agonist activity. *Nature* 258, 577.

Nakanishi, S., A. Inoue, T. Kita, A.C.Y. Chang, S. Cohen, and S. Numa (1979). Nucleotide sequence of cloned cDNA for bovine corticotropin—lipotropin precursor. *Nature* 257, 238–240.

Nicoll, R. A., G. R. Siggins, N. Ling, F. E. Bloom, and R. Guillemin (1977). Neuronal actions of endorphins and enkephalins among brain regions: a comparative microiontophoretic study. *Proc. Natl. Acad. Sci. U.S.A.* 74, 2584.

Reinscheid, R. K., A. Ardati, F. J. Monsma, Jr., and O. Civelli (1996). Structure–activity relationship studies on the novel neuropeptide orphanin FQ. *J. Biol. Chem.* 271, 14163–14168.

Trujillo, K. A. and H. Akil (1994). Inhibition of opiate tolerance by noncompetitive *N*-methyl-D-aspartate receptor antagonists. *Brain Res.* 633, 178–188.

Wagner, J. J. and C. Chavkin (1994). Neuropharmacology of endogenous opioid peptides. In *Psychopharmacology: The Fourth Generation of Progress* (F. E. Bloom and D. J. Kupfer, eds.). Raven Press, New York, pp. 519–529.

Somatostatin

DeLecea, L., J. R. Criado, O. Propsero-Garcia, K. M. Gautvik, P. Schweitzer, P. E. Danielson, C. L. M. Dunlop, G. R. Siggins, S. J. Henriksen, and J. G. Sutcliffe (1996). A cortical neuropeptide with neuronal depressant and sleep-modulating properties. *Nature* 381, 242–245.

de Lima, A. D. and J. H. Morrison (1989). An ultrastructural analysis of somatostatin-immunoreactive neurons and synapses in the temporal and occipital cortex of the macaque monkey. *J. Comp. Neurol.* 283, 212–227.

Effendic, S. and R. Luft (1980). Somatostatin: a classical hormone, a locally active polypeptide, and a neurotransmitter. *Ann. Clin. Med.* 12, 8794.

Moore, S. D., S. G. Madamba, M. Joels, and G. R. Siggins (1988). Somatostatin augments the M-current in hippocampal neurons. *Science* 239, 278–280.

Patel, Y. C. (1999). Somatostatin and its receptor family. *Front. Neuroendocrinol.* 20, 157–198.

Selmer, I., M. Schindler, J. P. Allen, P. P. Humphrey, and P. C. Emson (2000). Advances in understanding neuronal somatostatin receptors. *Regul. Pept.* 90, 1–18.

Vezzani, A. and D. Hoyer (1999). Brain somatostatin: a candidate inhibitory role in seizures and epileptogenesis. *Eur. J. Neurosci.* 11, 3767–3776.

Cholecystokinin

Baile, C. A., C. L. McLaughlin, and F.M.A. Della (1986). Role of cholecystokinin and opioid peptides in control of food intake. *Physiol. Rev.* 66, 172–234.

Chambers, M. S. and S. R. Fletcher (2000). CCK-B antagonists in the control of anxiety and gastric acid secretion. *Prog. Med. Chem.* 37, 45–81.

Fink, H., A. Rex, M. Voits, and J. P. Voigt (1998). Major biological actions of CCK—a critical evaluation of research findings. *Exp. Brain Res.* 123, 77–83.

Lotti, Y. J., R. G. Pendleton, R. J. Gould, H. M. Hanson, R. S. Chang, and B. V. Clineschmidt (1987). In vivo pharmacology of L-364,718, a new potent nonpeptide peripheral cholecystokinin antagonist. *J. Pharmacol. Exp. Ther.* 241, 103–109.

Noble, F. and B. P. Roques (1999). CCK-B receptor: chemistry, molecular biology, biochemistry and pharmacology. *Prog. Neurobiol.* 58, 349–379.

White, F. J. and R. Wang (1984). Interactions of cholecystokinin octapeptide and dopamine on nucleus accumbens neurons. *Brain Res.* 300, 161–166.

Wiesenfeld-Hallin, Z., G. de Araujo Lucas, P. Alster, X. J. Xu, and T. Hokfelt (1999). Cholecystokinin/opioid interactions. *Brain Res.* 848, 78–89.

Corticotropin-releasing hormone

Behan, D. P., E. B. De Souza, P. J. Lowry, E. Potter, P. Sawchenko, and W. W. Vale (1995). Corticotropin releasing factor (CRF) binding protein: a novel regulator of CRF and related peptides. *Front. Neuroendocrinol.* 16, 362–382.

Hernandez, J. F., W. Kornreich, C. Rivier, A. Miranda, G. Yamamoto, J. Andrews, Y. Tache, W. Vale, and J. Rivier (1993). Synthesis and relative potencies of new constrained CRF antagonists. *J. Med. Chem.* 36, 2860–2867.

Koob, G. F. and S. C. Heinrichs (1999). A role for corticotropin releasing factor and urocortin in behavioral responses to stressors. *Brain. Res.* 848, 151–162.

Perrin, M. H. and W. W. Vale (1999). Corticotropin releasing factor receptors and their ligand family. *Ann. N.Y. Acad. Sci.* 885, 312–328.

Stenzel-Poore, M. P., S. C. Heinrichs, S. Rivest, G. F. Koob, and W. W. Vale (1994). Overproduction of corticotropin-releasing factor in transgenic mice: a genetic model of anxiogenic behavior. *J. Neurosci.* 14, 2579–2584.

Vale, W., J. Spiess, J. Rivier, and C. Rivier (1981). Characterization of a 41-residue ovine hypothalamic peptide that stimulates secretion of corticotropin and beta-endorphin. *Science* 213, 1394–1397.

12

Cellular Mechanisms in Learning and Memory

Virtually every neurotransmitter system presented in this book is associated with some specific behavioral effects elicited by its agonists, antagonists, and often genetic manipulation of its receptors. A far more difficult question is just what role any given transmitter system plays in initiating, maintaining, or terminating the behavior. In this chapter, we consider an issue that represents one of the most complex behavioral achievements of nervous systems, the ability to learn and remember.

Substantial progress has been achieved over the last two decades in understanding the molecular and cellular basis of neuronal interactions, in particular the role of specific neurotransmitters engaged in the regulation of sleep and waking, appetite, and dependence on drugs of abuse. The same sorts of knowledge are being achieved with regard to the circuitry and molecular mechanisms underlying learning and remembering. Although we cannot yet provide inquiring minds with certified maps to today's Holy Grail in drug development, the cognitive enhancer designed to delay or reverse the ravages of Alzheimer's disease, there are now definite leads.

To develop a convincing basis for the neuropharmacology of learning and memory requires data establishing that a change in a specific neurotransmitter action at a specific synaptic location is necessary and sufficient for a behavioral change in a living organism. Until relatively recently, such data were either cellular or behavioral but it was not possible to study the same synaptic connections in awake, behaving experimental animals. However, that is no longer the case. Moreover, a wide range of observations combining behavioral, molecular, and cellular information in invertebrates and vertebrates has revealed that each species, despite obvious differences in neuronal circuitry, shares some cellular and molecular features in its memory storage functions.

Formulating Experiments in Memory and Learning

Biological analysis of memory and learning typically requires two parallel lines of work: *(1)* a behavioral component in which the subject is trained, given a training recess of variable intervals, and then tested for retention of the trained response and *(2)* a functional, structural, or biochemical component intended to define the essential changes, sites, or molecular sequelae underlying these events. Neuropharmacological probes offer excellent tools for the manipulation of molecular events and their behavioral outcomes.

Scientists working in this field commonly categorize learning into "nonassociative" and "associative" forms. *Nonassociative learning* describes the functional changes that ensue when an organism interacts with a single form of stimulus. For example, the decreased response that occurs when the same sensory stimulus is repeated sequentially, termed *habituation*, is a nonassociative form of learning, as is the reversal of the habituated response, termed *sensitization*, by a different intensity or quality of sensory stimulus. *Associative learning*, as the name implies, is the form of learning in which the subject associates a previously neutral stimulus with a response normally generated by another cue, the sort made famous by Pavlov's classic experiments, which triggered dog salivation with a bell instead of raw meat.

An exciting prospect for such work has been opened by the development of experimental test systems in invertebrates, and more recently in vertebrates, to show that the behavioral performance of the organism depends on synaptic events in specific locations and then to define the changes that occur at these locations as the organism learns and performs. Such possibilities for molecular and cellular explanations exist, it should be recognized, only because of the extensive data that have been obtained over this same interval.

The great neurocytologist Ramón y Cajal held strongly to the view that learning was a synaptic event. He regarded the synapse as the primary site of interneuronal communication. The subsequent advances in biochemistry, anatomy, and physiology that are emphasized in this text led to the notion that drugs should regulate the critical sites of synaptic transmission. Given this premise, the manipulations of memory processes by specific drugs used as tools for determining the roles played by specific transmitters and defined sites should not only illuminate the underlying process but also lead to eventual therapeutic interventions.

Cellular Models of Learning in Invertebrate Nervous Systems

Studies carried out in the visceral ganglia of invertebrates, particularly the marine mollusc *Aplysia*, have provided information on the molecular mechanisms of the cellular and synaptic changes that accompany simple forms of behavior modification, such as habituation and sensitization. *Sensitization* is the ability of nonspecific but strongly arousing sensory stimulation to enhance the effectiveness of synaptic transmission in other neuronal pathways. When the gill-withdrawal reflex is studied for this effect, strong electrical stimulation of the connections between the visceral ganglion and the head ganglion facilitates the withdrawal. The strong stimulation also diminishes habituation of the reflex if tested with identical repetitive sensory stimuli. As judged by analysis of postsynaptic potentials, the facilitating stimulation increases the amount of transmitter released by the afferent limb of the withdrawal reflex (i.e., the sensory nerves). Once initiated in the *Aplysia* ganglion, the effects of sensitization can last from several minutes to hours.

Further pursuit of the presynaptic mechanism of the enhanced release became possible when it was found that the sensitizing stimulation also increased the ganglionic content of cAMP and that exposure of the ganglion both to serotonin (known to activate cAMP production in the ganglion) and to cAMP (applied to the whole ganglion or injected intracellularly into the sensory nerve cell) replicated the effects of the sensitizing stimulation. In this case, the molecular mediation sequence of the second-messenger hypothesis (see Chapter 6) suggests that a serotonin-secreting interneuron is activated by the sensitizing stimulation, leading to increased presynaptic levels of cAMP and consequent enhancement of the sensory transmitter release.

Subsequent studies have shown that the short-term memory for sensitization of the *Aplysia* gill- and siphon-withdrawal reflex is distributed across at least four sites of circuit modification. Each of these invokes a slightly different circuitry and a different type of synaptic modification (presynaptic facilitation, presynaptic inhibition, posttetanic potentiation, and increased tonic firing rate), all of which result in facilitation. Nevertheless, all four of these mechanisms seem to be mediated by a common modulatory transmitter (serotonin), to utilize the same cAMP-mediated second-messenger system, and to require a phosphorylation event (affecting either intravesicular proteins or ion channels) to enhance transmitter release. These short-term changes, lasting from minutes to hours, do not require protein synthesis.

Kandel and associates have extended their definition of the molecular mechanisms involved in the long-term changes in the plasticity underlying the gill-withdrawal reflex. Having established that such long-term functional changes are indeed dependent on gene transcription and protein synthesis (while the short-term facilitative changes are not), several of the early and late genes involved specifically in the synaptic events have been defined. As with the short-term changes, cAMP generation appears to be a key step in long-term facilitation, leading to both protein and structural synaptic changes. In the most recent phases of this work, attention has been focused on the cAMP response element–binding protein (CREB), in which it can be shown by gene transfer experiments on cultured neurons that cumulative responses to 5-hydroxytryptamine (5-HT) lead to cAMP synthesis, activation of the cAMP-dependent protein kinase (PKA), translocation of the activated enzyme to the nucleus, and phosphorylation of serine-119 on CREB. By reducing the preparation to pairs of interconnected neurons in culture, it has been possible to determine that as few as five pulses of 5-HT are sufficient to engage the long-term changes.

The phosphorylated CREB presumably activates the transcription of immediate early genes, one of which has been specifically incriminated as a key player. Both 5-HT and cAMP rapidly induce the *Aplysia* CCAAT enhancer-binding protein (ApC/EBP), even in the presence of protein synthesis inhibitors. Furthermore, immunologically blocking the function of ApC/EBP blocks long-term facilitation selectively without affecting short-term facilitation. The switch from short- to long-term facilitation requires not only activation of CREB1 but also reversal of the tonic repression of of the *Aplysia* form of a second CREB, known as ApCREB2, as well as induction of ApC/EBP. A transcription activator, *Aplysia* A Factor, which is stimulated by PKA has been identified. This protein can dimerize with both ApC/EBP and ApCREB2 and appears to be necessary for the long-term facilitation that can be induced by as little as five pulses of serotonin. Present evidence suggests that this effect is mediated by either activation of CREB1 or derepression of ApCREB2. Overexpression of ApAF enhances the long-term facilitation further. Thus, ApAF is a candidate memory enhancer gene downstream from both CREB1 and ApCREB2.

These two themes, namely, that the short-term and long-term changes rely on different consequences of the 5-HT activation of cAMP synthesis and that the functional consequences of specific immediate early gene systems are important, also epitomize recent work in mammalian models of learning and plasticity.

Some General Principles

Kandel, Tauc, Gershenfeld, Strumwasser, Alkon, Levitan, Kaczmarek, and other explorers of invertebrate psychobiology launched the scientific search for the cellular and molecular bases of learning. They thought that these smaller nervous systems could be fathomed, and they set themselves the task of working out the rules of connectivity by which these systems adapt to environmental conditions. Their meticulous descriptive work has given us a logical, consistent, and compelling account of the cellular changes accompanying behavioral modification and the molecular basis of these changes.

The monoamine hypothesis of memory and learning in mammals advanced by Kety in the late 1960s could now be viewed as anticipating, in a more primitive mechanistic way, the serotonin and cAMP-mediated presynaptic facilitation model derived from the studies of invertebrate sensitization. Kety's model called for a brain system that could mediate the ability of arousal to consolidate experiences into adaptive behavioral mechanisms. To reconcile the two situations, we must first substitute the mammalian transmitter norepinephrine (NE) for the role played by 5-HT in the invertebrate. The central NE neurons and their synapses have now been well mapped, and electrophysiological recordings in behaviorally responding rats and monkeys have established that the NE cells fire when novel sensory events occur in the external environment. These data are logical and internally consistent with a role of NE neurons in mediating arousal. However, there are clearly many more neurotransmitter systems at work in mammalian mechanisms of learning and memory, especially when one begins to focus on discrete synaptic changes within precisely constrained regions of the brain in vitro or in vivo. We will examine some of these specific synaptic sites below. Learning and memory events in mammalian brain function—as in the invertebrate—are probably too critical to rely on only one transmitter system. In fact, if we substitute dopamine for NE in the above discussion, we have two-thirds of the work that won the Nobel Prize for Physiology or Medicine in 2000.

The Rabbit Nictitating Membrane Reflex and Associative Learning

The rabbit's nictitating membrane has provided another molecular and cellular model for memory and associative learning. This mammalian neuronal system permits the direct observation of an animal's performance during tests of learning and memory and the direct determination of exactly where in the circuitry the learning events have occurred.

The basic studies, largely done independently by John Harvey's and Richard Thompson's groups, showed that the rabbit blink reflex could be associatively conditioned. Rabbits will normally blink (i.e., cover their cornea with the nictitating membrane) every time a puff of air is directed at the cornea. Repeated presentation of a loud tone just before the air puff conditioned the rabbits to blink in response to the tone. Eventually, the rabbits blinked every time they heard the tone, with no air puff being required.

At this point, chemical and electrolytic lesions as well as micro-stimulations were used to define the pathways that relay the unconditional (air puff–induced) blink and the conditional blink triggered by the tone (see Fig. 12–1). The air puff stimulus travels through the sensory fibers of the trigeminal nerve to activate motor neurons of the abducens nucleus, causing the eye to retract into the orbit and then the nictitating membrane to extend over the eye. The acoustic stimulus travels through the cochlear nucleus to the pontine nucleus and seems quite separate at its early stages from the corneal sensory pathway. However, the researchers also noted that even before the rabbits learned to associate the sound with the blinking, the acoustic stimulus, which was rather loud, could often increase the response amplitude to the normal air puff stimulus. This suggested that the acoustic stimulus could also sensitize the unassociated air puff stimulus.

Based on comparisons of rabbits in which discrete lesions were made along the known anatomical paths within the two systems of sensory processing,

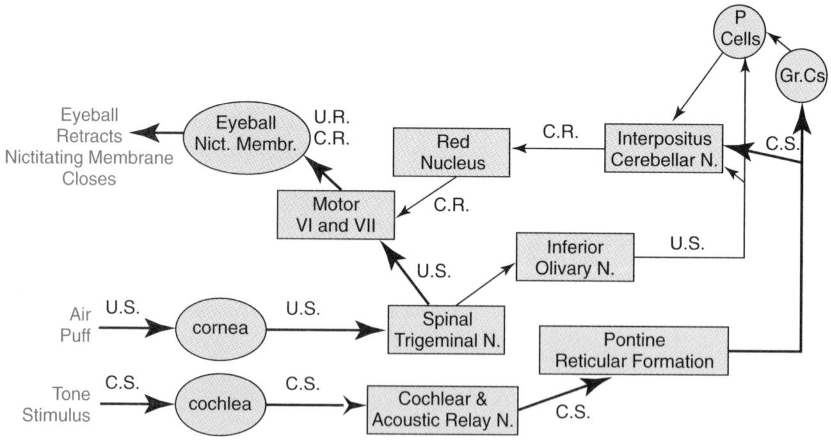

FIGURE 12–1. Schematic diagram of the sensory and motor circuitry for the rabbit nictitating membrane model of cellular mechanisms in associative learning. Pathways marked as U.S. participate in the elicitation of the unconditional reflex response (U.R. components) in which a puff of air to the cornea will always provoke a blink by closing the nictitating membrane. The auditory pathways that carry the tone information that can become associated with the air puff–induced blink are indicated as the C.S. (conditional stimulus) or C.R. (conditional response) components. When the animal has been well trained, the tone C.S. will evoke the C.R. even without a U.S.

the critical site for producing the association seems to be located in the interpositus nucleus, one of the deep cerebellar nuclei. The efferent limb of the motor reflex then courses through the red nucleus directly to the abducens nerve, where it activates the eyeball retraction reflex (see Fig. 12–1).

Since most of the transmitters operating in these hierarchical controlling chains have not yet been defined, it would seem that the model was not quite ready for transmitter-specific pharmacology to be a useful analytical tool. The only sites that have been probed so far are the synapses within the interpositus and red nuclei; here, the γ-aminobutyric acid (GABA) antagonists bicuculline and picrotoxin block the conditional reflex. Chemical lesions of the inferior olivary nucleus (the source of the climbing fiber projection to cerebellar Purkinje neurons) completely disrupt the acquisition and retention of the conditional reflex. Since glutamate is presumed to be the transmitter of this pathway, it is implicated as the transmitter for this final step. Inactivation of the deep cerebellar nuclei by local anesthetics abolished eyeblink conditioned responses whether the conditional stimulus was a tone or stimulation in the lateral reticular nucleus. Inactivation of the lateral pontine nucleus prevented only the acquisition and retention of tone-evoked eyeblink conditioned responses. Using multiple-unit recordings in the lateral pontine nucleus, Thompson and colleagues demonstrated that when lateral reticular nucleus stimulation was used as the conditional stimulus, inactivation of the interpositus nucleus abolished learning-related neuronal activity, whereas inactivation of the pontine nucleus had little effect on similar activity in the interpositus nucleus. They concluded that the learning-induced neuronal activity in the lateral pontine nucleus was most likely driven by the cerebellar interpositus nucleus.

Harvey and colleagues have used the associative eye blink model to show that low doses of amphetamine as well as lysergic acid diethylamide (LSD) enhance this form of learning, while a wide range of drugs from pimozide to atropine and scopolamine, as well as agonists for κ and σ opioid receptors, decrease acquisition at doses that do not affect the basic reflex. Because these latter drugs were given systemically, however, it was not clear where (within the cranial nerve–cerebellar circuitry or elsewhere) they acted to produce these effects.

In subsequent experiments, forebrain regions traditionally viewed as participating in memory and learning functions have also been implicated in the nictitating membrane associational reflex. For these experiments, rabbits received "classic conditioning" of the nictitating membrane response; the tone stimulus and the air puff were separated by a pause (or "trace") of 500 milliseconds. The short interruption made the simple learning task a bit more complicated. The extra time delay allowed the investigators to distinguish nonspecific arousal, possibly sensitizing the reflex, from more specific interactions.

Thompson and colleagues found that lesions of the hippocampus or cingulate/retrosplenial cortex disrupted acquisition of the interrupted conditioned response but that neocortical lesions did not. Neither lesion affected acquisition of the noninterrupted pairing of tone and air puff. When animals with hippocampal or cingulate/retrosplenial cortex lesions were switched to a standard delay paradigm in which the conditional stimulus and the unconditional stimulus were given at the same time, rabbits acquired the association in about the same number of trials as naive animals.

In the interrupted protocol, however, macroelectrode recordings showed substantial increases of multiunit activity in the CA1 region of the hippocampus, which began during the tone and persisted through the trace interval, even before the rabbit showed consistent association. As the rabbit's performance in responding to the tone stimulus improved, the hippocampal activity shifted to later in the trace interval. Although the transmitter has not been identified for this activation, it is a safe presumption that GLU is involved; in fact, changes in AMPA subtype receptor binding, but not in NMDA receptor kinetics, have been reported to accompany the learning.

These data thus directly relate behavioral demonstration of associational learning with discrete neuronal circuitry. Such an experimental system should eventually lend itself to pharmacological and electrophysiological analyses of the changes within the critical neurons that alter synaptic function. Student volunteers are welcome.

Long-Term Potentiation: A Vertebrate Model of Synaptic Plasticity

Next, we turn to an increasingly popular form of mammalian forebrain synaptic plasticity, in which electrophysiological changes at the synaptic level have been probed extensively by drugs to identify the transmitter systems responsible for discrete transductive mechanisms. The term *long-term potentiation* (LTP) refers to long-lasting enhancement of synaptic transmission (10 minutes to days, depending critically on the conditions used to evoke the response and where it is tested). The effect is measured as increased amplitudes of excitatory postsynaptic potentials (or the currents generated by these potentials) in specific circuits after high-frequency, high-intensity activation of the same circuits or other discrete paths. Because the phenomenon was originally described with macroelectrode recordings in the hippocampus of intact animals, it has been a candidate to link learning and cellular changes in vivo. However, the most intense studies have more often involved analysis of the pertinent connections in vitro in the now classic slice preparations of the hippocampus or neocortex in normal animals or transgenic animals with specific genes mutated.

This special form of enhanced synaptic transmission was first demonstrated by brief, high-frequency stimulation of the entorhinal cortex, through the perforant path (see Fig. 12–2), to enhance activation of the granule cells of the hippocampal dentate gyrus by subsequent single stimulation of the per-

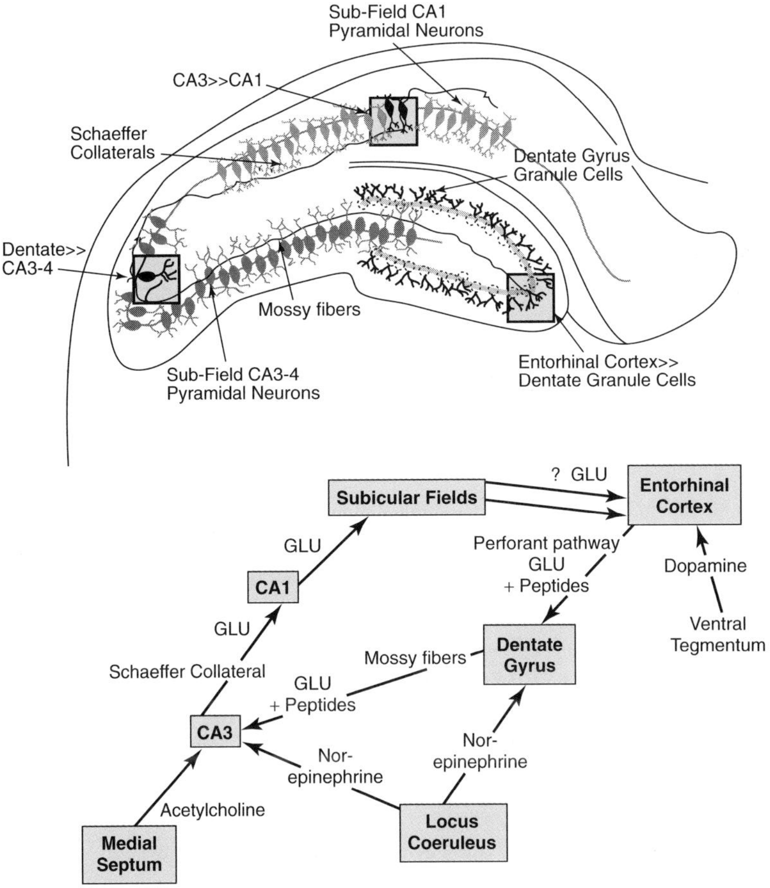

FIGURE 12–2. The synaptic circuits of the rodent hippocampal slice preparation in which long-term potentiation (LTP) can be elicited are indicated in the upper panel, while the neurotransmitters associated with these circuits or with the target cells capable of expressing the LTP response are shown in the lower panel. Each of the three major hippocampal circuits (entorhinal cortex to dentate, dentate to CA3, and CA3 to CA1) shows a generally similar LTP, but the pharmacology of the modification varies with the specific location and pathways. Glutamate (GLU)–mediated synaptic events figure prominently, as indicated in the text.

forant path. In awake cats, guinea pigs, and rats, this enhanced transmission could be seen for periods of days to weeks. The transmitter for this pathway is now thought to be glutamate (Glu) or possibly aspartate (Asp), although a role of at least one family of coexisting peptides, such as the proenkephalin-derived opioid peptides, has been reported (see Chapter 11).

LTP has also been observed at two other sites within the hippocampal formation: *(1)* at the synapses between the dentate granule cells and the targets of their mossy fiber synapses, the CA3 pyramidal neurons, and *(2)* between the Schaffer collaterals of CA3 pyramidal neurons, as well as associational fibers from the contralateral hippocampus, and the CA1 pyramidal neurons (see Fig. 11–3). The transmitter for these two pathways is thought to be Asp, Glu, or possibly homocysteate; coexisting peptides (prodynorphin-derived as well as cholecystokinin) remain candidates for some part of the action here, at least for the mossy fiber connection to CA3 pyramidal neurons. Although the basic phenomenon of LTP after a high-frequency, high-intensity period of afferent pathway activation is similar at all three synaptic sites, the relative importance of specific transmitters and their responsible transductive mechanisms differ on a site-by-site basis.

The essential pharmacological advance came with the recognition that the NMDA subtype of excitatory amino acid (EAA) receptor (see Chapter 7) was responsible for the induction phase of the long-lasting synaptic enhancement. Antagonists of this form of EAA receptor (e.g., aminophosphovaleric acid) prevent LTP induction without interfering with previously potentiated transmission or normal low-frequency transmission in the pathway under study. The critical link established by several groups is that the required high-frequency, high-intensity stimulation acts via the NMDA receptor to couple the depolarization with increased Ca^{2+} entry. This combined response is then thought to depolarize the small dendritic domains that are the site of convergent afferents in the pathway under study and to increase their subsequent transmission. In the perforant path-to-dentate circuit, drugs that block μ-opioid receptors or β-adrenergic receptors also block LTP. The latter antagonists are also very effective at blocking LTP of CA3 neurons, although not that of CA1 neurons. These drug effects coincide nicely with the regional patterns of monoamine innervation of the hippocampal formation (see Fig. 11–2).

The NMDA receptor has been accepted as a critical step in initiating LTP, as well as in a related intrahippocampal process, long-term depression, that has so far received much less scientific examination. The current evidence suggests that the NMDA receptor must be in the appropriate state for LTP to be induced and that a molecular switch triggered by the metabotropic Glu receptor is required for the proper setting. A minor glitch in validating this evolving consensus emerged when disruption of the NMDA receptor

NMDAR1 led to early postnatal death of transgenic mice, with grossly maldeveloped brains. However, subsequent efforts to knock out only one of the NMDA subunits (the NR2 subunit family) were more successful, leading to normal brain cellular structure and normal growth and development but significantly reduced LTP in the hippocampal CA1 subfield and reduced (albeit moderately so) spatial learning. Disrupted LTP induction in the transgenic mouse mutant *fyn*, in which one tyrosine kinase isoform is disrupted (but not three others), and in mice lacking the protein kinase Cγ isoform indicates that very specific protein phosphorylation can regulate the thresholds for LTP induction. Despite the lack of LTP in the latter mice, however, spatial learning deficits can be seen only under rather extreme test conditions.

The NMDA-regulated step alone does not seem to be adequate for the long-term maintenance of the enhanced transmission. Given the critical role established in the *Aplysia* and other invertebrate models for a cAMP-regulated process in longer-term plasticity, that process has also been explored in the hippocampal model. Again, transgenic mice were employed, in this case expressing mutated knockout constructs of the CREB α and δ isoforms. These mice responded normally to short-term conditional fear associative learning but were unable to retain the learned fear for more than 2 hours (which the wild-type mice could readily do). Interestingly, LTP induction was also normal in the CREB knockout mice but returned to baseline in 2 hours as well. Lest the reader think it is now safe to memorize these schemata, if not the underlying facts, we must deliver a cautionary note. The CREB had been previously knocked out in a couple of alternate ways, one leading to dwarf mice when the CREB could not be phosphorylated and another with full CREB knockout leading to no deficits at all. Attempts to reconcile these observations led to the recognition that CREB belongs to a family of transcription factor–regulatory proteins, which can often compensate for each other. Thus, there is an alternative interpretation of the LTP and memory problems of the CREB knockout mice (pointed out to me by a clever young molecular neurobiologist, who prefers anonymity until this idea is tested): the knockout CREB should have been compensated for by another of those transcription factors, and it is perhaps that failure to compensate which causes the problems. Life gets confusing, doesn't it?

In allowing the world to spin a few times between editions, we may have spared the reader at least one apparently abortive chapter in the LTP saga: nitric oxide (NO) found one of its many overnight stardom roles as a potential retrograde transmitter leading to long-term increases in presynaptic transmitter release, based in part on the ability of nitric oxide synthase (NOS) inhibitors to block LTP, at least at certain nonphysiological temperatures. However, when LTP was found to be unchanged in mice in which neuronal NOS had been knocked out, enzymatic activity was virtually immeasurable, and LTP could still be blocked by NOS inhibitors, covert side effects of those blockers pulled NO from center stage. You can bet this is not over by a long shot. Thus, much remains to be accomplished with this system, including the demonstration that such events within single pathways of the hippocampal formation are necessary and sufficient for the animal or the hippocampus to "learn" something.

Readers intrigued by these memory model synaptic systems should consult the selected references for other models in which the circuitry has been spelled out and awaits the application of creative pharmacological analysis: the olfactory memory pathways, long-term depression of hippocampal and cerebellar units, and the amygdaloid complex.

Are There Any Natural Memory Drugs?

Back in the third edition, we noted the growing literature on the ability of natural hormones such as vasopressin and corticotropin, as well as "endocrinologically inert" fragments derived from them, either to repair learning deficiencies in hypophysectomized rats or to delay or accelerate the extinction of a previously learned performance. Unfortunately, as the pathways containing these peptides were more clearly defined in their projections to targets other than the posterior pituitary and the known barriers to diffusion of these peptides from the bloodstream into the brain were shown to apply to all of them, this once-promising area became a source of contention. However, this body of research remains an important case study for scholars of the neuropharmacology of behavior. Extended analyses of the behavioral responses regulated by systemic peptide hormones have demonstrated that visceral afferent nerves can influence the brain even if the peptides are excluded by the blood–brain barrier. Furthermore, there is a strong possibility that secrets leading to the development of memory-enhancing drugs may be discovered among the intracerebral pharmacological effects of what were once considered to be exclusively endocrine peptides.

Therefore, despite the murky nature of this particular research strategy, students should persist. For example, several studies have shown that low doses of subcutaneously administered vasopressin can enhance acquisition of a behavioral response, such as maze running or extinction of active avoidance in a shuttle box. Since vasopressin was named for its potent ability to increase blood pressure, it is not surprising that, while low in dose weight, the behaviorally active systemic doses of vasopressin also cause systemic hypertension. Moreover, vasopressin analogues that block the peripheral vascular receptors will also block the memory performance effects of the hormone. All behavioral effects of vasopressin have been blocked by antagonists of the vascular receptor, and the apparent central effect is likely to be indirectly mediated by an arousal secondary to the inappropriate hypertension. However, substituted and truncated versions of vasopressin lacking any vascular activity can still alter memory performance. Since the latter demonstrations of effectiveness have been obtained with memory tasks driven by aversive stimuli, their interpretation is still open.

Regardless of the arguments as to how or where the peripherally injected vasopressin acts, still smaller doses of vasopressin (1/1000 the systemic dose) have similar effects on memory performance when given intracerebroventricularly or intracerebrally. These low-dose (nanogram or picomolar) central effects are also reduced or prevented by central administration of the peripheral vascular vasopressin antagonist. Furthermore, opposite effects (i.e., memory performance impairment) have been reported in some tasks after central administration of vasopressin antagonists that are thought to act at vascular receptors. This suggests that somewhere within the central vasopressin-containing pathways is a circuit whose release of vasopressin is critical in eliciting proper performance and which has a receptor highly similar to those on blood vessels. Whether this means that the vascular receptor is also present on the central targets or that the central targets are actually on the blood side of the blood–brain barrier is an unanswered question. In fact, without directly testing a role for vasopressin, Thompson and colleagues have reported that water deprivation increased the magnitude of hippocampal LTP (perforant path to dentate) and, in parallel experiments, facilitated contextual fear conditioning.

The concept of vasopressin as a simple natural memory hormone would seem to be seriously challenged by studies of the Brattleboro rat, whose genetic defect produces a defective mRNA that cannot be translated into a secretable vasopressin. Although initially reported as being memory-deficient, the Brattleboro rat has supporters who regard its memory ability as normal, especially when salt and water balance are maintained by oral or systemic loading. Furthermore, humans with diabetes insipidus (the human disease of vasopressin deficiency) are not reported to have significant memory impairment.

Without some rigorous identification of the precise cellular sites and specific receptor transductive mechanisms on which either blood-borne or cerebrospinal fluid-borne vasopressin acts, this line of work faces difficulties. The possibility that vasopressin-derived subfragments (whose informational value has been suggested but never defined) can influence neuronal operations critical to the acquisition or retention of the memory tasks that have been studied remains ripe for further study. Future research will probably establish the superficiality of such interpretations as the following: *(1)* vasopressin acts directly on memory processes and *(2)* vasopressin can be an aversive hormone that, when given at nonphysiological doses, arouses the animal, which then learns better. We await eagerly the answers to this mind–drug–behavior puzzle, but they may not be found in the next edition, either.

Approaching the Neuropharmacology of Human Memory

In the selected animal models reviewed here, several key ingredients were required for the partial resolution of how selective drugs could modify the process of memory and learning: well-defined and relatively simple memory tasks, well-defined neuronal circuits whose transmitters and transmitter receptors were identified, and drugs that could be administered in ways that restricted their access to the sites being observed. An ultimate goal of such studies would be to develop a specific neuropharmacology to improve human memory function, either in normal subjects or in those in the early stages of dementing illnesses such as Alzheimer's disease or acquired immunodeficiency syndrome. Until recently, almost none of the essential facts were in place. Furthermore, the prospects of inferring from noninvasive techniques which sites in a human's brain participated in memory-related functions or how systemically administered drugs with multiple sites of interaction affected these events seemed dim not all that long ago.

However, dealing with memory processes in humans or even nonhuman primates has become far more attractive, partly because of the ability of methods such as positron emission tomography to reveal very brief shifts in blood flow within the brain while humans perform complex cognitive processing, thereby indicating which locations were active with specific kinds of higher mental activity. These methods, combined with experimental neurosurgical approaches to the rapidly unfolding description of cortical circuits in the monkey brain, have helped clarify a long-studied area of human memory research, namely, the amnesias that follow discrete toxic or traumatic lesions of the human brain. These combined strategies have allowed investigators to return to a point made earlier in the chapter (namely, that memory functions in more complex brains are likely to involve multiple systems), to ask how the primate brain is organized for memory functions, and to determine which neuronal systems are involved.

As exciting as this work is, an interesting paradox has been recognized: while very discrete lesions of the human hippocampus can lead to profound anterograde amnesia, the hippocampus is not among the many cortical sites in which functional activation (i.e., blood flow or glucose utilization) occurs when normal humans are performing complex memory-forming or memory-testing tasks. However, selected regions of the frontal cortex are activated in these tasks.

From this body of work has emerged the hypothesis that information processing is partly tied to the specific processing areas of the brain that are engaged during learning and that these memory events are stored within the same neural systems that also participate in the processes of perceiving, analyzing, and processing sensory information. Thus, in the visual system, the temporal cortical region involved in analyzing complex visual patterns (like faces) seems also to be involved in processing the information and in storing it. These and other examples of complex sensory systems suggest that some aspects of learning can be localized to specific subtasks but that many parts of the brain are involved with the overall process through intricate, parallel neuronal circuits.

The effort to bolster short-term memory in aged rhesus monkeys is an indication of future prospects. Memory capacity was assessed by a moderately complex task that required the animal to remember information over short time intervals and to update this information on every trial. Through impressive structure activity comparisons of a series of α_2-adrenergic agonists, two forms of memory-related action could be defined: hypotensive, sedating, and memory-impairing effects at one type of site and memory-enhancing effects at another α_2-receptor site. This same drug was also found to improve performance of complex behaviors in young monkeys as well. In low doses, the α_2-adrenergic agonist guanfacine improved memory without inducing hypotension or sedation, offering hope for the treatment of at least some memory disorders in humans. Guanfacine has also been reported to improve the behavioral and learning problems of children with attention-deficit hyperactivity disorder.

Selected References

Abel, T., K. C. Martin, D. Bartsch, and E. R. Kandel (1998). Memory suppressor genes: inhibitory constraints on the storage of long-term memory. *Science* 279, 338–341.

Alkon, D. L. (1988). *Memory Traces in the Brain.* Cambridge University Press, Cambridge.

Arnsten, A. F. (1993). Catecholamine mechanisms in age-related cognitive decline. *Neurobiol. Aging* 14, 639–641.

Bao, S., L. Chen, and R. F. Thompson (2000). Learning- and cerebellum-dependent neuronal activity in the lateral pontine nucleus. *Behav. Neurosci.* 114, 254–261.

Bartsch, D., M. Ghirardi, A. Casadio, M. Giustetto, K. A. Karl, H. Zhu, and E. R. Kandel (2000). Enhancement of memory-related long-term facilitation by ApAF, a novel transcription factor that acts downstream from both CREB1 and CREB2. *Cell* 103, 595–608.

Baudry, M. and G. Lynch (1988). Properties and substrates of mammalian memory systems. In *Psychopharmacology: The Third Generation of Progress* (H. Y. Meltzer, ed.). Raven Press, New York, pp. 449–462.

Bear, M. F. and R. C. Malenka (1994). Synaptic plasticity: LTP and LTD. *Curr. Opin. Neurobiol.* 4, 389–399.

Bortolotto, Z. A., Z. I. Bashir, C. H. Davies, and G. L. Collingridge (1994). A molecular switch activated by metabotropic glutamate receptors regulates induction of long-term potentiation. *Nature* 368, 740–743.

Gallagher, M. and P. C. Holland (1994). The amygdala complex: multiple roles in associative learning. *Proc. Natl. Acad. Sci. U.S.A.* 91, 11771–11776.

Haley, D. A., R. F. Thompson, and J. I. V. Madden (1988). Pharmacological analysis of the magnocellular red nucleus during classical conditioning of the rabbit nictitating membrane response. *Brain Res.* 454, 131–143.

Harvey, J. A. (1987). Effects of drugs on associative learning. In *Psychopharmacology: The Third Generation of Progress* (H. Y. Meltzer, ed.). Raven Press, New York, pp. 1485–1491.

Ito, M. (1987). Long term depression as memory process in the cerebellum. In *Synaptic Function* (G. M. Edelman, W. E. Gall, and W. M. Cowan, eds.). Wiley-Interscience, New York, pp. 431–447.

King, D. A., D. J. Krupa, M. R. Foy, and R. F. Thompson (2001). Mechanisms of neuronal conditioning. *Int. Rev. Neurobiol.* 45, 313–337.

Markowitsch, H. J. and E. Tulving (1994). Cognitive processes and cerebral cortical fundi: findings from positron-emission tomography studies. *Proc. Natl. Acad. Sci. U.S.A.* 91, 10507–10511.

Mayford, M. and E. R. Kandel (1999). Genetic approaches to memory storage. *Trends Genet.* 15, 463–470.

Mintz, M., D. G. Lavond, A. A. Zhang, Y. Yun, and R. F. Thompson (1994). Unilateral inferior olive NMDA lesion leads to unilateral deficit in acquisition and retention of eyelid classical conditioning. *Behav. Neural Biol.* 61, 218–224.

Rampon, C. Y.-P. Tang, J. Goodhouse, E. Shimizu, M. Kyin, and J. Z. Tsien (2000). Enrichment induces structural changes and recovery from non-spatial memory deficits in CA1 NMDAR1-knockout mice. *Nat. Neurosci.* 3, 238–245.

Richter-Levin, G., M. L. Errington, H. Maegawa, and T. V. Bliss (1994). Activation of metabotropic glutamate receptors is necessary for long-term potentiation in the dentate gyrus and for spatial learning. *Neuropharmacology* 33, 853–857.

Scahill, L., P. B. Chappell, Y. S. Kim, R. T. Schultz, L. Katsovich, E. Shepherd, A. F. Arnsten, D. J. Cohen, and J. F. Leckman (2001). A placebo-controlled study of guanfacine in the treatment of children with tic disorders and attention deficit hyperactivity disorder. *Am. J. Psychiatry* 158, 1067–1074.

Schacher, S., E. R. Kandel, and P. Montarolo (1993). cAMP and arachidonic acid stimulate long-term structural and functional changes produced by neurotransmitters in *Aplysia* sensory neurons. *Neuron* 10, 1079–1088.

Tang, Y. P., E. Shimizu, G. R. Dube, C. Rampon, G. A. Kerchner, M. Zhuo, G. Liu, and J. Z. Tsien (1999). Genetic enhancement of learning and memory in mice. *Nature* 401, 63–69.

Williams, J. H., Y. G. Li, A. Nayak, M. L. Errington, K. P. Murphy, and T. V. Bliss (1993). The suppression of long-term potentiation in rat hippocampus by inhibitors of nitric oxide synthase is temperature and age dependent. *Neuron* 11, 877–884.

13

Treating Neurological and Psychiatric Diseases

By this stage, the reader will almost certainly have detected a not-so-subtle underlying theme, namely, that a full understanding of the molecular mechanisms by which healthy neurons communicate and are sustained constitutes the starting point for developing drugs to treat neurological and psychiatric disorders. Except for a few passing references, however, we have not said very much about the nature of the brain's diseases or indicated those for which treatments are now effective and why. In this final chapter, we provide a brief introduction to those important neuropharmacological motivations.

The modern neuropsychiatric therapeutic armamentarium contains some powerful, selective, and highly effective drugs that often save lives and restore function. With the few exceptions noted below, however, most of these treatments came about by refining drugs that were initially developed for dubious disease hypotheses because astute clinicians observed potentially useful central nervous system (CNS) actions initially considered to be side effects. Many times, the predecessors for today's drugs told us more about the nature of the brain's normal operations by revealing drug-sensitive functions that had not been previously recognized. Moreover, several of the most effective drugs (e.g., those used to treat complex diseases of thinking, emotion, movement, or appetite) work for reasons that are still not very clear. In such cases, pursuit of molecular explanations of the pathophysiology of the disease continues to drive research. The corollary is that the diseases that we understand the most clearly in terms of causative mechanisms are not yet very treatable.

One last confession before we begin: New drugs will unquestionably emerge from the mountains of molecular information now being compiled in the push to sequence the entire human genome (see below). Still, the goal remains to gain far more insightful views of pathogenesis and then to exploit that knowledge for ways to treat and prevent disease.

When It's All In The Genes

In Chapter 3, we went through some of the fundamentals of molecular genetics in order to define the basic properties of the neurons and glia and their functions. To understand the genetic basis of CNS disorders when there is one, we need to revisit those concepts. The *genome* is the complete genetic set of a given individual, established at the moment of conception when the half-set of genes in a sperm combine with the half-set in the lucky ovum. In humans, the cells of the embryo contain 22 pairs of homologous autosomes and one pair of sex chromosomes that are heterologous (XY) for males and homologous (XX) for females. Each chromosome contains many genes, each with its own specific location (its *locus*) along the chromosome. Since each chromosome in the pair has one copy (or *allele*) of each gene, a person can be either *homozygous* (if both genes are the same) or *heterozygous* for any given gene. One of the early surprises from the human genome sequencing effort is that humans and other mammals may have far fewer than the 50,000–100,000 genes that had been previously conjectured, perhaps only twice as many as the fruit fly, the most complex organism of the previously sequenced genomes. If this conclusion holds, one interpretation is that in order to create organisms as complex as humans and other mammals from relatively few genes probably means that the richness of the required proteins is based on their modifications, either during transcription of the gene by much more intense splicing events or after translation of the mRNA into the protein.

Small, nonfunctional differences in nucleotide sequences that arise by mutation allow for some genes to contain polymorphisms. Because these differences also change the vulnerability of the DNA to cleavage by specific restriction endonucleases, a given digestion protocol can produce DNA fragments of different lengths in the alleles for a given gene. These varying sequence patterns and their differential cleavage patterns led to the method of restriction fragment length polymorphisms (RFLP) analysis to follow the inheritability of a particular allele by its telltale-sized fragments across family pedigrees. Often, the polymorphisms are the result of single nucleotide mutations that may or may not change the structure of the protein that the gene encodes. Polymorphisms can also occur within introns, where they can influence gene expression or splicing. The most recent molecular trends to emerge from the tracking of single-nucleotide polymorphisms (SNPs) and disease frequencies or morbidities is that clusters of SNPs, occurring within distinct genes, may form the inherited altered vulnerability for the pathology.

With that background, we can now categorize disease-producing genetic alterations in individuals in three ways:

1. *Single-gene mutations*, which, because they are inherited in recognizable patterns within families, are often referred to as *mendelian* mutations. These can be further categorized as *autosomal dominant* (when having one disease-causing allele is all it takes, as with Huntington's disease or neurofibromatosis) or *autosomal recessive* (when the disease may be carried across generations and not be detected unless the individual inherits the gene mutation from both mother and father and thereby becomes homozygous). Genomic information has already had an enormous impact on the medical sciences by providing absolute diagnostic markers for the highly inheritable neurodegenerative disorders that have generally borne the names of dead neurologists (like Huntington's disease), including the rare forms of some diseases that have traditionally not been regarded as inheritable (like Parkinson's and Alzheimer's). Special forms of recessive disorder arise from the X chromosome. Because the Y chromosome carries no somatic traits and codes only for gender-determining factors, a single X-linked disease-producing allele will cause the disease in males. However, nonsymptomatic females will pass it on to half of their sons but none of their daughters. Familial dysautonomia is an autosomal recessive condition affecting Ashkenazi Jews that can be diagnosed by a supersensitivity of the iris to methacholine. Because the gene for phenylalanine hydroxylase is subject to several different forms of inheritable mutation, even heterozygous carriers can produce offspring vulnerable to absence of the functional enzyme, leading to phenylketonuria and progressive mental retardation unless dietary phenylalanine is severely restricted well into early adolescence. Other inheritable metabolic disorders leading to mental retardation are shown in Table 13–1.
2. *Polygenic diseases*, in which multiple genetic factors are responsible for the appearance of the disorder. Many of the most prevalent brain diseases, such as depression, schizophrenia, multiple sclerosis, and epilepsy, are complex multigenic diseases. That is, for an inheritable vulnerability to eventuate in the expression of the disease, sev-

TABLE 13–1. Hereditary Diseases Associated with Cerebral Impairment

Disorder	*Metabolic Defect*
Amino acid metabolism	
Arginosuccinic aciduria	Arginosuccinase
Citrullinemia	Arginosuccinic acid synthetase
Cystathionuria	Cystathionine-cleaving enzyme
Hartnup disease	Trytophan transport
Histidinemia	Histidase
Homocystinuria	Cystathionine synthetase
Hydroxyprolinemia	Hydroxyproline oxidase
Hyperammonemia	Ornithine transcarbamylase
Hyperprolinemia	Proline oxidase
Maple syrup urine disease	Valine, leucine, and isoleucine decarboxylation
Phenylketonuria	Phenylalanine hydroxylase
Lipid metabolism	
A,β-lipoproteinemia (acanthocytosis)	β-Lipoproteins
Cerebrotendonous xanthomatosis	Cholesterol
Gaucher's disease	Cerebrosides
Juvenile amaurotic idiocy	Gangliosides
Krabbe's globoid dystrophy	Cerebrosides
Kufs' disease	Gangliosides
Metachromatic leukodystrophy	Sulfatides
Niemann-Pick disease	Gangliosides and sphingomyelins
Refsum's disease	3,7,11,15-Tetramethylhexadecanoic acid
Tay-Sachs disease	Gangliosides
Carbohydrate metabolism	
Galactosemia	Galactose-1-phosphate uridyl transferase
Glycogen storage disease (Type 2)	α-Glucosidase
Hurler's disease	Chondroitin sulfuric acid gangliosides
Unverricht's myoclonus epilepsy	Polysaccharides
Lafora's disease	Polyglucosan

(*continued*)

Table 13–1. *(Continued)*

Disorder	*Metabolic Defect*
Miscellaneous	
Cretinism	Thyroid hormone
Hallervorden-Spatz disease	Iron deposition in basal ganglia
Intermittent acute porphyria	δ-Aminolevulinic acid
Lesch-Nyhan syndrome	Hypoxanthine-guanine Phosphoribosyltransferase
Methylmalonic acidemia	Methylmalonyl CoA mutase
Wilson's disease	Ceruloplasmin

eral genetic events must coexist, perhaps brought on by environmental influences, such as a particularly intense life event. One consequence of developing drugs for such complex, multigenic diseases is that, even with today's molecular wizardry, scientists cannot create a pertinent transgenic animal model for complex genetic diseases in which to evaluate new drugs, as they can with those much more rare brain diseases caused by mutations in a single gene. On the other hand, another way to look at complex genetic diseases, arising from studies of monozygotic twins, is that the concordance for these diseases is far less than the expected 100%, indicating that factors other than genetic must be able to modify the vulnerability or resistance to the disease. Pursuit of that line of reasoning may lead to medications that enhance disease resistance rather than treat the disease after its emergence.

3. *Physically abnormal chromosomes* can arise during embryogenesis as cells divide, and the structural abnormality of the chromosomes can be detected by microscopy. *Trisomy 21* (a triplication of all or parts of chromosome 21, better known as Down syndrome), fragmentation and deletions of chromosome 7 resulting in Williams syndrome, and the fragile X form of mental retardation are examples of this kind of genetic problem.

Inheritable Metabolic Errors and Brain Development

Most of the diseases listed in Table 13–1 are single-gene defects leading to mental retardation or other forms of behavioral abnormality. Almost all of them are untreatable, so the forward-looking emphasis has been on genetic screening. In Lesch-Nyhan syndrome (an X-linked recessive disorder), affected males are both physically and mentally retarded and exhibit a characteristic self-mutilating behavior. Although the enzyme is expressed in many other cellular systems, it is basically only the formation and function of the brain in which the problems arise, for reasons that are not clear. Mice lacking this enzyme show no behavioral phenotype, a lesson for all those seeking to make transgenic mouse models of human disease.

Cretinism, a severe neonatal hypothyroidism, may arise from loss of thyroxin-producing enzymes and can be treated by replacement therapy as soon as it is recognized. The same is true of deficiencies of growth hormone and of gonadotropin-releasing hormone. The latter hormone is lost in males by a mutation (Kallmann's syndrome) in a specific neuronal adhesion molecule required to provide the trail by which the gonadotropin-releasing hormone neurons migrate from their birthplace in the olfactory bulb to their intended final location in the ventral hypothalamus. The porphyrias are a group of mostly autosomal dominant mutations of enzymes that synthesize the blood pigment heme; in one form, acute intermittent porphyria, an otherwise asymptomatic case can be triggered by infections, dehydration, or other unknown factors into a variety of incapacitating neurological and behavioral symptoms, including depression, mania, and hallucinations. King George III had this disease with severe manifestations during the American Revolution.

When Genes are Influenced by Environment

In many diseases, several unknown genes clearly play a role but in a manner that is both environmentally and behaviorally influenced. The extent to which specific gene products are expressed in the brain is a matter of the demands placed on the neuron or glial cell by the incoming information, as we saw in Chapter 3. The environment, both physical and social, is the source of much of this external information. Eventually, the information is converted into synaptic signals, which are in turn transduced into intracellular second messengers and ultimately into altered cytoplasmic and nuclear signals that, through transcriptional regulation, can determine which genes are turned on or kept off. Thus, the degree to which our genotype is reflected in our functional form or phenotype comes under environmental influence. The effects of the environment are likely to be cumulative and perhaps vary with the stages of development, and these influences on early brain function may not be apparent until much later in life (see Fig. 13–1).

Therefore, the cumulative effects of external events will continually modify gene expression as the cellular systems of the brain adapt to the adverse conditions imposed by environmental challenges. So long as the neurons and their supporting glia can adapt, the system will appear to be "healthy." However, when the environmental demands exceed in magnitude or duration the ability of the brain's adaptations to maintain normal function, the dysfunctional state that emerges is recognizable through the signs and symptoms of a neurological or psychiatric disease. If learned behaviors have their roots in changes in synaptic efficacy and response to specific external sensory signals, then early "personality-like" behaviors could appear to be inheritable when they are instead learned. For example, Meaney and colleagues reported that some rat mothers (dams) are much more active in grooming and tending to their pups than others and that the female pups grow up to follow their dam's style of maternal duty. However, when pups are cross-fostered from high-

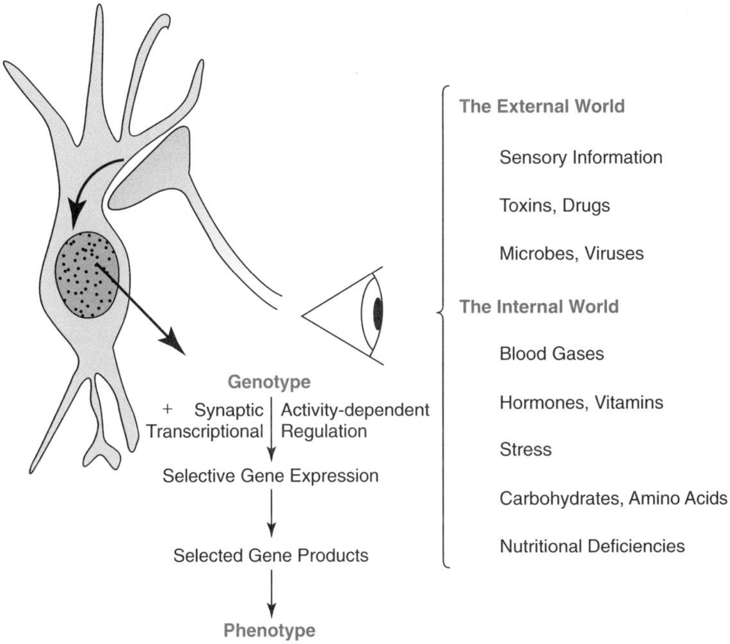

FIGURE 13–1. A neuron sees the world through the information received by its synaptic afferents. The cumulative results of synaptic activity modify gene expression by means of the nuclear actions of metabolic end products (intracellular second messengers, ions, and immediate early genes leading to transcriptional regulation) triggered by synaptic events. Changes in gene expression, resulting from the influence of external events or from changes in the internal environment of the brain, alter the neuron's phenotype (e.g., the rate of utilization of its own transmitter or receptors) and, hence, the operation of the circuits in which it participates.

to low-intrusive dams and vice versa, the daughters still grow up to exhibit maternal behaviors like the dam who raised them regardless of their biological dam.

The three examples we will now examine are relatively common disorders, and many powerful medications have been developed to provide significant alleviation for some, but not all, of those affected.

Depression

The family of psychiatric diseases epitomized by changes in mood (emotion, affect) is called the *affective disorders*, most commonly depression, and they are very common. Between 10% and 20% of the population will have a serious (i.e., clinically significant) episode over their lifetime. Depression and other serious affective manifestations may account for as many as 50% of those hospitalized for psychiatric reasons. There are two major forms, with different gender and genetic characteristics. *Bipolar depression* is characterized by recurrent, wide swings in mood from depression to extreme elation, occurs slightly more often in women than men, shows strong familial patterns of inheritability, and is primarily found in young adults aged 18–45. *Unipolar* or *major depression* is characterized by one or more prolonged episodes of depression (measured in weeks), occurs far more frequently in the population than the bipolar form, and is nearly three times as common in women as in men. Depression can also occur in childhood and early adolescence when the genders are nearly equal in vulnerability, perhaps an important clue to etiology. Studies of those with depression epitomize the possibility for successful biological approaches to the treatment of psychiatric disorders and exemplify the current thinking about the study of animal models of human brain diseases.

Catecholamine Theory of Affective Disorder

Animal studies indicate that catecholamines play a role in the periphery with regard to stress and emotional behavior. The catecholamine hypothesis of affective disorder states that, in general, behavioral depression may be related to a deficiency of catecholamine (usually norepinephrine) at functionally important central adrenergic receptors, while mania results from excess catecholamine. While substantial experimental work supports this proposal, most of the experiments on which this hypothesis is based derive from studies on the brains of "normal" animals.

The original impetus for developing the catecholamine hypothesis was the finding in patients being treated for tuberculosis that various monoamine oxidase inhibitors (MAO-Is) administered with other goals in mind, notably iproniazid, acted clinically as mood elevators or antidepressants. Shortly thereafter, this class of compounds also produced marked increases in brain amine levels. By the same token, reserpine, a potent tranquilizer and effective antihypertensive, depletes brain amines (serotonin as well as the catecholamines) and produces a serious depressed state (clinically indistinguishable from endogenous depression) in about 20% of those treated, and even suicide attempts in some people. The similarity in overt behaviors of "reserpinized" rodents and humans is one of the foundations of neuropharmacology and generated biochemical insight into the underlying disease.

Both drugs (MAO-Is and reserpine) alter brain levels of catecholamines and 5-HT quite equally. However, the fact that a precursor of catecholamine biosynthesis, L-DOPA, can reverse most of the reserpine-induced symptomatology in animals has tended to bias many researchers in favor of the catecholamine theory. In fact, by no means does the available evidence for the involvement of norepinephrine rule out the participation of dopamine, 5-HT, epinephrine, or other putative transmitters in similar events. The three general classes of drugs used to treat depressive disorders are the MAO-Is, the tricyclic antidepressants, and the psychomotor stimulants such as amphetamine. All of these pharmacological agents appear to interact with catecholamines in a way that is consistent with the catecholamine hypothesis. Thus, all four of the suggested modes of action of amphetamines (partial agonist, inhibitor of catecholamine reuptake, competitive inhibitor of MAO, and displacer of presynaptic norepinephrine and dopamine) would be expected to increase catecholamines temporarily at their receptors. Administration of long-term or high-dose amphetamine produces an eventual depletion of brain norepinephrine and dopamine and inhibition of neuronal activity in catecholamine neurons. This chronic depletion of transmitter and the prolonged inactivity of catecholamine neurons may account for the clinical observation of amphetamine tolerance or the well-known poststimulation depression or fatigue observed after chronic administration of this class of drugs.

Well after they were recognized as effective antidepressants, studies of norepinephrine conservation revealed that these drugs shared an important action, the ability to block the monoamine transporters. All of the tricyclic antidepressants inhibit both the catecholamine and 5-HT presynaptic plasma membrane transporters. Other reuptake inhibitors, not necessarily of the tricyclic structural motif, have been made selective for one or another monoamine transporter. For example, mazindol and bupropion are dopamine-selective reuptake inhibitors, and fluoxetine (Prozac) was the first of the serotonin-selective reuptake inhibitors (SSRIs). These drugs are far more selective than was the original tricyclic imipramine, which is now recognized as a mixed norepinephrine/5-HT transporter inhibitor. Although they have proven to be effective in treating subsets of depressed patients, some depressed patients, especially those with onset in childhood or adolescence, are quite resistant to such treatment.

The action of the MAO-Is also supports the catecholamine hypothesis. All of these agents inhibit an enzyme responsible for the metabolism of norepinephrine and various other amines (5-HT, dopamine, tyramine, tryptamine). This inhibition results in a marked increase in intraneuronal levels of norepinephrine. According to the conceptual model of the adrenergic neuron presented above, this intraneuronal norepinephrine might eventually diffuse out of the neuron and reach receptor cells, thereby overcoming the presumed deficiency. A similar mechanism may also explain the antagonism of reserpine-induced sedation with MAO-Is, since there will be an initial replenishment of the norepinephrine deficiency caused by reserpine.

Lithium, one of the main agents used to treat manic episodes in bipolar disease, has also been studied with regard to its action on the life cycle of the catecholamines. Interestingly, pretreatment with lithium blocks the stimulus-induced release of norepinephrine from rat brain slices. Other investigators have suggested that lithium may facilitate reuptake of norepinephrine. If the inhibition of release observed in stimulated brain slices is due to a facilitated recapture mechanism, then the mechanism of action of lithium is the exact opposite of that of the antidepressant drugs, as would be expected according to the catecholamine hypothesis. Although a single injection of lithium may antagonize responses to norepinephrine, chronic treatment of rats leads to modest supersensitive norepinephrine responses.

On closer scrutiny of the clinical and experimental pharmacological data cited in support of the catecholamine theory of affective disorder, however, even the keenest enthusiasts recognize significant inconsistencies. *(1)* Cocaine is a very potent inhibitor of catecholamine reuptake and thus, like the tricyclic antidepressants, should increase the availability of norepinephrine at central synapses; but cocaine does not possess any significant antidepressant activity. *(2)* Iprandole, a tricyclic compound without any significant effect on catecholamine uptake or any influence on central noradrenergic neurons, is an effective antidepressant. Furthermore, desensitization of postsynaptic β-receptors after chronic administration of tricyclic antidepressant drugs is not entirely compatible with the proposed theory of a synaptic catecholamine deficit. *(3)* In laboratory studies, the ability of tricyclic antidepressants to inhibit catecholamine uptake and the ability of MAO-Is to block MAO and elevate brain catecholamines are apparent soon after administration. Clinically, however, antidepressants must be given for several days (10–14) to produce therapeutic effects. Perhaps some novel genes must be turned on or off, or enhanced catecholamine action at recurrent axons onto monoamine cell bodies may further reduce their firing such that even less catecholamine is released at synapses. *(4)* On the surface, the pharmacological and clinical effects of the tricyclics and lithium seem to fit nicely with the catecholamine theory of affective disorder. These agents produce opposite effects on norepinephrine disposition, and lithium is effective at treating mania while the tricyclics are useful in treating depression. However, lithium is also effective in the treatment of bipolar depressed patients, where it dampens the emotional oscillations into either mania or depression.

The fact that several questions have been raised about the pharmacological data used to support the catecholamine theory of affective disorder has prompted many investigators to seek direct evidence of the involvement of norepinephrine systems in affective disorders. The most extensive studies have involved analysis of urinary excretion patterns of catecholamine metabolites in patients with affective disorders. The rationale has been that the urinary excretion of a particular metabolite, like MHPG, may be a useful reflection of central catecholaminergic processes. Despite the inherent problems in urinary catecholamine metabolite measurement, such as complications because of large contributions from peripheral sources, some interesting findings have emerged.

Findings from several clinical studies now indicate that *(1)* depressed patients as a group excrete less than normal quantities of MHPG, *(2)* diagnostic subgroups of depressed patients are particularly likely to have low urinary MHPG values, *(3)* bipolar patients who switch from a depressive to a euthymic or hypomanic state show a corresponding increase in urinary MHPG, and *(4)* pretreatment urinary MHPG values are predictive of the type of therapeutic response obtained with catecholamine-directed reuptake inhibitors. Although provocative and essentially consistent with the norepinephrine theory of affective disorder, these clinical findings hinge on the issue of the degree to which urinary MHPG reflects central norepinephrine metabolism.

For the above reasons, the original hypothesis that antidepressant drugs increase the availability of monoamines in the brain has been updated to include the effects of long-term antidepressant treatment on monoamine receptor sensitivity. A wide array of effects of long-term treatment with antidepressants has been reported in several monoamine systems. The most consistent findings following chronic administration of most of the clinically effective antidepressant drugs to experimental animals are *(1)* a reduction in the number of β adrenoceptors and downregulation of β-adrenoceptor functioning; *(2)* an increase in the sensitivity of central α adrenoceptors, suggesting up-regulation in central α-adrenoceptor functioning; and *(3)* similarly, both behavioral and electrophysiological studies also point to upregulation in the sensitivity of central 5-HT receptors. These changes require time to be accomplished, and perhaps that is why the delay in antidepressant action is observed. However, all of the monoaminergic neurons have a rich recurrent innervation, and enhancement of this autoinhibition would also be expected from the use of re-uptake inhibitors. If the monoamine neurons are thus reduced in their excitability, the actual amount of transmitter released from distal axons may be initially reduced, even with reuptake inhibition. Therefore, another part of the lag in response to treatment may be the requirement for reequilibration of perikaryal excitability and restoration of baseline synaptic function before the reuptake inhibition per se can enhance postsynaptic receptor durations of action. Alternatively, perhaps new gene products must be made and time is required to ship them in sufficient amounts to the synapses.

Throughout this analysis of antidepressant drug actions, we have noted frequent intrusions by 5-HT-related effects. In fact, 5-HT deficiency deserves attention as an independent causative factor in some forms of depression. Diminished plasma levels of the rate-limiting 5-HT precursor L-tryptophan (L-TRP) have been seen in several series of depressed patients and linked to either reduced intestinal absorption or increased hepatic catabolism of absorbed L-TRP stimulated by pyrrolase; dietary restriction of L-TRP may also precipitate depression in recently remitted patients. Cerebrospinal fluid levels of the 5-HT catabolite 5-hydroxyindoleacetic acid (5-HIAA) have been studied extensively in depressed subjects, but the best correlations here are with attempts to commit suicide and with impulsive, violent behavioral patterns. Some postmortem studies of depressed patients have shown increased numbers of certain 5-HT receptors, especially in frontal cortical regions. Positron emission tomographic (PET) studies of patients with depression show decreased glucose utilization in the frontal cortex. Other pharmacological challenges to depressed subjects, based on changes in sleep electroencephalographic (EEG) patterns and neuroendocrine secretion patterns, further support the concept of a central 5-HT deficiency and postsynaptic receptor upregulation. Perhaps the greatest support for this concept arose with the major, and to some extent unexpected, antidepressant success of fluoxetine, sertraline, and paroxetine, the SSRIs. An interesting polymorphism in the transcriptional regulatory domain of the 5-HT transporter has been found to be a potential marker for patients who will more rapidly respond to SSRIs. In this polymorphism, a block of 40 nucleotides may be present or absent, but the "normal" form is not clear. Those with the fragment missing seem to make less of the transporter, perhaps suggesting why they show greater sensitivity to its inhibition. However, it is not a good diagnostic marker for depression.

More direct studies are necessary, however, before it is possible to conclude whether any or all of the therapeutic actions of antidepressant drugs are indications that the emotional disorder was truly caused by functional deficiencies of one or another monoamine. These studies await the development of appropriate models for monitoring central neurotransmitter functioning in humans, especially those at high risk of developing depression based on their family pedigrees. An interesting development, yet to be confirmed, is the potential for other classes of drugs, in particular antagonists of the neuropeptides substance P and corticotropin-releasing hormone, to work as antidepressants in both rodent models and patients with depression

Dopamine Hypothesis of Schizophrenia

Schizophrenia is a chronic disorganization of mental function that affects thinking (paranoid ideas, inability to maintain a focused thought, easily distracted, loose associations between thoughts), feeling (typically referred to as *blunted affect* and inappropriate responses to social situations), and movement (from hyperactivity and excitement to persistent inactivity to the point of maintaining bizarre postures for long periods of time). The disease is most commonly recognized in very young adults, particularly when confronted with severely stressful life events, although evidence of social withdrawal and disorganized thinking may have been noted before the episode. About one-third of those having a single episode may fully recover. Like bipolar depression, schizophrenia shows strong familial patterns of inheritance in some cases, but no mendelian pattern; thus, it is best relegated to the category of complex genetic disorder. Available medications can be very beneficial, but for most affected individuals, the blunted affect, apathy, lack of volition, and social withdrawal (the "negative" symptoms) may be intractable even though the medications dampen the hallucinations and aggressive behaviors (the "positive" symptoms).

In analogy to the catecholamine hypothesis of depression, a biochemical explanation for schizophrenia arose from observations that the only consistent feature among the antipsychotic drugs used to treat the disease was their ability to antagonize D_2 dopamine receptors and that chronic administration of amphetamine and other (mainly) dopamine-mediated psychostimulants could produce a psychotic state loosely resembling some aspects of schizophrenia. This hypothesis in its simplest form states that schizophrenia may be related to a relative excess of central dopaminergic neuronal activity. Attempts to validate this hypothesis in clinical studies have been intense but inconclusive. While D_2 receptors are consistently increased in most postmortem studies of schizophrenic brains, the influence of prior antipsychotic treatments confounds the interpretation, and no direct evidence of increased dopamine synthesis, turnover, or function has been obtained, nor has noninvasive imaging of dopamine receptor densities in previously unmedicated young schizophrenics produced any consistent evidence for an increase in D_2 receptors within the striatum. However, the D_4 receptor subtype, functionally akin to the D_2 receptors in inhibiting adenylate cyclase, has greater affinity for the atypical antipsychotic drug clozapine (these drugs were called "atypical" precisely because they were not good D_2 antagonists) and to be increased specifically in postmortem schizophrenic brains.

Experiments in animals have generated substantial support for the idea that antipsychotic drugs are effective blockers of dopamine receptors, but most of these animal studies (whether behavioral, biochemical, or electrophysiological) have been carried out after acute drug administration. This is a very serious drawback since, as with the antidepressants, the clinical effects of antipsychotic drugs (both antipsychotic and neurological) take days, weeks, or even months to develop. In nonhuman primates, chronic treatment with antipsychotic drugs causes tolerance to the homovanillic acid (HVA) increase normally observed in the putamen, caudate, and olfactory cortex after a challenge dose of an antipsychotic drug. In cingulate, dorsal frontal, and orbital frontal cortex, however, increased levels of HVA are maintained throughout the time course of chronic treatment. Similar observations have also been made in studies carried out on autopsied human brain specimens. If patients with the diagnosis of schizophrenia are chronically treated with antipsychotic drugs, a significant increase in HVA is found in the cingulate and frontal cortex but not in the putamen and nucleus accumbens, suggesting a locus for the therapeutic action of these drugs and providing the first direct experimental evidence that antipsychotic drug treatment increases the metabolism of dopamine in the human brain in a regionally specific manner. Establishing with certainty that the elevation in dopamine receptor density is part of the disease process would be very important for both etiological and diagnostic purposes and would serve as a possible basis for treatment strategies. The use of new noninvasive techniques, such as PET and single-photon emission computed tomography (SPECT), to examine dopamine receptor distribution and density in schizophrenia holds promise that this may be accomplished soon.

Unlike depression, where there is no obvious neuropathology, postmortem microscopic examination of the brains of patients with schizophrenia has revealed pathological evidence for abnormalities of neuronal density but without frank neurodegeneration. Together with other findings (enlarged ventricles, thin cortex) that are stable over years of observation, this has led to the interpretation of schizophrenia as a neurodevelopmental disorder. The negative symptoms of cognitive disruption and affect have been correlated with reduced brain metabolic activity in the frontal lobes, and this may be an adaptive downregulation of dopaminergic activity in this region of cortex. The atypical antipsychotic drugs have received increasing attention for three reasons: *(1)* they can reduce symptoms in patients resistant to other drugs, *(2)* they produce fewer side effects, and *(3)* their pharmacology may extend the neurotransmitter base for schizophrenia to 5-HT as well as dopamine. Other evidence points to disruption of corticostriatal glutamatergic circuits in the disease. In part, this is based on the observation that the hallucinatory side effects of so-called dissociative anesthetics, such as ketamine (named for their apparent opposing effects on the EEG and behavior) are also antagonists of the *N*-methyl-D-aspartate (NMDA) and glutamate receptors. Animal models of altered brain dopamine neurochemistry and behaviors have been proposed in both rodents and primates by acute or chronic injection of phencyclidine, another NMDA antagonist and hallucinogen. The area is being intensively explored (see Fig. 13–2).

Drug Abuse

The continued, compulsive obsession with obtaining, consuming, and experiencing self-administered drugs is a major social and medical problem throughout the world. Specific drugs of abuse have specific patterns of use and dependence. Seven families of drugs have been recognized to be obsessively self-administered by humans. In order of prevalence, they are caffeine (as in coffee or tea), nicotine, alcohol (grouped with benzodiazepines and barbiturates), marijuana (and the congeners hashish and tetrahydrocannabinol), psychostimulants (cocaine and amphetamines), opiates (morphine, heroin, and other agonists), and the hallucinogenic drugs [LSD], phencyclidine, and MDMA [3,4-dioxymethylene methamphetamine], otherwise known as "ecstasy"). Note that the most widely abused substances, caffeine, nicotine, and alcohol, are legal in the United States and that both the federal and state governments collect substantial "sin taxes" on the latter two, while attempts to place taxes on caffeine helped spark the American Revolution. Recreational use of alcohol or nicotine, however, may serve as the gateway to illicit and powerful drugs of abuse.

The neurobiological substrate for the acute rewarding effects of the four major classes of abused drugs (alcohol, nicotine, opiates, and psychostimulants) has focused on the area of the ventral forebrain that surrounds the nucleus accumbens. This nucleus, with neurons of several forms and efferents, is heavily innervated by pontine monoamine neurons and is considered to be in a continuum with the nearby neurons of the amygdaloid complex. Increased release of dopamine within the region of the nucleus accumbens has

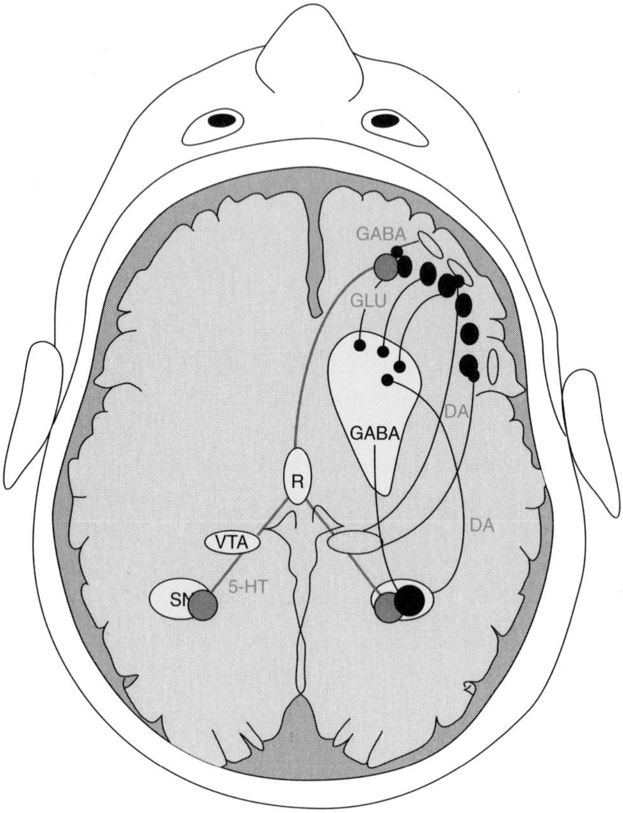

Figure 13–2. Schematic diagram of the main neuronal circuits and transmitters dysregulated in schizophrenia. The frontal neocortical projections to the striatum and other subcortical structures are primarily glutamatergic (GLU) efferents. Within the cortex, these units are regulated by γ-aminobutyric acid (GABA)–containing interneurons (neuropeptides of those GABA neurons not shown) and by dopamine (DA) projections from the ventral tegmental area (VTA) and serotonergic (5-HT) projections from the raphe (R) nuclei. The latter monoaminergic neurons may project to either the cortical interneurons or to the cortical efferent neurons. The raphe also projects to the substantia nigra (SN), whose DA axons innervate the basal ganglia but not the cortex. Current evidence favors decreased levels of operation in the nigrostriatal projections but overactivity in the mesolimbic DA systems. "Atypical" antipsychotic drugs may work in part through their capacity to inhibit cortically enriched DA receptors (such as the D_4) as well as 5-HT receptors. (Based on circuitry described by Maes and Meltzer, 1994; see Meltzer, 1999.)

been directly observed by microdialysis in animals trained to self-administer alcohol and nicotine. In the case of cocaine, antagonists of the D_3 dopamine receptor selectively reduce the rewarding effects of the drug, as shown by decreases in self-administering behavior. However, while the key site for eliciting opiate reinforcement also seems to be within the nucleus accumbens, the dopamine afferents are not required for maintenance of opiate self-administration.

Certain forms of alcoholism are thought to arise from unknown but genetically transmittable factors that increase vulnerability to alcohol dependence, especially in the male offspring of male alcoholics, who may also show antisocial personality disorder. Animal models of genetically transmittable alcohol preference have been achieved several times by constant cross-breeding of the few rats that initially show modest alcohol preference. Later generations develop an almost universal predilection to consume alcohol for its pharmacological (as opposed to nutritional or thermal) properties. In the alcohol-preferring animals developed in the United States and in Sardinia, there is diminished dopamine and 5-HT innervation of the nucleus accumbens and increased GABA innervation of that site. Among other manipulations, 5-HT-specific reuptake inhibitors will reduce alcohol self-administration in these rats. Alcohol-accepting rats bred in Finland, however, do not show these neurochemical differences.

Alcoholism vulnerability in Asian human populations is actively suppressed by a prevalent genetic mutation leading to reduced capacity to oxidize the ethanol catabolite acetaldehyde. Those carrying this mutation (half of these populations) react to alcohol as though they were receiving the drug antabuse, which directly inhibits acetaldehyde dehydrogenase. After a brief flurry of excitement over the possibility that certain alleles of certain dopamine receptors might be associated with alcoholism and other forms of dependence, the search for the responsible genes has resumed. While there are major differences in the sensitivity of certain inbred rat strains to the reinforcing actions of opiates and psychostimulants, the evidence for such inherited vulnerability has not been well studied in humans, largely because these drugs are universally highly reinforcing once their recreational use has been initiated. Nonetheless, these high- and low-sensitivity animal models of drug response may prove useful in determining the genetic basis for vulnerability to drug dependence in humans.

Efforts to deal with drug abuse have focused on reducing the supplies of illicit drugs by law-enforcement strategies aimed at the prevention of drug smuggling. After spending hundreds of millions of dollars on "supply reduction" only to see the suppliers able to sell more products at lower prices, agencies have devised other strategies directed toward demand reduction. For many years, methadone, a mixed agonist–antagonist for morphine with reduced euphorigenic properties, and some of its long-acting congeners have been known to reduce opiate dependence, increasing job performance and reducing criminal behavior; but its widespread adoption was delayed because of the belief in some quarters that it was merely substituting one drug for another. Only recently has methadone been approved for the long-term treatment of opiate abuse. In animal studies, the opiate antagonist naloxone, also reduces some of the rewarding effects of alcohol self-administration for reasons that are not yet clear. Naltrexone, the orally active congener of naloxone, has been approved for the long-term treatment of alcoholism. Specially reformulated long-acting naltrexone is under investigation as a means to improve compliance by removing the daily decision to take the medication.

When Knowing the Genes Does Not Help

Lastly, we consider a group of diseases that fall under the general heading *neurodegenerative disorders*, most of which remain untreatable, although there are animal models that pose striking opportunities for pharmacological development in this area. Some of the more creative avenues to restoring brain function, from transplantation of neurons and neuronal stem cells to the transfer of autosomal genes, are explored here.

Alzheimer's Disease

This chronic, progressive, degenerative disorder, which is in some rare cases familial, is becoming a greater health-care problem as the population survives to older ages. The nearly linear incidence of Alzheimer's detection with age makes this a very prevalent disorder, perhaps being detected in as many as 50% of those older than age 85, which is no longer a unique occurrence in our society. Alzheimer's disease is characterized behaviorally by a severe impairment in cognitive function, including memory, the ability to recognize objects, and the ability to orient oneself in time and space. Neuropathologically, the disease is epitomized by the appearance of *neuritic plaques* (amorphous extracellular deposits of proteinaceous matter) and *neurofibrillary tangles* (intracellular whorls of paired helical filaments), generally restricted to specific cortical regions (hippocampus, frontal and temporal lobes more than parietal and far more than occipital lobes) and specific cortical layers (especially prominent in layers III, V, and VI, where the large neurons are located). The pathology is also seen within a few of the subcortical structures (ventral pallidum, locus ceruleus) projecting to the cortical areas of involvement.

A cholinergic hypothesis of cortical dysfunction has been implicated in Alzheimer's disease based on the findings that *(1)* administration of centrally acting muscarinic-blocking agents to normal individuals induces a loss of recent memory; *(2)* in the cerebral cortex and hippocampus of patients with Alzheimer's disease, there is a dramatic reduction of ACh, choline acetyltransferase, and high-affinity choline uptake; *(3)* in Alzheimer's disease patients, there is a severe reduction of neurons in the nucleus basalis of Meynert, the primary cholinergic input to the cortex; and *(4)* in some, but not all, studies, a decrease in muscarinic and nicotinic receptors has been noted. However, patients with Alzheimer's disease also have decreased levels of somatostatin, neuropeptide Y, substance P, and corticotropin-releasing factor, increased numbers of corticotropin-releasing factor and NMDA receptors; as well as reduced numbers of locus ceruleus neurons. More recent detailed comparisons of the rate at which synapses in general and cholinergic markers specifically are lost suggest that many of the other neurotransmitters are affected before the cholinergic afferents, indicating that the cholinergic decline may be secondary to the loss of cortical neurons. A decrease in the neuronal content of the nucleus basalis of Meynert has also been observed in some patients with Down syndrome (trisomy 21, the only other condition known to express neuritic plaques) and with Parkinson's disease.

Thus, the stormy marriage of Alzheimer's disease to a cholinergic dysfunction may involve some extramarital relationships. Little success has been achieved to date by treating patients with choline, lecithin, physostigmine, or the muscarinic agonist arecoline. Current efforts focus on the development of drugs that can block acetylcholinesterase centrally without hepatoxicity and peripheral autonomic side effects. Future therapy had been thought to focus on specific M_2-muscarinic antagonists, nicotinic agonists, or the transplantation of fetal cholinergic tissue. However, a potentially more revealing strategy over the past decade has been a direct assault on the causes of the disease, made possible by the isolation, purification, and sequencing of the proteinaceous matter that has been the hallmark of the pathology, the amyloid plaques. Once the aggregated forms had been sequenced, scientists isolated and sequenced the entire gene and identified its natural product as a large protein of unknown function, named the *amyloid precursor protein* (APP). The structure of normal APP, combined with studies of relatively rare families with greatly increased frequencies of Alzheimer's with especially early onset, has led investigators to focus on errors in APP processing that could generate the short fragments of the amyloid plaques. Although no such proteases were previously known, concentrated efforts to find enzymes capable of such proteolysis (termed β- and γ-*secretases*) were soon successful and will provide important targets for drug development and perhaps for the known effects of mutations in the proteins presenilin-1 and -2.

The creation of transgenic mice overexpressing one form of the β-*APP* gene in cortical neurons and producing plaques within 18 months offered the possibility of screening for drugs that may reduce amyloid depositing and perhaps even of revealing the reasons why the aged brain overproduces this and other proteins in the first place. In addition, it has been possible to reverse the extracellular deposits by immunizing the mice with the fragment of APP that forms the aggregates (A-β_{1-42}). It remains unclear how other recognized genetic predispositions (e.g., inheritance of the apoliprotein E-4 allele) also interact with environmental conditions or events such as head trauma to initiate the process of pathological change. The recognition that APP may be a receptor for the light chain of kinesin-1, the motor molecule underlying axonal transport, suggests a functional role for APP; but as of now, there is still much to be resolved. Nevertheless, at least in mice, it appears that combined pathology in the genes of APP and tau (the major protein of microtubules) is required to get the full pathological picture of plaques and tangles.

Parkinson's Disease

Parkinson's disease (PD) is a progressive neurodegenerative disorder of the basal ganglia characterized by tremor, muscular rigidity, difficulty in initiating motor activity, and loss of postural reflexes. It is observed in approximately 1% of the population over age 55. For over 75 years it has been known that PD is characterized pathologically by loss of pigmented cells in the substantia nigra, but only since 1960 has the substantial loss of dopamine in the striatum been documented. It is now clear that PD can be defined in biochemical terms as primarily a dopamine-deficiency state resulting from degeneration or injury to dopamine neurons. The most striking degenerative loss of dopamine neurons is observed in the nigrostriatal system. Even in patients with mild symptoms, a striatal dopamine loss of 70%–80% occurs, while severely impaired subjects have striatal dopamine depletions in excess of 90%. Since the dopamine transporter is heavily expressed in the terminals of dopamine neurons that are lost in PD, it is not surprising that striatal binding of agents that label this site (cocaine, nomifensine, and mazindol) is lost in the parkinsonian striatum. This alteration corresponds well with the loss of functional dopamine uptake visualized in vivo by PET, using ^{18}F-L-DOPA uptake or ^{18}F-nomifensine, or by SPECT, using other presynaptic dopamine labels. Although striatal dopamine loss represents the primary neurochemical abnormality in the PD brain, typical parkinsonism is accompanied by loss of other dopamine systems and other monoamine neurons. Some degeneration of dopamine-containing neurons is also apparent in the mesolimbic, mesocortical, and hypothalamic systems, as is loss of norepinephrine-containing neurons in the locus ceruleus and of serotonin neurons. Nonmonoamine systems are also affected, with depletions observed in somatostatin, neurotensin, substance P, enkephalin, and cholecystokinin-8. Since many of these nondopamine systems indirectly interact with mesotelencephalic dopamine systems, changes in some of them are bound to influence the function of dopamine neurons in a complex way. Before it was known that dopamine is severely depleted, for example, treatments with anticholinergic drugs had been viewed as moderately effective.

Since the earliest and most substantial neurochemical abnormality in PD is the loss of dopamine, the modern strategy for treatment has concentrated on restoring the dopamine deficit. The theoretical strategies here include substrate supplementation, direct and indirect dopamine agonists, metabolic inhibitors (MAO-Is) and uptake inhibitors. The most successful treatment has been the use of L-DOPA, recognized by the Nobel Prize for Physiology or Medicine in 2000. Currently, L-DOPA treatment is usually combined with an inhibitor of DOPA decarboxylase that acts only outside the brain. This enhances the amount of the absorbed L-DOPA that can enter the brain and alleviates the gastrointestinal symptoms that arose from the higher doses required previously. As the degeneration of dopamine (and other) neurons progresses, the requirements for L-DOPA increase and are accompanied by interruptions of its effectiveness that are poorly understood ("wearing-off" and "on/off" responses). Direct dopamine agonists have some benefit for patients whose responsiveness to L-DOPA is greatly reduced or erratic. So far, the only direct-acting dopamine agonist that has found extensive use in PD is bromocriptine, primarily a D_2 agonist. However, selective D_1 agonists, such as ropinirole and cabergoline, have been developed which exhibit longer durations of action without the absorption or blood–brain barrier permeation problems that have been suggested as causes for the variations in L-DOPA effectiveness. Other agents belonging to this class will no doubt prove useful in the future as supplements or alternatives to L-DOPA.

In view of the behavioral and electrophysiological studies that suggest that D_1 receptor activation is necessary for the effects of D_2 receptor stimulation to be maximally expressed in normal animals as well as in animals with supersensitive dopamine receptors, the functional interaction between D_1 and D_2 receptors could have important implications in PD, where stimulation of postsynaptic dopamine receptors confers symptomatic benefit. Knowledge of the optimal ratio of relative drug activity at D_1 and D_2 receptors that is required to elicit effective stimulation of dopamine-mediated function may provide a basis for the design of new drugs. Also, more knowledge of the distribution and function of various dopamine receptor subtypes should facilitate the development of new agents to treat dopamine-deficiency states. Fetal neurons and neuronal stem cells are likely to be used in model systems to treat the earliest symptoms of PD since the motor effects are an excellent index of functional recovery. As with Alzheimer's, the recognition of rare familially transmitted disease has led to the isolation of an abnormal protein, parkin, in autosomal recessive juvenile parkinsonism and of α-synuclein mutations in adult forms, apparently related to abnormal accumulations of protein in aging dopamine neurons, known as Lewy bodies. A special form of dementing illness in aging that affects frontal lobe function is epitomized by cortical Lewy bodies whose composition and relation to nigral neuron Lewy bodies is under intense investigation.

Primate Model of Parkinson's Disease

In 1983, researchers at Stanford University identified a contaminant in locally produced "synthetic heroin" that induced a PD-like syndrome in some individuals who self-administered this street drug. The contaminant identified in this preparation, 1-methyl-4-phenyl-1, 2,3,6-tetrahydropyridine (MPTP), exhibits a high degree of anatomical and species-specific toxicity. MPTP administered systemically in low doses to nonhuman primates produces parkinsonian symptoms and destroys nigrostriatal dopamine neurons while sparing several other brain dopamine systems. The discovery of the selective neurotoxic properties of MPTP and the development of a primate model of parkinsonism stimulated a strong resurgence of inquiry into the causes and treatment of PD. Still, the mechanism responsible for the selective dopamine neurotoxic features of MPTP in primates has not been conclusively established.

MPTP appears to act as a protoxin, and MAO-B–mediated bioactivation of MPTP to 1-methyl-4-phenylpyridinium (MPP^+) plays a critical role in the ultimate neurotoxic action. Administration of MAO-B inhibitors, including L-deprenyl, affords full protection against the neurotoxic action of MPTP. Since dopamine uptake inhibitors such as mazindol and GBR-12909 also protect against MPTP-induced neurotoxicity, it has been suggested that MPTP is oxidized to MPP^+ outside the dopamine neurons, perhaps by MAO in astrocytes. MPP^+ is then transported and concentrated by the dopamine uptake system in dopamine neurons. Once inside the dopamine neuron, MPP^+ exerts its neurotoxic effect by acting as a mitochondrial poison through inhibition of respiration at site i of the electron transport chain. This postulated mechanism for MPTP's neurotoxicity in dopamine neurons is illustrated in Figure 13–3.

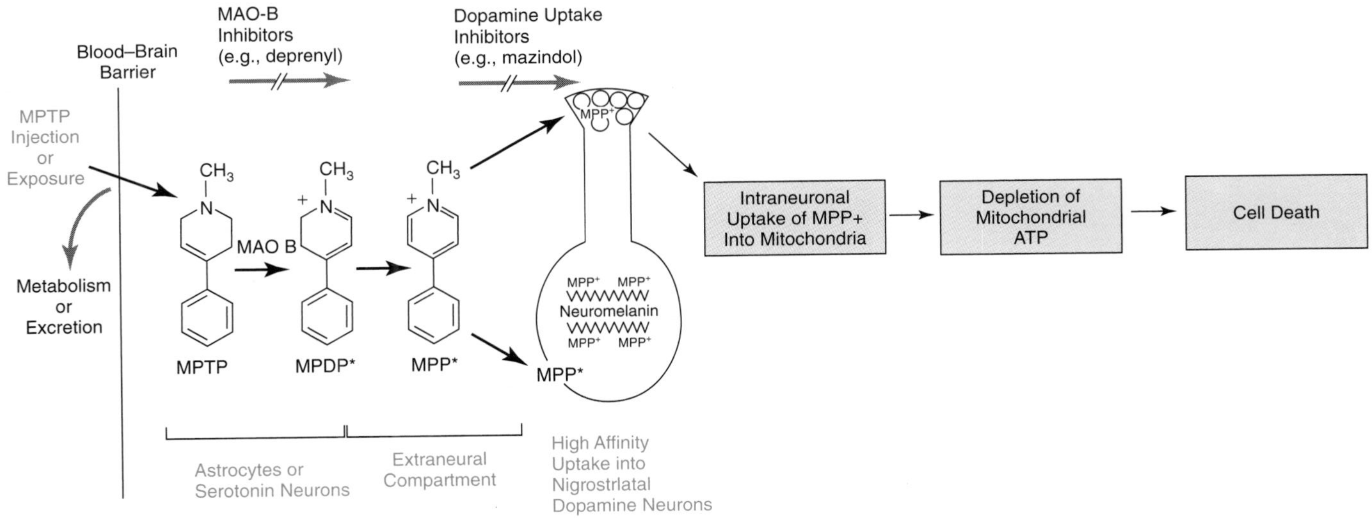

FIGURE 13–3. Hypothesized mechanisms of neurotoxicity of 1-methyl-4-phenyl-1,2,3,6-tetrahydropyridine MAO-B, monoamine oxidase type B; ATP, adenosine triphosphate; MPP^+, 1-methyl-4-phenylpyridinium (MPTP), which produces parkinsonism in primate species.

These animal data on the mechanisms of MPTP killing of dopamine neurons have been applied in the effort to delay the progression of early PD: large clinical trials combining MAO-Is and antioxidants have been interpreted as showing some effectiveness. Future treatments may also profit from studies now actively being pursued in the animal model: surgical transplantation of fetal substantia nigra neurons, transplantation of autologous (i.e., from the same patient) adrenal medullary cells or skin fibroblasts genetically transfected to express tyrosine hydroxylase, and even special virus constructs to transfer the tyrosine hydroxylase gene into surviving striatal neurons, with or without some of the neurotrophic factors which may be required.

Future Prospects

There is no shortage of treatment opportunities for neuropharmacologists, and we have surveyed only some of the diseases in which causes and treatments are under intensive investigation. The interested reader may find additional targets of opportunity in the literature of other neurological or psychiatric disorders that follow the dictum of the moment: that treatment success seems to be inversely proportional to the molecular understanding of the etiology or pathophysiology. Along this spectrum from treatable, not understood to untreatable, almost-understood diseases, we find the following:

The hypertensions: A large collection of disorders identified by the medical finding, often without symptoms, of elevated systolic and diastolic blood pressures. Depending on cause, a variety of medications may restore blood pressure to normal levels: diuretics, peripheral sympatholytics, ganglionic blocking agents, central α_2-adrenergic agonists (clonidine) and α_1 antagonists (prazosin), and quite unexpectedly drugs that inhibit angiotensin III, either by inhibiting its formation through molecules engineered as angiotensin-converting enzyme inhibitors or by receptor antagonism.

Obsessive-compulsive disorder and anxiety disorder: Highly prevalent conditions characterized by chronic, unrealistic anxiety and compulsive behavior patterns, in some cases chemically elicitable, exhibiting semispecific cerebral metabolic alterations. For unknown reasons, these disorders often respond well to certain medications such as fluoxetine and other SSRIs; inbred animal model systems may provide clues someday.

The epilepsies: Convulsive disorders arising from trauma, infections, tumors, or unknown causes. Postencephalitic or posttraumatic scarring and acute fever are known causes, but many others are suspected. Several drugs are partially effective for reasons that are poorly understood, and the only useful animal models are those that simulate the disease condition (electroshock or drug treatments). Again, studies of the rare familial forms of epilepsy have led to isolation of mutations in ion channels, which is not surprising in view of the problems with excitability.

Multiple sclerosis: Chronic or recurrent demyelination leads to progressive, focal neurodegenerative loss of sensory and motor function, often accompanied by depression and cognitive deterioration. Drugs that slow down the immune system (interferon-b) or reduce cell division in the macrophage–monocyte lineage (2-Cl-deoxyadenosine) have shown some potential.

Human immunodeficiency virus (HIV)–associated cognitive and motor dysfunction: HIV, known not to infect neurons, nevertheless produces progressive neurodegeneration, perhaps mediated by either inflammatory cytokines or neurotoxins of either central (e.g., NO or Glu) or peripheral (e.g., quinolinic acid) origin. No current treatments are effective. The high-activity antiretroviral treatment cocktails that have been successful in many HIV-infected patients at great financial cost seem not to penetrate the blood–brain barrier, leaving the infected brain as a sanctuary for the virus. Primate models of HIV have been found to reflect the slow inflammatory changes of the early illness and may be a model for CNS treatment development.

Amyotrophic lateral sclerosis: An untreatable progressive neurodegenerative disorder restricted to the motor neurons of the spinal cord, corticospinal tract, and medulla and linked to a mutation in the enzyme Cu–Zn superoxide dismutase (SOD). In a mutant mouse model, this enzyme has been overexpressed, and motorneurons do die, leading advocates of the free radical toxicity to load up their vitamin E stores to ward off the ravages of the degenerative process; but that may not be an explanation for the human disease at all. Either eliminating natural SOD or overexpressing the natural SOD has no effect on the toxicity seen when the human disease's mutated form is expressed, suggesting that there is something toxic about the molecule itself; and pathological aggregates have been reported.

Before leaving student readers to find their own path forward into this daunting field of major untreated diseases, let us make one final observation. The present-day armamentarium of drugs useful in the treatment of CNS disorders arose, for the most part, without good insight into the nature of the diseases, by making fortuitously clever structural modifications of drugs often developed initially for the wrong reasons and by refining animal models that only very loosely simulated the human condition or predicted the results of structure–activity modifications in the leading compound. Future treatments may be accomplished most directly when one or more disease-causing genes are understood and then replaced by gene transfer technology. Several promising animal models have been reported in which cells transfected with gene constructs are stereotaxically inserted into the brain or a cerebral ventricle; these cells then express the desired gene product (e.g., tyrosine hydroxylase or a neurotrophic growth factor) and provide enduring replacement at least in experimental PD or Alzheimer's disease models. Much work is being done to develop viruses as packaging vectors to deliver next-generation gene constructs to specific target neurons, without the need for neurosurgical manipulation to transfer the genes, although this remains an unrealized goal.

It is not surprising that drug development in this field is a multibillion dollar annual investment since untreatable chronic dysfunctions of the brain represent one of the largest pockets of cost in health care. Investors in this process are firmly convinced that the "tools of modern drug discovery," as exemplified by the angiotensin-converting enzyme inhibitors and the angiotensin III antagonists, will successfully guide logical drug development in the future. These tools are *(1)* the genetic understanding of diseases and their environmentally sensitive regulatory factors, *(2)* animal models that closely simulate the pathophysiology, *(3)* cellular models in which to express cloned transmitter receptors, and *(4)* detailed three-dimensional, atomic-level understanding of the target molecules where the drugs are intended to act.

A recent example may illustrate how a previously lethal neurodegenerative disease can go from untreatable to potentially treatable through such a sequence of insights. Huntington's disease, an autosomal dominant disease, was the first inheritable brain disease to have its mutation mapped and its sequence identified; the normal sequence was named "huntingtin." One of the many surprises of this mutated protein was that it represented the tip of the iceberg of a series of related genetic disorders in which a trinucleotide repeat, CAG (coding for the amino acid glutamine), expands from the normal 10–35 to more than 41 and is a near certain predictor of the disease having been inherited. (The spinocerebellar ataxias share a similar trinucleotide expansion but much greater genetic and clinical variation.) The expansion of CAGs and glutamines was shown to lead to a gain of function in protein–protein binding, and the most recent hot trail was considered to be death of the basal ganglion neurons, which indicates Huntington's disease signs and postmortem pathology through some consequence of the enhanced binding, or failure to dissociate from a protein partner. However, a totally different picture, loss of a beneficial effect of the normal protein, has emerged from recognition that normal huntingtin can upregulate in vitro the expression of brain-derived neurotrophic factor by cortical neurons known to project to the striatum, a function that is lost by the mutations to huntingtin. If brain-derived neurotrophic factor could be delivered to striatal neurons in adequate amounts, perhaps they need not die in the future.

Given these tools, the imaginative drug design team will then rely on the capacity of synthetic chemistry to compute the potential steric requirements of receptor pockets for manipulation and to create molecular dynamic simulations of putative drug candidates. In the next-to-final step before selecting the color of the tablets, the ideal candidate will then be synthesized within the computer work station from chiral molecular building blocks in the archives and then matched against the world's database of previously made molecules for initial validation of activity. After that, it is off to the patent office, the Food and Drug Administration, and a few million dollars' worth of clinical trials. Piece of cake? Maybe there are better ways, but no one knows them yet. Students, start your engines.

Selected References

Bard, F., C. Cannon, R. Barbour, R. L. Burke, D. Games, H. Grajeda, T. Guido, K. Hu, J. Huang, K. Johnson-Wood, K. Khan, D. Kholodenko, M. Lee, L. I. Ieberburg, R. Motter, M. Nguyen, F. Soriano, N. Vasquez, K. Weiss, B. Welch, P. Seubert, D. Schenk, and T. Yednock (2000). Peripherally administered antibodies against amyloid beta-peptide enter the central nervous system and reduce pathology in a mouse model of Alzheimer disease. *Nat. Med.* 6, 916–919.

Bruijn, L. I., M. K. Houseweart, S. Kato, K. L. Anderson, S. D. Anderson, E. Ohama, A. G. Reaume, R. W. Scott, and D. W. Cleveland (1998). Aggregation and motor neuron toxicity of an ALS-linked *SOD1* mutant independent from wild-type SOD1. *Science* 281, 1851–1854.

Davis, K. L., R. C. Mohs, D. B. Marin, D. P. Purohit, D. P. Perl, M. Lantz, G. Austin, V. Haroutunian, and A. D. Korczyn (1999). Neuropeptide abnormalities in patients with early Alzheimer disease. *Arch. Gen. Psychiatry* 56, 981–987.

Esler, W. P. and M. S. Wolfe (2001). A portrait of Alzheimer secretases—new features and familiar faces. *Science* 293, 1449–1454.

Fox, H. S., M. R. Weed, S. Huitron-Resendiz, J. Baig, T. F. Horn, P. J. Dailey, N. Bischofberger, and S. J. Henriksen (2000). Antiviral treatment normalizes neurophysiological but not movement abnormalities in simian immunodeficiency virus-infected monkeys. *J. Clin. Invest.* 106, 37–45.

Francis, D., J. Diorio, D. Liu, and M. J. Meaney (1999). Nongenomic transmission across generations of maternal behavior and stress responses in the rat. *Science* 286, 1155–1158.

Jentsch, J. D., D. E. Redmond, Jr., J. D. Elsworth, J. R. Taylor, K. D. Youngren, and R. H. Roth (1997). Enduring cognitive deficits and cortical dopamine dysfunction in monkeys after long-term administration of phencyclidine. *Science* 277, 953–955.

Jones, J. E., and R. B. Johnston, Jr., eds. (2001). *Multiple Sclerosis: Current Status and Strategies for the Future*. National Academy Press, Washington, D.C.

Kamal, A., G. B. Stokin, Z. Yang, C.-H Xia, and L. S. B. Goldstein (2000). Axonal transport of amyloid precursor protein is mediated by direct binding to the kinesin light chain subunit of kinesin-1. *Neuron* 28, 449–459.

Koob, G. F. and M. LeMoal (2001). Drug addiction, dysregulation of reward, and allostasis. *Neuropsychopharmacology* 24, 97–129.

Kramer, M. S., N. Cutler, J. Feighner, R. Shrivastava, J. Carman, J. J. Sramek, S. A. Reines, G. Liu, D. Snavely, E. Wyatt-Knowles, J. J. Hale, S. G. Mills, M. MacCoss, C. J. Swain, T. Harrison, R. G. Hill, F. Hefti, E. M. Scolnick, M. A. Cascieri, G. G. Chicchi, S. Sadowski, A. R. Williams, L. Hewson, D. E. J. Carlson, R. J. Hargreaves, and N. M. J. Rupniak (1998). Distinct mechanism for antidepressant activity by blockade of central substance P receptors. *Science* 281, 1640–1645.

Kupfer, D. J. and E. Frank (2001). The interaction of drug- and psychotherapy in the long-term treatment of depression. *J. Affect. Disord.* 62, 131–137.

Lewis, J., D. W. Dickson, W.-L. Lin, L. Chisholm, A. Corral, G. Jones, S.-H. Yen, N. Sahara, L. Skipper, D. Yager, C. Eckman, J. Hardy, M. Hutton, and E. McGowan (2001). Enhanced neurofibrillary degeneration in transgenic mice expressing mutant tau and APP. *Science* 293, 1487–1491.

Marenco, S., and D. R. Weinberger (2000). The neurodevelopmental hypothesis of schizophrenia: following a trail of evidence from cradle to grave. *Dev. Psychopathol.* 12, 501–527.

Meltzer, H. Y. (1999). Treatment of schizophrenia and spectrum disorders: pharmacotherapy, psychosocial treatments, and neurotransmitter interactions. *Biol. Psychiatry* 46, 1321–1327.

Nestler, E. J. and G. K. Aghajanian (1997). Molecular and cellular basis of addiction. *Science* 278, 58–63.

Phillips, M. I., E. A. Speakman, and B. Kimura (1993). Levels of angiotensin and molecular biology of the tissue renin angiotensin systems. *Regul. Pept.* 43, 1–20.

Plaitakis, A. and P. Shashidharan (1994). Amyotrophic lateral sclerosis, glutamate, and oxidative stress. In *Psychopharmacology: The Fourth Generation of Progress* (F. E. Bloom and D. J. Kupfer, eds.). Raven Press, New York, pp. 1531–1544.

Price, D. L. S. S. Sisodia, and D. R. Borchelt (1998). Genetic neurodegenerative diseases: the human illness and transgenic models. *Science* 282, 1079–1083.

Sipe, J. C., J. S. Romine, J. A. Koziol, R. McMillan, J. Zyroff, and E. Beutler (1994). Cladribine in treatment of chronic progressive multiple sclerosis [see comments]. *Lancet* 344, 9–13.

Tallman, J. F. and S. G. Dahl (1994). New drug design in psychopharmacology: the impact of molecular biology. In *Psychopharmacology: The Fourth Generation of Progress* (F. E. Bloom and D. J. Kupfer, eds.). Raven Press, New York, pp. 1861–1874.

Tiraboschi, P., L. A. Hansen, M. Alford, E. Masliah, L. J. Thal, and J. Corey-Bloom (2000). The decline in synapses and cholinergic activity is asynchronous in Alzheimer's disease. *Neurology* 55, 1278–1283.

Weinberger, D. R. (1994). Neurodevelopmental perspectives on schizophrenia. In *Psychopharmacology: The Fourth Generation of Progress* (F. E. Bloom and D. J. Kupfer, eds.). Raven Press, New York, pp. 1171–1184.

Wirdefeldt, K., N. Bogdanovic, L. Westerberg, H. Payami, M. Schalling, and G. Murdoch (2001). Expression of alpha-synuclein in the human brain: relation to Lewy body disease. *Brain Res. Mol. Brain Res.* 92, 58–65.

Index

Abused drugs, seven major classes of, 387
Acetylcholine (ACh), 151–179
  acetylcholinesterase, 156–161
  assays, 151–152
  cellular effect, 165–167
  cholinergic pathways, 164–165
  cholinergic receptors, 167–175
  choline transport, 154–155
  in disease states, 175
  synthesis, 152–153
  uptake, synthesis and release, 161–164
ACPD receptor, 139–140. *See also* Metabotropic receptor
Adenosine, 102, 317–318
Adenosine receptors, 317
Adrenergic neurons, 182–183
α- and β-Adrenergic receptors, 199–205
  desensitization, 204–205
  pharmacology, 201–204
  regulation, 202–204
  subtypes, 201
Agmatine, 105
AJ-76, 244
Alzheimer's disease, 175, 390–392
γ-Aminobutyric acid, 106–127. *See also* GABA
γ-Aminobutyrobetaine, 112
γ-Aminobutyrylcholine, 112
γ-Aminobutyrylhistidine (homocarnosine), 112
γ-Aminobutyrl-lysine, 112
γ-Aminohydroxybutyric acid (GABOB), 112
Aminoxyacetic acid, 111
AMPA receptor, 139–140, 143–144
  distribution and function, 143–144
  pharmacology, 140, 143–144
Amphetamine, 216, 218, 263–264
Antidepressants, 218
Apomorphine, 264
AP4 receptor, 139
AP-5, 133, 140
AP-7, 140
Autoreceptors, 193, 215–217, 243–244, 310–311
  transmitter release, 193–194

Baclofen, 119–120
βARK, 202–203
Benzodiazepines, 116–118
Bicuculline, 123, 124
Blood–brain barrier, 11
BMAA, 147
BOAA, 148
Bufotenine, 272

Calcineurin, 89
Carbon monoxide, 101
Catecholamines, 181–221
  and affective disorders, 380–385
  biosynthesis, 184–189
  coexistence, 212–213
  metabolism, 194–196
  plasma membrane transporter release, 192–193
  reuptake, 196–198
  storage, 191–192
  synthesis regulation, 189–191
  systems in CNS, 206–212
Catecholamine Theory of Affective Disorder, 380–385
Catechol-O-methyltransferase (COMT), 194–197
β-CFT, 232, 240
Choline acetyltransferase, 155–156
Cholinergic receptors, 167–170
7-Chlorokynurenic acid, 140
Cholescystokinin, 347–348
Chromaffin granules, 191
Cimetidine, 313

β-CIT, 232, 240
Clonidine, 216–219
Cloning, 48–52
Clorgyline, 194
Clozapine, 250, 254, 262–263
CNQX, 140
Cocaine, 145, 232
Compound 48/80, 307
Corticotropin releasing factor, 349
Cretinism, 378

DARPP-32, 90, 247, 248
Deprenyl, 194
Depression, 380
Desipramine, 218
Diethyldithiocarbamate, 188
Dihydrexidine, 242
Dihydropteridine reductase, 186
Dihydroxyphenylacetic acid (DOPAC), 235–236
Dimaprit, 313
DNQX, 140
Domoic acid, 148
Domperidone, 242
DOPA decarboxylase, 187
Dopamine, 225–270
  anatomy, 227–228
  autoreceptors, 243–245
  β-hydroxylase, 187–188
  and cyclic AMP, distribution, 241–242
  DARPP-32, 247–248
  imaging, 266
  molecular biology, 249–251
  pharmacology, 253–261
  receptors, 239–255
  receptor mRNA distribution, 251–253
  release, 231
  schizophrenia, dopamine hypothesis of, 268, 385–387
  signal transduction, 245–249
  synthesis, 229
  synthesis regulation, 229–231
  transporter (DAT), 231–235
  uptake, 231–235
Dopamine hypothesis of schizophrenia, 268, 385–387
Dopaminergic neuronal systems, 226–228
  incertohypothalamic neurons, 226–227
  interplexiform neurons, 226–227
  medullary periventricular neurons, 226–227
  mesocortical neurons, 226–227, 256–261
  mesolimbic neurons, 226–227
  nigrostriatal neurons, 226–227
  periglomerular neurons, 226–227
  pharmacology, 257–258
  tuberohypophysial neurons, 226–227, 261–262
Drugs, 262–265. *See also specific drugs*
  antipsychotic, 254, 262–265
  stimulants, 263–265
Drug abuse, 387–390
Drug transport, 3–4
D-serine, 141

Eicosanoids, 96–99
Eliprodil, 142
EMD-23-448, 242, 244
Epinephrine neurons, 210–212
  anatomy, 211
  pharmacology, 212
Excitatory amino acids (EAAs), 138–148
  agonists, 140
  AMPA receptor, 143–144, 147
  antagonist, 140
  AP-4 receptor, 139
  kainate receptor, 143–144, 147
  and long-term potentiation, 146, 148
  metabotropic receptors, 144–148
  NMDA receptors ionophore complex, 139–143, 147
  non-NMDA receptors, 143–44

Familial dysautonomia, 175
FLA-63, 188
Fluorescence microscopy
  of catecholamines, 181–184
Fluoxetine, 290, 302

GABA, 106–127
  agonists, 120, 125–126
  alternative metabolic pathways, 111–113
  antagonists, 120–124
  distribution, 106–107
  endogenous modulators, 126–127
  glutamic acid decarboxylase, 108–109
  metabolism, 107, 108
  neurotransmitter role, 106
  pharmacology, 122–127
  receptors, 114–122
  release and reuptake, 113–114
  storage, 113
  synthesis, 108, 109
  shunt, 107
  succinic semialdehyde dehydroganase, 110, 111

-transaminase (GABA-T), 109, 110
vesicular GABA transporter (GAT), 113
$GABA_A$ receptor, 115–118
endogenous modulators, 116–118
molecular biology, 116–118
pharmacology, 116–118
$GABA_B$ receptor, 118–122
pharmacology, 120–122
GBR, 231, 232
Genetic code, 46
Genome, 341, 374
Glia, 10
Glutamic acid (glutamate), 132–138
and long-term potentiation, 146
release, 134–135
storage, 135
synthesis metabolism, 133–134
transporter, 136–137
Glutamic acid decarboxylase (GAD), 108–109
Glycine, 127–132
distribution, 127, 128
metabolism, 128–130
modulator of NMDA receptors, 131–132
neurotransmitter role, 129–130
receptors, 129
transporter, 128–129

HA-966, 141, 261
Haloperidol, 254, 262–264
Hallucinogens, 299–301. *See also* LSD
Hemicholinium-3, 154–175
Histamine, 304–317
agonists, 311–315
antagonists, 311–316
catabolism, 305–307
containing cells, 307–309
pharmacology, 314–316
receptors, 309–317
synthesis of, 305–307
Histidine decarboxylase, 305–306
inhibitors of, 305
HIV, 397
Homovanillic acid (HVA), 235–236
γ-Hydroxybutyric acid (GHB), 111–113
5-Hydroxytryptamine, 271–304. *See also* Serotonin
5-Hydroxytryptophan decarboxylase, 275–276

IBMX, 218
Idazoxane, 218–219
Ifenprodil, 142
Iodotyrosine, 186
Ion channels, 18, 23, 59–61
Ion pumps, 16
Iplandule, 32

Kainate receptor, 139–140
Kainic acid, 139
Ketamine, 141
Kynurenate, 142

Lambert-Eaton syndrome, 175
Learning and memory, 357–372
associative learning, 361
cellular models, 359
long-term potentiation, 364
vasopressin, 368
Lesch-Nyhan syndrome, 378
Locus ceruleus, 208–209
anatomy, 208–209
functional hypothesis, 210, 220
LSD, action of, 299–301

Marijuana, 387
Mast cells, 307
Mazindol, 232
Median eminence, 226
Melatonin, 272
Metabotropic receptor, 144–147
function, 146–147
pharmacology, 140, 146
signal transduction, 144–145
3-Methoxy-4-hydroxyphenethyleneglycol (MHPG), 213–214, 216–217
Methylenedioxymethamphetamine (MDMA), 283, 387
Methylhistamine, 305–306
α-Methyl-5-hydroxytryptophan, 186
α-Methyl-3-iodotyrosine, 186
α-Methyl-norepinephrine, 188
α-Methyl-*p*-tyrosine (AMPT), 186
Methylxanthines, 218. *See also* IBMX
MHPG, 213–214, 216–217
Modulation, 85–104
definiton, 85–88
presynaptic, 86
postsynaptic, 87
second messengers, 89–102
protein phosphorylation, 89–90
Molecular interactions, 44
Molecular strategies, 50
Monoamine oxidase inhibitors, 194–196
Morphine, 218, 219

MPTP(1-methyl-4-phenyl-tetrahydropyidine), 268
Multiple sclerosis, 397
Muscimol, 120, 124
Myasthenia gravis, 173

Naloxonc, 218, 219
NBQX, 140, 143
Neuroactive peptides, 321–356
Neurohormone, definition of, 3
Neuromodulator, definition of, 3
Neurosteroids, 116–117, 127
Neurotensin, 348
Neurotransmitter, definition of, 2
Neurotransmitters, false, 188
Nicotine, 173, 387
Nisoxetine, 199
Nitric oxide, 99–101
NMDA receptor, 139–143
  function, 139
  pharmacology, 140–141
  regulation, 141–143
Nomifensine, 232
Non-NMDA receptors, 143–148
  function, 147
  pharmacology, 146–148
Noradrenergic
  locus ceruleus, 208
  neurons, the lateral tegmental, 210
  neurons, peripheral, 182–183
  systems in CNS, 209–210
Norepinephrine, 181–220
  biosynthesis, 184–189
  coexistence, 212
  metabolism, 194–196
  pharmacology, 214–217
  physiology, 218–220
  release, 192–194
  storage, 191–192
  synthesis regulation, 189–191
  transporter (NET), 197–199
  uptake, 196–199

Opioid peptides, 338–344
Orphan receptors, 65
Oxytocin, 330–333

Pargyline, 216
Parkinson's disease, 266–268, 392–396
Paroxetine, 290
PCP (phencyclidine), 140–141, 387
Perphenazine, 264
Phaclofen, 121
Phenotype, 41
Phenylethanolamine-*N*-methyltransferase, 188–189
Phosphoinositide hydrolysis, 90–96
Physostigmine (eserine), 175
Picrotoxin, 120, 123
Pineal body, 278–279
Piperoxane, 218–219
Polyamines
  modulators of NMDA receptors, 142
Polygenic diseases, 375
Polymixin B, 307
Postsynaptic potentials, 20–23
3-PPP, 244
Prostaglandin, and norepinephrine release, 193
Proteomics, 143
Psilocin, 272

Quisqualic acid, 140
Quinolinic acid, 142–143

Receptors, 67–84
  assays, 67–69
  definition, 67
  G protein–coupled receptors, 76–77
  identification, 69–70
  kinetics and theories of drug action, 71–76
  ligand-gated channels, 76
  purinoreceptors, 79–81
  steroid hormone receptors, 78–79
  tyrosine kinase receptors, 79
Reserpine, 216, 264, 302

Schizophrenia, 385
Serotonin, 271–304. *See also* 5-hydroxytryptamine
  adaptive regulation, 296
  behavioral aspects, 290
  biosynthesis and metabolism, 272, 277
  cellular effects, 283–284
  clinical disorders and drug actions, 298–299
  cytochemistry of, 279–283
  electrophysiology, 297–298
  hallucinogenic drugs, 299–301
  indolealkylamines, structure, 272
  molecular biology, 285–290
  pathways, 279–283

pharmacology, 302
physiology, 296–297
pineal body, 278–279
receptors, 284–285
receptor subtypes, 285–290
regulation of synthesis, 272–275
signal transduction pathways, 293–296
transporters, 297–298, 290
Sertraline, 290
Single-gene mutations, 375
Somatostatin, 345–347
Steps in transmission, 34
Strychnine, 129
Succinic semialdehyde dehydrogenase, 110–111
Sulpiride, 250
Synapse, 8

Tacrine, 175
Tachykinin peptides, 331–334
Tetrabenazine, 216, 264, 302
Tetrahydrobiopternin (BH_4), 229
Tomoxetine, 199
Transmitter identification, 31
Transport proteins, 81–83
Transporter. *See individual types*
plasma membrane, 136–137, 197–199, 231–235, 277–278
vesicular, 191–192, 200, 230, 295
Tropolone, 186
Tryptophan hydroxylase, 273–275
Tubero mammillary nucleus, 308
Tyrosine, 184–185
Tyrosine hydroxylase, 184–191
inhibitors of, 184
phosphorylation, 189–190
regulation of, 189–191

UH-232, 244

Varicosities, 182–183, 207
Vasopressin, 330
Vesamicol, 174
Vesicular monoamine transporter (VMAT), 191–192, 200, 230, 295
Vesicle proteins, 62–64
VIP, 335
von Euler, Ulf, 181

Yohimbine, 218–219